Enlightened Absolutism
Reform and Reformers in Later Eighteenth-Century Europe

PROBLEMS IN FOCUS SERIES

Each volume in the 'Problems in Focus' series is designed to make available to students important new work on key historical problems and periods that they encounter in their courses. Each volume is devoted to a central topic or theme, and the most important aspects of this are dealt with by specially commissioned essays from scholars in the relevant field. The editorial Introduction reviews the problem or period as a whole, and each essay provides an assessment of the particular aspect, pointing out the areas of development and controversy, and indicating where conclusions can be drawn or where further work is necessary.

Europe's Balance of Power 1815–1848
edited by Alan Sked

FURTHER TITLES ARE IN PREPARATION

Series Standing Order

If you would like to receive future titles in this series as they are published, you can make use of our standing order facility. To place a standing order please contact your bookseller or, in case of difficulty, write to us at the address below with your name and address and the name of the series. Please state with which title you wish to begin your standing order. (If you live outside the United Kingdom we may not have the rights for your area, in which case we will forward your order to the publisher concerned.)

Customer Services Department, Macmillan Distribution Ltd, Houndmills, Basingstoke, Hampshire, RG21 2XS, England.

Enlightened Absolutism

Reform and Reformers in Later Eighteenth-Century Europe

EDITED BY
H. M. SCOTT

MACMILLAN

First published 1990

Published by
MACMILLAN EDUCATION LTD
Houndmills, Basingstoke, Hampshire RG21 2XS
and London
Companies and representatives
throughout the world

Typeset by Wessex Typesetters
(Division of The Eastern Press Ltd)
Frome, Somerset

Printed in Hong Kong

British Library Cataloguing in Publication Data
Enlightened absolutism: reform and reformers
in later eighteenth-century Europe—
(Problems in focus series)
1. Europe. Enlightened despotism, 1760–1790
I. Scott, H. M. (Hamish M.), 1946–
II. Series
321.6
ISBN 0–333–43960–0 (hardcover)
ISBN 0–333–43961–9 (paperback)

Contents

List of Maps

List of Tables

Preface

The editing of this volume has been a more prolonged process than anticipated, and I am deeply grateful to the publishers and to all the contributors, not only for their unfailing co-operation but for their encouragement and forbearance during this delay. The withdrawal of the scholar who was to have surveyed the Austrian Habsburgs deprived the volume of a key chapter and in the circumstances I decided to produce this article myself. I am particularly indebted to Dr Robert Evans, who agreed to become a contributor at an advanced stage and who produced his own articles to a particularly tight deadline, and to Professor Derek Beales and Dr Derek McKay for their helpful and fortifying comments on a draft of my chapter on the Habsburgs. I am also grateful to Ms Sabina Arnot-Krause for her invaluable assistance with the research for the introduction. This was paid for by the Committee on Research in Arts and Divinity at the University of St Andrews, whose support I am glad to acknowledge. The project was initially suggested by Ms Vanessa Couchman, who provided essential support, advice and encouragement in the planning stages. The completion of the volume has been facilitated by the skill and understanding of her successor at Macmillan, Ms Vanessa Graham. I am grateful to Graeme Sandeman, who drew the maps, and to Dr McKay and Dr Janet Hartley for their help with the proofs.

H.M.S.

Introduction: The Problem of Enlightened Absolutism

H. M. SCOTT

I

FEW historical concepts have had their obituaries written more frequently than enlightened absolutism, yet so obstinately refuse to die. In its classical form, the theory of enlightened absolutism asserted that during the second half of the eighteenth century the domestic policies of most European states were influenced and even dictated by the ideas of the Enlightenment and were therefore sharply distinguished from what had gone before. Government became a systematic and rational attempt to apply the best recent knowledge to the task of ruling, while the main aim of internal policy came to be the improvement of educational opportunities, social conditions and economic life.

The principal enlightened monarchs were the King of Prussia, Frederick the Great (1740–86); the Russian Empress, Catherine the Great (1762–96); and the Habsburg Emperor, Joseph II (co-regent 1765–80, sole ruler 1780–90); along with his brother Leopold, Grand Duke of Tuscany (1765–90); the Swedish ruler Gustav III (1771–92); and the Spanish monarch Charles III (1759–88). To these can be added several minor German rulers, above all Frederick II of Hesse-Cassel and Charles Frederick of Baden, and some enlightened ministers, notably the Marquês de Pombal in Portugal (1750–77), Johann Friedrich Struensee in Denmark (1770–72) and subsequently the government in Copenhagen headed by A. P. Bernstorff after 1784, and Bernardo Tanucci in the Kingdom of the Two Sicilies (1734–76). Together, these rulers and their influential advisers carried through a wide-ranging series of reforms, the principal inspiration for which was the political philosophy of the Enlightenment. Administrations and legal and fiscal systems were modernised; commercial and economic development was encouraged; efforts were made to improve agriculture and even to abolish the institution of serfdom; state control over the Catholic Church was extended and a determined attempt was made to channel some of its wealth into increased and improved pastoral activity; teaching in universities was brought up to date; and the provision of secondary and primary education was vastly increased. These reforms, and the amount of success achieved, were remarkable in the context of the later eighteenth century, and the sense of social responsibility which lay behind many of these measures was also novel and striking. Though the precise policies adopted varied from country to country, a common explanation was found in the movement of ideas known as the European Enlightenment, which was at its peak exactly when the policies of enlightened absolutism were being pursued: the middle decades of the eighteenth century until the outbreak of the French Revolution in 1789.

During the past two generations, the historical reputation of enlightened absolutism has undergone remarkable vicissitudes, above all in Anglo-American historical writing about later eighteenth-century Europe. It first rose to prominence during the 1930s, and for a generation thereafter was widely accepted among historians.[1] By the 1960s, however, the tide was beginning to turn and the sceptics, who saw only the conventional aims of state-building in these policies and regarded the enlightened professions which accompanied them as mere window-dressing, were gaining the ascendant. The critics of enlightened absolutism argued that few monarchs, and certainly none of the major rulers, could afford to adhere blindly to an enlightened blueprint and ignore, or even neglect, the demands of self-defence. The competitive states-system within which the great powers operated required large and powerful armies and administrative and fiscal systems to support them. These priorities pre-empted much of the scope for enlightened reform. Above all, critics of the notion of enlightened absolutism argued that the success of these initiatives was often incomplete and could, at times, be very limited indeed.

Many of these criticisms were marshalled by M. S. Anderson in a textbook published in 1961.[2] Professor Anderson mounted a frontal attack on the whole idea of enlightened absolutism, denying its practical impact and questioning its theoretical validity, and his arguments were clearly influential. The relevant volume of *The New Cambridge Modern History*, when it appeared four years later, contained a fairly watered-down version of the theory in the editorial introduction.[3] Even this proved to be too strong a brew for some contributors, three of whom found it necessary to deny the validity of enlightened absolutism outright.[4] And in 1968, François Bluche, in what is still the only full-scale modern survey, added his voice to the chorus of scepticism, finding much cynicism and little genuine enlightenment in these rulers' policies.[5]

By the 1970s, the notion of enlightened absolutism seemed in full retreat.[6] In 1975, C. B. A. Behrens, in a notably unguarded phrase, went as far as to declare that the term had 'no value as a historical category' and 'has . . . now outlived its usefulness'.[7] A further severe blow appeared to come later the same year when Derek Beales, through some remarkably skilful detective work, demonstrated that many of the celebrated quotations used to support the cause of Joseph II, one of the few remaining enlightened absolutists, were, in fact, forgeries.[8] Matters were not helped by the late Albert Soboul's bizarre and confusing attempt to portray it as a social movement, the ideological superstructure of the 'second serfdom' in eastern Europe.[9] A rather more sophisticated explanation was advanced by Helen Liebel-Weckowicz in 1970, but its thrust was also essentially negative.[10] Though rightly emphasising the importance of recurring social problems and economic difficulties throughout the period 1763–89, she then reduced enlightened absolutism to an attempt by the ruling class to promote stability in the face of the resulting unrest.

These and similar obituaries, however, have been accompanied by clear evidence that they may be premature and probably unnecessary. During the 1970s, the widespread assumption that enlightened absolutism was a fiction came to be challenged by a growing number of monographic studies of particular countries which make clear that the reforms of this period were influenced by ideas and were not simply another stage in the growth of the absolutist state. This trend has increased during the 1980s. It was symptomatic of the changing emphasis that when Professor Anderson produced a second edition of his standard textbook in 1976, he was more willing to admit the enlightened intentions, if not the achievements, of many rulers.[11] During the 1970s and 1980s it has become clear that enlightened absolutism is not an idea in the mind of historians. The remarkable extent and the considerable success of the reforming initiatives is now established beyond question. This collection re-examines the generation of reform at the national level on the basis of the best recent research. Its starting point is the conviction that the past criticisms of the whole idea of enlightened absolutism are, in fact, criticisms of the terms in which the theory has hitherto been formulated by historians, not of the reality of reform in later eighteenth-century Europe and of the importance of ideas in bringing this about. Many and perhaps most of the objections are critical of the way the concept was first expressed and has subsequently developed.

II

There are essential distinctions between enlightened absolutism – or the more usual but less satisfactory form in Anglo-American historical writing, enlightened despotism – as a term, as a concept, and as a theory and historiographical category. The problem here is the way in which terms employed in the eighteenth century subsequently acquired more precise – and sometimes different – meanings in the hands of later historians. The category and theory of enlightened absolutism/despotism were created by scholars writing in the nineteenth and twentieth centuries, though many of the basic ideas were familiar during the Enlightenment.

The term 'absolutism' was an invention of the first half of the nineteenth century[12] and the phrase 'enlightened absolutism' was obviously not employed at the time. Indeed, it only became current after 1847, when it was coined by Wilhelm Roscher (see below, pp. 5–6). The phrases 'enlightened despot' and 'enlightened despotism' were occasionally used during the second half of the eighteenth century.[13] This is, at first sight, surprising, given the overtones of tyranny and even dictatorship which the word 'despotism' possessed in the eighteenth century when it was equated with authority not subject to the restraint of any laws,[14] and its use certainly excited lively controversy. The first recorded – and apparently isolated – use of 'despote . . . éclairé' would seem to be in 1758 by Baron Grimm in his *Correspondance littéraire*, but enlightened was only to be one of his qualities. Grimm's ideal ruler was also to be just, vigilant and well-intentioned. Nine years later, Grimm employed the precise term, in response to the appearance of books by two prominent physiocrats, François Quesnay and Paul-Pierre Le Mercier de la

Rivière, and this is the real origin of the term 'enlightened despot'. In 1767, Le Mercier de la Rivière published his celebrated *L'ordre naturel et essentiel des sociétés politiques*, which has been widely regarded as a handbook for an enlightened ruler and which declared the best government to be that of a *'despote patrimonial et légal'* who would rule by discovering and applying established fundamental laws.

The debate which followed revealed the deep divisions within the Enlightenment. Jean-Jacques Rousseau immediately denounced 'legal despotism' as a contradiction in terms[15] since it united two words which when placed together meant nothing, and his scepticism was widely shared. No coherent theory of enlightened despotism was produced at the time.[16] Indeed, some influential scholars have denied that the concept received much support, at least from among the ranks of the *philosophes*.[17] This seems altogether too negative. The term, at least, was established and the concept was becoming familiar by the closing decades of the eighteenth century. Many of the physiocrats, that influential group of French economic theorists whose importance was at its peak during the 1760s, supported the ideal of an enlightened ruler and, particularly after 1770, many of the *philosophes* came to favour direct involvement in government, as a minister or official, as a means of securing a better society.[18] Outside France, there was significantly more support among later eighteenth-century thinkers for the kind of reform-from-above which enlightened despotism embodied. The full elaboration of this idea, however, was not the work of contemporaries but of later historians.

The classical theory of enlightened absolutism had two sources which, to an extent that is surprising, have remained distinct. The first was the writings of certain German historians during the second half of the nineteenth and early twentieth centuries and principally those of the distinguished scholar of *National-ökonomik*, Wilhelm Roscher, who is the most plausible father of the modern historical concept and who first invented the term. In 1847, in the course of a wide-ranging survey of government and its theoretical sources during the early modern period, Roscher distinguished three stages of development, each represen-ting a stronger form of monarchy and produced by the natural laws of social evolution.[19] The first was a phase of 'confessional absolutism', the aim of which was embodied in the celebrated formula *cuius regio, eius religio* and which lasted from the Reformation until the end of the Thirty Years War with Philip II of Spain as the prime example. This was followed after 1648 by a period of 'courtly absolutism' dominated by Louis XIV of France under the banner '*l'état c'est moi*'. Finally, this gave way during the middle decades of the eighteenth century to a phase of 'enlightened absolutism', the principal exponents of which according to Roscher were Frederick the Great of Prussia and Joseph II in the Habsburg Monarchy.

Roscher's ideas were widely influential, both in their own day and subsequently. Though passing mention was made of other rulers or ministers who fitted the category of enlightened absolutist, they were primarily inspired by and applied to developments in the German-speaking lands: it is significant that one of the key articles for the development of this notion was written by the celebrated biographer of Frederick the Great, Reinhold Koser. Roscher and those who followed him were confident about the beneficent role of the state and saw enlightened absolutism primarily in terms of the growth of its power. Indeed, for patriotic and, increasingly, authoritarian historians such as Heinrich von Treitschke, it was nothing less than 'a new idea of the state'.[20] Enlightened absolutism enabled the ruler, in the name of *raison d'état*, to exercise wide and unprecedented authority to mobilise his country's resources. Under the banner that 'the King was the first servant of his people', Frederick the Great (who employed the phrase and embodied the ideal) carried out a series of economic and judicial reforms, the main aim of which was to strengthen Prussia at home and abroad. The sources for these policies were, firstly, the ideas of the German Cameralists (Roscher was a noted historian of their ideas and influence) and, secondly, the Enlightenment, broadly defined. The links between such theories and the internal policies actually pursued were never systematically investigated and, to a significant extent, enlightened absolutism remained a label not a theory. In this view, the Enlightenment was simply a means of strengthening the state and enlightened absolutism the most powerful form of monarchy.

Roscher's ideas were not accepted in their entirety. The most important refinement and development of them came from Koser in an article published in 1889.[21] While upholding Roscher's main arguments, this questioned his view that enlightened absolutism was the most powerful form of monarchy. Koser instead pointed out that, since later eighteenth-century rulers placed the interests of the state above their own personal concerns, this in itself imposed certain restrictions on their rights and freedom of action. He instead saw a crucial and novel element in enlightened absolutism to be the acceptance by the monarch of a social contract, imposing obligations in return for the obedience and support derived from the population at large. This was accompanied by a limited extension of the geographical focus of enlightened absolutism to include rulers and ministers such as Catherine II in Russia, Leopold in Tuscany, Tanucci in Naples and Bernstorff in Denmark. Yet in all important respects, Koser endorsed the idea first put forward by Roscher.

Within Germany in particular, the concept of enlightened absolutism developed during the second half of the nineteenth century has proved remarkably tenacious. The celebrated survey written by Fritz Hartung during the 1950s, and the influential collection of essays edited by Karl Otmar Freiherr von Aretin in 1974, both stand in a direct line of descent from Wilhelm Roscher.[22] Influential as this has been within German historiography, however, its impact outside Germany and on Anglo-American scholarship in particular has been surprisingly limited,[23] and it has contributed far less to the development of the classical theory of enlightened absolutism than the work of the International Commission of Historical Sciences [ICHS] during the later 1920s and 1930s.

The enquiry into government during the second half of the eighteenth century sponsored by the ICHS was the second and far more important source of the concept. One principal – and unhappy – contribution was to the terminology. The form employed by nineteenth-century German scholars, 'enlightened absolutism', was ignored in favour of 'enlightened despotism' which, as will be seen (below, pp. 9–10), is much less satisfactory. This typified the extent to which the earlier work of Roscher, Koser and other German scholars was ignored – and perhaps not known – to the members of the ICHS, who were visibly surprised when it was drawn to their attention once their own enquiry was already far advanced.[24] The concept of enlightened absolutism/despotism which evolved within the English-speaking historical world is primarily the work of the ICHS and the enquiry which it organised between 1928 and 1937.[25]

The context of this enquiry is all-important, though too often overlooked. The ICHS had held several meetings before the First World War, but these only became regular during the 1920s, when it acquired a more formal organisation. It was inspired by a laudable desire to encourage international understanding and thereby increase harmony between nations: the shadow of the battlefields of the Great War at times fell heavily across its deliberations.[26] The enquiry into enlightened despotism was one of a series which aimed to identify common developments and shared experiences in the past, and thereby to increase mutual understanding in the present.[27] It was launched in 1928. The moving spirit behind this investigation was the influential French historian, Michel L'Héritier, who first proposed the enquiry and subsequently shaped the form it took. Two points have to be made about L'Héritier, who was the first Secretary of the ICHS. He was an exceedingly successful academic politician who made use of the opportunity provided by this enquiry to move into positions of considerable power within the ICHS, ending up as its influential Secretary-General. These ambitions are of some enduring importance, and partly explain the bland tone and the vagueness of the final, much-cited report produced in 1937, though the spirit of international co-operation also contributed to this.[28]

In a second, more important respect, L'Héritier decisively shaped the concept of enlightened despotism conceived and nurtured by the ICHS. His own research lay principally in the administrative history of eighteenth-century France, particularly at the provincial level, and in questions of local government. He had become interested in the influence of the *philosophes* on certain reforming French intendants during the second half of the eighteenth century and he published an article on this the year after the ICHS enquiry was launched.[29] This was to be of considerable significance for the concept of enlightened despotism which now took shape. It is not too much to say that the work of the ICHS was an extension of L'Héritier's own research onto the international stage and into comparative history.[30] His demonstration of the importance of the *philosophes* in French government was to be widened into an exploration of the influence of French thought on the internal policies of other European states. There is a sense in which, beneath the rhetoric of international scholarly co-operation, L'Héritier was a strong French nationalist. Under his direction, the ICHS enquiry became a search for the impact of French ideas, and especially the economic teachings of the physiocrats, on European governments, particularly during the generation after the ending of the Seven Years War in 1763.

The project was launched at the 6th Congress in Oslo in 1928, and two years later a special international commission was established to co-ordinate the enquiry. National committees of historians produced bibliographies; reports were written on enlightened despotism in every major European state and quite a few smaller countries as well; plenary sessions were held to debate the reports produced by teams of historians from each nation; and, finally, in 1937, L'Héritier – by now Secretary-General – produced the final survey, significantly entitled 'Enlightened Despotism, from Frederick II to the French Revolution'. This enquiry, and particularly the final report, is the real origin of the modern historical concept of enlightened absolutism. In the course of the next ten or twenty years, the notion entered the textbooks in the terms formulated by the ICHS and became a historical commonplace. Simultaneously, however, a reaction set in against this view of later eighteenth-century government.

This began after 1945, in the starker atmosphere created by the Second World War which put idealism at a disadvantage and made historians even more sceptical of lofty motives in the past. It was, above all, a rejection of the theory of enlightened absolutism developed by L'Héritier and his fellow-scholars. So many potent objections were put forward that it is difficult to know where to begin. A relatively minor source of criticism was the chronological and geographical excesses of the ICHS. During its discussions, the period and the physical location of enlightened absolutism (c. 1750–89 in Europe) were vastly extended, sometimes in a way that was patently ludicrous. Several Roman Emperors (Marcus Aurelius above all) and certain of the iconoclastic Emperors of Byzantium were canvassed in all seriousness; so too were some of the princes of Renaissance Italy and a couple of Ming emperors from China; rather more plausibly, Napoleon I and Napoleon III were brought forward. Even Solomon was advanced as a Biblical enlightened absolutist, though this interjection was probably ironical. Too much should not be made of this, though the extension of the idea far beyond later-eighteenth-century Europe was certainly seized on by critics, who argued that a concept so elastic in temporal and geographical terms was next to useless.

More seriously, the very term 'enlightened despotism' was objected to as absurd and contradictory. It was pointed out that 'enlightened despotism' was an oxymoron and that to suggest a ruler could be both despotic and enlightened was ludicrous, illogical and a glaring contradiction in terms. The German variant, 'enlightened absolutism', was certainly preferable and, for this reason, is generally employed throughout this volume. But again it was argued that the practices of absolute monarchy and the ideas of the Enlightenment were equally antithetical and contradictory. It was pointed out with some force that, in the major states at least, the requirements of great power status and international rivalry, and in particular the resulting military and fiscal priorities, were likely to be incompatible with the kind of reforming programme envisaged by enlightened absolutism.[31] This remains an important limitation on any too-mechanical belief in the strict applicability of a theory of enlightened absolutism. Most historians – and certainly the contributors to this volume – would now argue that the need to defend and maintain a state's international position could set limits to the kind of reforming measures attempted in the major powers at least. But though it placed constraints on a ruler's freedom of action, it did not preclude measures of enlightened reform. It did make certain kinds of initiatives difficult and perhaps impossible. Above all, the celebrated failure of Frederick the Great and Catherine the Great to attack the institution of serfdom, and Joseph II's opposition to such reform in the later 1770s, is to be explained largely in terms of their fear of the social and fiscal dislocation which might follow, with its threat to their military power and international position (see below, pp. 273–4, 307–10 and 180–1).

The third difficulty presented by the idea which emerged from the deliberations of the ICHS is that it incorporated a view of the Enlightenment as an international movement dominated by a handful of major, and usually French, thinkers. This accorded with the prevailing view, as evident in Ernst Cassirer's influential *The Philosophy of the Enlightenment*, first published in German in 1932.[32] For Cassirer, as for most of his contemporaries, the study of the Enlightenment was the study of its philosophical giants: Kant above all, but mainly the celebrated French thinkers, Diderot, Montesquieu, Rousseau, Voltaire, Condillac and so on. In one sense the ICHS enquiry, with its search for the Enlightenment's contemporary and practical influence, went directly against the established and prevailing trend of studying it primarily as an autonomous movement of ideas which was the source of later philosophical and political doctrines. Until comparatively recently, the Enlightenment was seen less as a force in eighteenth-century history than as a seminal phase in the history of ideas which looked forward to the subsequent development of liberalism and democracy. During recent decades, the emphasis on the central place of French thinkers and on the intellectual and philosophical giants has given way to a more rounded approach to the Enlightenment and this, as will be seen, has been important for the revival of enlightened absolutism (see below, pp. 13–14).

Finally, a particularly serious objection to the ICHS view was that it failed to establish a clear and persuasive link between theory and practice, or to demonstrate convincingly how ideas influenced actions. As the critics of enlightened absolutism were quick to point out, it was remarkably difficult to identify actual policies which were clearly inspired by a particular doctrine of the Enlightenment. The relationship of thought to action is, of course, a particularly complex problem in the history of ideas in general. But it has to be said that the members of the ICHS solved it by pretending that it did not exist. What is very striking is the general failure to address the question of exactly how the ideas of the *philosophes* in particular influenced the reforms attempted and the policies pursued. The reforming monarchs of the later eighteenth century were declared to be inspired by the Enlightenment, but the precise extent of this influence and the actual way in which it was manifest were never systematically explored. This proved to be one of the weakest points in the classical theory when it came to be attacked during the two decades after 1945. Critics were quick to point out that very few actual reforms can be 'proved' to have been inspired by a particular Enlightenment doctrine. Searching for a linear relationship between thought and action, an elementary 'idea A caused reform X' linkage, they were rapidly disillusioned. The one initiative which could confidently be identified as being dictated by the ideas of the Enlightenment was the attempt by the Margrave Charles Frederick of Baden to introduce a single tax on land in 1770. This was the famous *impôt unique* advocated by the physiocrats, and even its partial introduction was a failure.[33]

Indeed, it was not merely impossible to establish a linkage between thought and action, but it seemed that many later-eighteenth-century rulers were acting in a way that was actually opposed to the doctrines of the Enlightenment. Two particularly glaring examples of this fundamental contradiction were highlighted. The first was the notorious failure of Eastern European rulers to do anything about the existence of serfdom, the drawbacks of which they certainly recognised and the continuation of which could not easily be reconciled with the ideas of the Enlightenment and in particular the doctrine that was – perhaps wrongly – attributed to it: the notion that all men were free and equal. It has earlier been suggested (see above, p. 10) that the case of serfdom primarily demonstrates the practical limits on reform, particularly in states struggling to maintain their status as great powers; but it was seized on by critics in an attempt to discredit the whole notion of enlightened absolutism. The second contradiction highlighted was that between the Enlightenment's emphasis on peace and international harmony and the actual conduct of international relations at this time. It was demonstrated that all the major enlightened monarchs pursued notably aggressive and expansionist external policies. The cynical onslaught on Poland, partitioned out of existence between 1772 and 1795, could not easily be reconciled with their supposedly enlightened domestic policies.

By the 1960s, these and similar objections had done much to discredit the concept as it had been formulated by the ICHS, and influential textbooks were dismissing it out of hand.[34] Yet these criticisms are essentially objections to the work of the ICHS, and not to the validity of enlightened absolutism itself. They relate to the way the concept was formulated during the later 1920s and 1930s, and subsequently passed into the literature, not to the nature of government in later eighteenth-century Europe. During the past generation, even while the ICHS version was being discredited, a new and more convincing theory has been taking shape. The rebirth of enlightened absolutism overlaps chronologically with the discrediting of the ICHS version. It begins in the 1960s and has become influential during the past two decades. This collection is intended to advance its progress. The essays that follow are informed not by the older ICHS idea of enlightened absolutism but by the broader and more convincing approach to later-eighteenth-century reform that is now current.

Several reasons for this rebirth can be suggested. In the first place, enlightened absolutism has re-emerged primarily because of the sheer weight of evidence that the second half of the eighteenth century did see a series of remarkable reforms which were, to a significant extent, inspired by ideas current at the time. During the past generation, numerous detailed studies have appeared which explore the formulation and implementation of administrative, legal, religious, educational, agrarian and economic reforms and make clear the importance of the Enlightenment, broadly-defined, in the genesis of many of these measures.[35] It would require scepticism and even cynicism of a very high order for a historian of eighteenth-century Europe to deny that there was some connection between the intellectual currents of the time and the domestic policies of many states. The extent of that influence and the precise way in which ideas shaped actual policies remain subjects of debate and disagreement, but not the existence of a link between the intellectual context and the reforms attempted.

This has, in part, been made possible by fundamental changes in the way in which the Enlightenment is studied. In broad terms, the older approach exemplified, for example, by Cassirer's *The Philosophy of the Enlightenment*, was of a French-centred movement dominated by great thinkers and of enduring philosophical importance. This has given way to an appreciation of distinct national Enlightenments within a broader European movement and to a recognition that these must be located in their precise social and political contexts and that second- or even third-rate thinkers could be equally or even more important at the time and must be studied.[36] Underlying this shift in emphasis has been a search for the contemporary impact of the Enlightenment, and this has contributed to the revival of enlightened absolutism.

The influence of the distinguished Italian historian Franco Venturi has here been crucial.[37] Both through his own research and massive publication and through the scholarship he has inspired and co-ordinated, Venturi has contributed more than anyone else to the breaking of the mould within which the Enlightenment had long been studied. His rediscovery of the native Italian Enlightenment, with all its intellectual richness and diversity, provided the earliest and most persuasive example of a national Enlightenment. A second and even more important contribution has been his emphasis on the practical impact and contemporary importance of Italian reformist ideas and to the interest of thinkers in the peninsula in practical measures of reform. In an Italian context, he demonstrates exactly how actual policies were influenced, sometimes decisively, by the ideas of Italy's Enlightenment and also reveals the role of particular thinkers in planning and implementing these reforms. For Venturi, the Enlightenment in Italy and in Europe as a whole is primarily a movement of reform and renovation, and must be located precisely in its proper context.[38] Whatever minor reservations may be felt about Venturi's broader perspectives, his contribution has been enormous. His concept of reform extends far beyond the area of government activity and is much wider than that embraced by the idea of enlightened absolutism. Yet his overall approach and his emphasis on locating ideas in their context and in studying practical measures of reform, as well as his programmatic suggestion that we should think of a 'reforming eighteenth century',[39] have been quite fundamental for the revival of enlightened absolutism.

Before turning to the elements in any new theory of enlightened absolutism, two other approaches must be examined. The first is that embodied in recent Marxist historical writing, particularly in present-day Eastern Europe. Though there are obviously important differences of interpretation and nuances of meaning, certain broad points can be made. The general Marxist approach has been to highlight the discrepancies between theory and practice and to argue that enlightened absolutism is a contradiction in terms.[40] Absolutism is seen as essentially aristocratic, being monarchy in the interests of and supported by the dominant social class; while the Enlightenment is assumed to be the ideology of the rising middle class.[41] Enlightened absolutism thus becomes a futile attempt to reconcile the resulting tension and to shore up the feudal system and the power of the nobility by modifying the policies of absolutism to take account of middle-class interests. In this way, the reforms of the later eighteenth century become a series of attempts to promote the interests of the bourgeoisie, but they were bound to fail because of the historic contradictions which they embodied. In the hands of such historians, enlightened absolutism becomes simply an attempt to modernise feudalism and an ideological superstructure, both of 'second serfdom' and of the repressive policies of the state. Crude, simplistic and schematic perspectives such as these are accompanied by an ignorance of eighteenth-century realities, by a general cynicism about the motivation of reform and by a refusal to accept that it produced many – sometimes any – positive results. Though some impressive detailed research has been produced by historians working within this tradition, it cannot be said that the overall framework is satisfactory and it has little to contribute to the study of later eighteenth-century government. The central contention of this approach – that 'enlightened' policies were in some mysterious way a response to 'social forces' – is effectively and decisively rebutted by Derek Beales in Chapter 1.

An altogether more convincing approach has been that of the distinguished historian of eighteenth-century Russia, Marc Raeff, though its implications are again ultimately negative. In an important and widely-influential book, Professor Raeff has recently found the key to enlightened absolutism, particularly in Russia, in the operation of a German-inspired *Polizeistaat* pursuing fundamentally cameralist policies.[42] Two points which he makes can be readily accepted: the extent of continuity between the period after 1750 and earlier generations, and the crucial import-ance of cameralism as a source of government policy. The problem with this approach is that it comes close to arguing – as orthodox Marxists would – that the Enlightenment was an ideological superstructure and that the policies of the reformers were simply the established prescriptions of cameralism. Professor Raeff reduces enlightened absolutism to the implementation of a programme (which he believes was provided by the *Polizeistaat*) whereas many historians, and certainly the contributors to this volume, would rather see it as a process, a matter of nervous interaction between ruler and government, and the problems confronting them. His views are important and must be integrated into any theory of enlightened absolutism, but they cannot in themselves be accepted as a total explanation of government in later eighteenth-century Europe.

III

The chapters which follow make clear the crucial importance of particular national circumstances and even provincial conditions in determining the kind of measures attempted and the degree of success with which they were implemented. They also reveal that the intellectual and political environment within which the reforms took shape varied significantly from one state to another. This is only to be expected. European countries were, in the later eighteenth century, at different stages of historical development and subject to diverse influences and pressures and no over-arching explanation, of the kind embraced by the ICHS enquiry, is possible. Yet several common themes emerge from the national chapters. The most obvious is that the idea of enlightened absolutism cannot itself provide a complete explanation for the reforming initiatives of this period. This is obvious and, it might be thought, self-evident, were it not for the exaggerations introduced by earlier scholarship. The ICHS enquiry in particular assumed that government actions could be explained simply in terms of enlightened absolutism. It believed, in other words, that the internal policies of most European states were guided solely by the Enlightenment. In fact, there were both important continuities and significant new directions in domestic policy during the second half of the eighteenth century. Enlightened absolutism was an additional layer of statecraft and the origin of radical and innovative measures, but there was a significant degree of continuity as well.[43] To recognise important innovations is not to deny that the policies of all states were, to some extent, influenced by traditional and established concerns. Above all, rulers and their ministers sought to recruit soldiers and to raise revenue to finance administration and the Court and, in the major states, to make possible ambitious foreign policies. These aims were well-established and continued to be important, but during the second half of the eighteenth century, they were supplemented by new objectives.

The best and certainly most celebrated example of this blend of continuities and new directions was Prussia, as T. C. W. Blanning makes clear (see below, pp. 266–8). Frederick the Great's domestic policies were to a significant extent those pursued by his father, the notably unattractive Frederick William I (1713–40), whose reign had been crucial for the administrative, military and social evolution of the Hohenzollern state. Yet this continuity was accompanied after 1740 by significant innovations which – as Dr Blanning makes clear – are principally to be explained by the new king's receptivity to the doctrines of the Enlightenment. This amalgam of the familiar and the novel can be seen in the internal policies of many European countries during the generation after 1750. In certain of the Italian states, in Spain and in Russia, some of the reforms introduced during the second half of the eighteenth century built on the achievements of earlier rulers.

An important corollary is that enlightened absolutism cannot anywhere provide a complete explanation of government policy. Though always an important element and frequently a dominant one, reform also proceeded from other, often pragmatic motives. There is an important distinction between the nature and content of the reforms, where the influence of the Enlightenment, broadly defined, was very real, and the reasons which led rulers to overhaul their domestic structures, which could be very different. The desire to strengthen a state internally, to make it more formidable on the international stage, was particularly important in the major powers, and above all in the Habsburg Monarchy and Prussia. The destructive impact of the Seven Years War (1756–63) was crucial and convinced several rulers that new initiatives were imperative: even Spain's brief and ill-starred intervention at the very end of the conflict was a significant stimulus to Charles III's reforms, as Charles C. Noel makes clear (see below, pp. 133–4).

The importance of the Seven Years War has sometimes been underestimated. Its length, scale and destructiveness made clear the urgent need for reconstruction and reform in all the belligerents. Its cost in material terms was everywhere very high. The Habsburg Monarchy, for example, was burdened with a state debt that in 1763 was seven or eight times annual revenue and a peacetime deficit for the foreseeable future.[44] In France, the fiscal and economic costs of the Seven Years War were also enormous.[45] This impact was not merely financial: everywhere the conflict exposed the shaky domestic foundations upon which states sought to maintain their international positions. In the Habsburg Monarchy, it highlighted the shortcomings of the administration and made clear the need for further changes, while in Prussia, across whose territories much of the fighting took place, the extent of material destruction was enormous. Though the Seven Years War provided an obvious impetus for reform, it did not dictate the kind of measures attempted. Enlightened absolutism, as Kenneth Maxwell astutely points out (see below, p. 116), was creative rather than merely reactive: it was the way individuals responded to problems and challenges, the manner in which they confronted the difficulties, that was crucial, and not the circumstances and problems themselves.

This, in turn, modifies the role of the Enlightenment in bringing about the reforms. Previous scholarship too often assumed a linear relationship between ideas and actions and searched for the influence of a particular doctrine on a certain policy. Historians now approach the problem of enlightened absolutism in a rather more sophisticated way and are more aware of its complexities. The Enlightenment is seen as providing the broad intellectual context within which reforms took shape more than the direct source of a particular measure. Viewed in this light, enlightened absolutism becomes a matter of mental attitudes, not of trying to plant physiocratic doctrines in foreign soils. Reformers were usually practical statesmen, aware of what was possible, rather than blinkered idealists embarking on doctrinaire policies. Rulers and their ministers aimed, above all, to govern their states according to the most up-to-date ideas, and they found these principally but not exclusively in the doctrines of the Enlightenment. During the past generation, this recognition has been facilitated by a growing awareness of the practical nature of much Enlightenment thinking and of its considerable impact in governing circles. Linked to this has been a realisation that the French Enlightenment was in some respects less important than other bodies of ideas, notably Italian reformist doctrines and German cameralism and natural law theories.

The contribution of the *philosophes* to enlightened absolutism is now seen to have been far less than once supposed. The distinctive economic theories of the physiocrats were certainly studied and sometimes emulated, and the new spirit of humanitarianism apparent in many reforms owed something to the idea of natural rights. In a similar way, certain of the more radical doctrines of the French Enlightenment can be seen to have contributed to the introduction of religious toleration. But France was a less important source of innovative ideas than the Italian peninsula and the German-speaking lands. The extent, variety and originality of Italy's Enlightenment is now widely recognised and its distinctive doctrines shaped reforms not only within the peninsula itself but in southern Europe and in the Habsburg Monarchy. Germany's contribution throughout the northern half of the continent was even greater. Many of the essays which follow make clear the crucial importance of cameralism. This was a distinctive body of economic ideas which emphasised the primacy of a state's wealth and the prosperity of its subjects. In many ways it resembled the kind of mercantilism associated with Louis XIV's economic minister, Colbert, though it gave more emphasis to the role of government. Its advocates believed that the best foundation for a wealthy and strong state was a happy and prosperous population, and this led rulers to regulate the lives of their subjects in considerable detail. Cameralism had originated during the second half of the seventeenth century and the interventionist policies which it produced were in some respects the forerunners for the reforms attempted after 1750.

But its ideas were only rigorously implemented during the second half of the eighteenth century when they became an important source of internal policy throughout Germany and in Russia, the Habsburg Monarchy and Denmark.

German natural law theories were scarcely less influential.[46] Associated above all with thinkers such as Samuel von Pufendorf, Christian Thomasius and Christian Wolff, they particularly fostered the idea of a social contract by which the people pledged obedience and loyalty in return for a ruler's protection and his efforts to advance their interests. This superseded the older conception of monarchical authority based on traditional patriarchal and religious assumptions and imposed duties on the ruler as well as conferring rights. The natural law theorists also supported the idea of religious toleration and advocated legal reforms in order to define the new concept of the state inherent in their doctrines. This was a radical change in the way in which royal authority was viewed and was both an essential preliminary to the new policies that were adopted after 1750 and a source of particular reforms. Influential in German universities as well as in the education of rulers such as Frederick the Great, natural law was an important source of enlightened absolutism throughout Germany.

This was part of a more fundamental transformation in the political philosophy which underlay government. Proprietary, divine-right monarchy gave way to the new sense of the duties and responsibilities of kingship, epitomised by Frederick the Great's famous aphorism that 'the king was the first servant of his people': a phrase that has often seemed the watchword of enlightened absolutism. The ruler's traditional sense of responsibility for his subjects, rooted in Christian belief and the paternalistic assumptions which had long been part of monarchy, was now accompanied and even superseded by the new doctrine of the social contract as the source of authority. The most radical statement of this was provided by the Grand Duke Leopold of Tuscany. Writing in his political testament, drawn up in 1790 – two years before his death – he declared that 'a ruler, even a hereditary ruler, is only a delegate, a servant of the people whose cares and troubles he must make his own'.[47] One of Leopold's leading advisers in Tuscany, Francesco Maria Gianni, defined the duties incumbent upon the monarch. 'Government is best', he wrote, 'where the Sovereign is best, who knows how to lead, with legislation and administration toward the happiness of his subjects'.[48] These two quotations embody the essence of enlightened absolutism. This was as much a matter of aims and style, of attitudes and intentions, as of policies and actual reforms. Rulers came to approach the craft of kingship in a novel and radical way. They confronted the problems they faced in a new spirit and ruled, or believed they were ruling, in the interests of all of the people. Their aim was to provide better government and to improve the material conditions and advance the prosperity of their subjects. This was the novel element in internal policy during the decades after 1750. It co-existed alongside the state's traditional fiscal, military, legal and administrative concerns.

The success of these reforms must be judged against what was possible in the context of the second half of the eighteenth century. At first sight, this is obvious, not to say a platitude. Historians would now agree on the necessity of a properly historicist approach to the study of the past. Central to the theory and practice of modern historical writing is the idea that every age should be viewed and interpreted in its own terms and not by reference either to some subsequent yardstick or to immutable and timeless standards, however admirable these may be. This has not always been done when examining the problem of enlightened absolutism. The remarkable nature of some of the reforms has been underestimated because they subsequently became a commonplace, whereas in the context of the later eighteenth century they were radical and sometimes startling measures. The efforts to make primary and secondary education more widely available, to weaken the entrenched position of the Catholic Church and to ameliorate the position of the East European serfs were remarkable measures which cannot simply be explained in terms of *raison d'état*.

A particularly good example is the matter of religious toleration, an issue which was pre-eminent for the Enlightenment. It is nowadays taken so much for granted that it is difficult to conceive of a time when this was not so. Yet in the eighteenth century, hatred and vilification of Jews was widespread, and co-existence between Catholics and Protestants far from common and perhaps even unusual. This makes the efforts of Catherine the Great and her leading minister, Nikita Panin, in the mid-1760s to stop the persecution of non-Catholics in Poland appear remarkable, however distasteful the methods of Russian intervention were in practice. In a similar way, the courage and the radicalism of Joseph II's edicts granting a measure of toleration to Protestants and Jews (see below, pp. 167–71) are only fully apparent in the context first of Maria Theresa's renowned Catholic bigotry and notorious anti-Semitism and also of the deep and vociferous public hostility which greeted the measures. In a similar way, we now take law codes so much for granted that there is a danger of underestimating the herculean labours that went into compiling them in the first place. This is also the case with freedom of expression and of the press, which are also axiomatic, at least in modern western democracies. Much has been made of Frederick the Great's restrictions on free speech: in effect, he forbade discussion of political and military matters. But in the context of the later eighteenth century, it is the extent of the debate which he permitted and encouraged that is remarkable (see below, pp. 287–8). This is also true of Catherine the Great, who stimulated and even participated in free intellectual exchange (see below, pp. 302–3).

The correct historical context is also crucial in assessing the impact of the reforms. One established criticism has been that the enlightened intentions of the rulers were not realised in practice and that their aims were seldom translated into actual changes. It can be admitted that the success of the reforms was often incomplete and in some places could be very limited: all the contributors make clear that there were in practice real limitations on the authority of central government in later eighteenth-century Europe. These arose from two particular circumstances: the structure of European states and the relatively small size of the administrations which sought to implement the changes. Eighteenth-century states were far from internally united or uniform structures. On the contrary, the process of national unification, and with it the authority of central government, was everywhere incomplete and sometimes very limited indeed. A host of corporate bodies – towns, guilds, universities, even whole provinces – retained and defended important privileges, which were usually legal and fiscal in nature, and these were a significant barrier to change and sometimes a target for reformers as well. Popular conservatism and apathy were also an obstacle to central government initiatives, particularly when these attacked traditional Catholic religious practices or established forms of social and economic organisation. Yet this is a comment on the limitations of all government authority in early modern Europe and not simply on enlightened absolutism. The chapters in this volume make clear the real success and the scale of the reforms, at least when judged against what was possible in the later eighteenth century. Where the power of traditional institutions was less well-established or simply less important, either in the colonial possessions of Spain, Portugal and France, or in newly-acquired territories inside Europe, the enlightened intentions of the governments were much clearer and their achievements rather more considerable.

The impact of innovative policies was also restricted by the limited number of officials in eighteenth-century administrations which, by present-day standards, were very small indeed. This was an obvious limitation on the success of enlightened absolutism. So, too, was the shortage of trained personnel: not merely bureaucrats to implement the changes but teachers for the newly-established schools and doctors for the protean health services. Everywhere the task facing reformers was massive and the basis of support for innovations dangerously narrow. Rulers and their ministers also encountered resistance in the ranks of their own officials: the unquestioned support of the administrative machines could certainly not be relied on, and the bureaucrats were at times reluctant to implement the reforms. This was particularly important in Portugal, where Pombal had to create his own 'party' among the administration in order to carry through his planned changes (see below, pp. 106–7). Yet the later seventeenth and eighteenth centuries saw a significant increase in the size of state administrations, and this was crucial for the success of the reforms undertaken. The role of bureaucracy in enlightened absolutism was usually important and at times crucial, as it was particularly in the Habsburg Monarchy and in the smaller states of Germany and the Italian peninsula. The dominance of central government by the nobility was broken and the extent of centralisation increased. The enlarged state administrations were increasingly staffed by men with a university or college education. This was often in practical subjects such as law, recent history and, within the German-speaking lands, the new science of government, cameralism. Many of these men were themselves committed reformers, and they provided not merely a cadre of trained bureaucrats to implement the changes but even some of the ideas and policies of enlightened absolutism.

This emphasis on the importance of state administrations qualifies rather than overturns the established emphasis on the crucial role of individual rulers. Enlightened absolutism was predominantly but not exclusively a monarchical phenomenon, as the chapters in this volume make clear. Some powerful ministers were able to carry out wide-ranging reforms with only the tacit support of the monarch: this was most notably the case in Portugal during Pombal's long ascendancy and in Denmark during the exceptional circumstances of Struensee's brief dominance and again after 1784, during the administration headed by A. P. Bernstorff (see below, Chapters 3 and 9). But royal support was everywhere important, and monarchical leadership often crucial, for the progress of reform. Rulers as different as Charles III of Spain, Leopold as Grand Duke of Tuscany, Catherine II in Russia and Joseph II in the Habsburg Monarchy all played a central role in drawing up and carrying through reform programmes.

One recurring theme in these chapters is the extent to which reform in later eighteenth-century Europe must be seen in an international and comparative perspective. This is sometimes disguised by the insularity of national historical traditions, but it is an important dimension of enlightened absolutism. Successful states had always been studied and copied by countries anxious to emulate them. The monarchy of Louis XIV had served as an exemplar in the decades around 1700, and Prussia's spectacular military success after 1740 made Frederick the Great's state a model for his contemporaries to admire and imitate: as, for example, it was in the Habsburg Monarchy (see below, pp. 150–1). The pragmatic, eclectic nature of many innovations was apparent from the widespread willingness of the reformers to study and learn from the experiences of more successful states. Foreign models were everywhere important and could be decisive. This was particularly evident in Southern Europe which now lagged behind the more dynamic north-western countries. A common theme in the Italian states, in Spain and in Portugal (see below, Chapters 2, 3 and 4), was the efforts of once-great countries, aware of the extent to which they had fallen behind, to catch up on more obviously successful and dynamic states. Britain's wealth and power excited most admiration and was studied by Pombal in his attempt to restore Portugal's economic prosperity, while the English 'Agricultural Revolution' of the eighteenth century was one inspiration for the agrarian reforms introduced in Denmark and, to a lesser extent, Sweden. Even more remarkably, the English judicial system and even the writings of Sir William Blackstone were one inspiration for certain of Catherine II's reforms of the Russian legal system (see below, pp. 294–6). The ideas of Adam Smith also exerted significant influence in Russia, Germany and Denmark.

This borrowing could occasionally be very direct and obvious: the overhaul of the Habsburg administration at the end of the 1740s was directly modelled on the Prussian system of government (see below, pp. 152–4), while Catherine II copied directly from Joseph II's educational reforms in her efforts to establish a Russian school system (see below, pp. 301–2). Such direct influence was unusual, as was the simple transplanting of foreign institutions and innovations into native soil. More usually, foreign models provided an inspiration and an example of successful changes, which could then be studied and applied where appropriate. The extent and importance of such interaction was considerable and is one key to the process of reform.

An international perspective is also important in appreciating the wide influence of certain theories. This is particularly true of Italian reformist ideas and of German cameralism, the crucial importance of which is apparent from the chapters that follow. In the Catholic states of Southern Europe and in the Habsburg Monarchy, the ideas of Ludovico Antonio Muratori were crucial in bringing about the notable religious reforms of this period.[49] The influence of the celebrated Italian penal reformer, Cesare Beccaria, was even more wide-ranging and direct. His *Dei delitti e delle pene (On Crimes and Punishments)* of 1764 was widely and immediately read, and it was probably the most influential single work, influencing legal reform throughout Europe. In Northern and Central Europe, the ideas of the German cameralists were scarcely less important. In the German-speaking lands, in Denmark and in Russia, many of the economic, agrarian, social and even educational initiatives of this period were inspired by their teachings. As a generalisation, it can be suggested that in Southern Europe, Italy was the single most important source of reforming ideas, while in the northern half of the continent, German cameralism was equally widely influential. Within this pattern, the place of the Habsburg Monarchy was unique. Its central geographical position and established links with Italy and with Germany ensured that reforming ideas from the peninsula and cameralist teaching from Germany were both widely influential within its borders.

This geographical pattern is reflected in the chapters that follow. Three chapters are devoted to the Southern European reformers, where common themes are provided by the efforts of once-great countries to remedy their backwardness and by the assault on the privileges of the Roman Catholic Church, partly inspired by the ideas and policies of the Italian reformers. The following three chapters examine the vicissitudes of reform in the Habsburg Monarchy, which receives such extensive treatment because of the remarkable extent of the changes introduced – Maria Theresa and Joseph II presided over the most radical series of reforms anywhere in Europe – and because of the sheer size and diversity of its territories, with the implications for the innovations attempted. The final four chapters examine the territories where one obvious source of unity is provided by the obvious influence both of German cameralism and of the ideas of the French Enlightenment. Pressure of space has imposed certain limits on the topics and territories covered, and something must briefly be said about three subjects which do not receive separate treatment: the Enlightenment's influence on international relations, and the impact of enlightened absolutism in Sweden and in France.

IV

An acceptance of the Enlightenment's importance in domestic policy immediately raises questions about its impact on diplomacy and international relations.[50] Rulers and statesmen who pursued progressive and humanitarian ends at home do not appear, at first sight, to have followed these same principles abroad. This discrepancy has, in the past, been seized on by critics of enlightened absolutism. By contrasting the moralistic professions of a Frederick the Great or a Catherine the Great with their unscrupulous diplomatic conduct, they have sought to discredit the whole idea. Any revival of the concept of enlightened absolutism must resolve this difficulty.

The decades when the Enlightenment was at its peak – broadly the half century after the 1740s – were not a noticeably peaceful era, nor was the level of international morality particularly high. The period opened with Frederick the Great's unprovoked invasion and annexation of the Habsburg province of Silesia in 1740 and ended with the second and third partitions of Poland (1793 and 1795) and the outbreak of war between the established European powers and the French Revolutionary régime. These decades saw not merely the actual partition of Poland but the planned partition of Prussia (by the Austrian-led coalition in the Seven Years War). They also witnessed substantial Russian territorial gains in south-eastern Europe by means of two successful wars against the Ottoman Empire and the forceful seizure of territory in peacetime, most notably Catherine's annexation of the Crimea in 1783. Overall, there is something to be said for the view that European diplomacy during the second half of the eighteenth century was characterised by new standards of unscrupulous and cynical behaviour, which were justified in terms of *Realpolitik*.

Many of the most prominent thinkers of the Enlightenment were highly critical of the contemporary practice of international relations.[51] There were clearly important differences between the views of different individuals, but certain common features can be recognised. The *philosophes* in particular found much to criticise in the conduct of diplomacy and in the premises which underlay it. They censured its duplicity, its dependence on military force, the frequent wars which it produced and, more generally, its cynicism and *Realpolitik*. Treaties, declared Rousseau, were no more than 'temporary armistices', alliances mere 'preparations for treason'. In the place of the bad old ways of the diplomats and of the chancelleries of Europe, the men of the Enlightenment proposed to substitute the rule of reason. The *philosophes* came to believe that, in a new political world governed by reason, there would be no need for diplomacy or for wars either. Such optimism was buttressed by certain of the ideas of the physiocrats, who argued that destructive wars weakened not only the vanquished but the victor as well. Rejecting the prevailing mercantilist view that states were competitors for fixed resources, they instead emphasised that countries were united by the economic interests they had in common and stressed the importance of free trade, which they believed would increase wealth. This would not merely remove any commercial motive for conflict, but made wars themselves appear unnecessary, since if states were bound together by trade rather than divided by dynastic and political rivalry, then competition and open hostilities were obviously futile. Enlightenment thinkers visualised a 'new diplomacy', a foreign policy based on moral laws and the power of reason. This, they believed, would usher in an era of peace and harmonious relations between states. Ultimately, diplomacy and diplomats would become redundant, as the need for an active foreign policy simply vanished. In this way the Enlightenment came to stand for peace and prosperity rather than war and destruction. It rejected the idea that international relations should be based on aggression, expansion and *raison d'état*, and instead proposed to substitute morality, justice and pure reason as the bases of external policy, with the eventual aim of abolishing diplomacy. The Enlightenment assumed that, as reason should determine relations between individuals, so it should generate rational laws to govern and civilise relations between states.

At first sight, such thinking had little immediate impact upon the conduct of diplomacy. The foreign policies of Russia and Prussia, and even Austria and Spain, were not noticeably 'enlightened'. The relations between the major European states continued to be guided by *Realpolitik*, aggression and armed might. Yet in two distinct, though not completely separate ways, the Enlightenment did exert a measure of influence on contemporary diplomacy. The first was the acceptance by a small number of powerful individuals of the idea that peace should be the principal aim of foreign policy and that aggression and the use of force should be abjured. Its adherents were to be found predominantly among the states of the second rank. One of the most striking and persuasive examples is the Habsburg Archduke Leopold, Grand Duke of Tuscany after 1765 and, following his brother Joseph II's death, briefly ruler of the entire Habsburg Monarchy (1790–92). Leopold as ruler of Tuscany followed the Enlightenment's dictates very closely in foreign as well as domestic policy (for his internal reforms, see below, pp. 65–8). He was a convinced physiocrat and believed absolutely in the value of free trade. Leopold made Livorno a free port, forswore any active foreign policy, disbanded his army and formally declared Tuscany neutral in all circumstances. His policies while Grand Duke were, to a surprising extent, guided by the doctrines of the Enlightenment. Even more remarkably, Leopold sought to pursue a policy based on the same guidelines when he succeeded his brother as Emperor. Habsburg foreign policy during Leopold's brief reign in Vienna was, to a considerable degree, based on the principles he had espoused as Grand Duke. He immediately ended Austria's involvement in the Russo-Turkish War which had begun in 1787 and he struggled, though in the end unsuccessfully, to prevent armed intervention against the French Revolution and in support of his own sister, Queen Marie Antoinette of France. Leopold's most authoritative modern biographer has argued that the principal theme in his conduct both as Grand Duke and as Emperor was a remarkably consistent adherence to the ideas of the Enlightenment.[52]

Danish foreign policy from the 1750s to the 1790s, under the guidance of the two Bernstorffs, provides a second example of enlightened diplomacy.[53] J. H. E. Bernstorff, foreign minister from 1751 to 1770, and his nephew A. P. Bernstorff, who held this office from 1773 to 1780 and from 1784 to 1797, were both influenced by their dislike of war. This was both moral and religious in origin. Both men accepted the physiocratic doctrine that free trade was essential for the development of states and both pursued pacific and neutralist foreign policies. During the Seven Years War, the elder Bernstorff vigorously defended Danish neutrality and kept Denmark at peace, even though Copenhagen was France's client and was expected to join the anti-Prussian coalition. His correspondence at this time contains several discussions of the moral objections to war and continually emphasises the advantages, both practical and philosophical, of peace. His nephew pursued an even more enlightened diplomacy. A. P. Bernstorff believed that external policy should be conducted in an open, honest and straightforward manner and that, above all, peace should be preserved. He thought that there were moral laws for all mankind which states should not violate, and he was also influenced by a cosmopolitan sense of international brotherhood: the phrase *Patria ubique* ('Fatherland is everywhere') recurs in his correspondence. This underlined his belief that all peoples were members of the same nation and that therefore conflict between them was absurd. Trade and progress, rather than war and destruction, were his aims in foreign policy. A. P. Bernstorff was, of course, the originator of the principles which found expression in the celebrated Armed Neutrality, the league of neutral commercial states which was formed under Catherine the Great's leadership in 1780. This was certainly claimed by the *philosophes* as their child, and it embodied the Enlightenment's preference for free trade and its wish to humanise warfare.

Denmark or Tuscany, as states of the second or even third rank, possessed greater freedom of action and fewer binding commitments than one of the major powers. To this extent the more enlightened diplomacy pursued by Leopold or by A. P. Bernstorff was always unlikely to be emulated by the ruler of one of the major states. There is an obvious analogy here with the limitations imposed by great power commitments on domestic reform in Prussia or Russia (see above, p. 10). Yet, at certain points, the diplomacy of one first-rank state at least was influenced by the teachings of the *philosophes*, and that was France. During the 1740s, in the middle years of the Austrian Succession war, the French foreign minister, the Marquis d'Argenson, attempted to renounce territorial conquests and to promote peace and international brotherhood. The moment for such an initiative was, to say the least, inopportune. France had entered the war with the intention of destroying Austrian Habsburg power once and for all, and d'Argenson's attempts to replace aggression with international co-operation and harmony were notably unsuccessful.

The Comte de Vergennes, who directed French foreign policy from 1774 until 1787, is a better example of the way in which the established motives of great power rivalry came to be supplemented and modified, though never replaced, by the newer doctrines of the Enlightenment.[54] In many ways, Vergennes epitomised the older diplomatic traditions. Yet in some respects he was very much the pupil of the *philosophes*. This can be seen in his conviction that France should not make further territorial conquests, in his recurring emphasis on the need for foreign policy to be based on justice, in his repeated denunciations of the use of force in international relations and in his highly moralistic denunciation of the first partition of Poland. Vergennes' distaste for war did not, however, prevent him taking France into the War of American Independence on the side of Britain's rebellious colonists. This intervention was carefully prepared in advance and exemplified the established *Realpolitik* of the eighteenth century. Vergennes' difficulty in reconciling his enlightened inclinations with the needs of France's great power position confirms the Enlightenment's limited direct impact on the relations of the major states.

Such ideas also had a second, indirect influence on the way foreign policy was conducted. During the second half of the eighteenth century, diplomatic calculations acquired a new precision, and this can be related to the ideas of the Enlightenment. The most celebrated exponent of this more systematic and mathematical approach to diplomacy, known as 'political algebra', was Wenzel Anton von Kaunitz, Austrian State Chancellor and controller of Habsburg foreign policy from 1753 until the early 1790s.[55] This embodied a more logical, rational approach to foreign affairs which became widespread during the decades after 1750. It was as characteristic of Frederick the Great's approach to diplomacy (see Dr Blanning's remarks, below, pp. 279–80) as of that of the Bernstorffs; it was also the basis of the policies of Choiseul (France's leading minister 1758–70) and of Vergennes. Kaunitz and the other exponents of 'political algebra' were convinced that policy could and should be determined by reason, by the power of rational analysis. Their fundamental premise was that the 'true interests of states' were absolute and should be the foundations of policy. Since these interests were comprehensible and amenable to calculation, they enabled the formulation of equally precise diplomatic strategies. An 'enlightened' statesman should ascertain not only the true interests of his own state, but of other powers as well, and alliances could in this way be perfected on the basis of self-interest through the application of reason. The Enlightenment, in the broadest sense, thus helped to make the conduct of diplomacy more rational, systematic, and precise; but only in one or two of the minor states did it also lead to policies designed to promote peace and harmony. Though its impact on diplomacy was always much more limited than on internal affairs, it did have some influence on the conduct of international relations.

V

Something must finally be said about two countries that, because of pressure on space, do not receive separate chapters: Sweden and France. The Swedish King Gustav III (1771–92) certainly possesses many of the credentials of an enlightened absolutist. His education and upbringing had steeped him in French culture and in the philosophy of the French Enlightenment, and he himself corresponded with some of France's leading reformers and visited that country in 1770–71. His tutor, C. F. Scheffer, was an admirer of the physiocrats and made the future king a pupil of their doctrines. Gustav also read Mercier de la Rivière's *L'ordre naturel et essentiel des sociétés politiques* immediately after its publication in 1767 and he became an enthusiastic supporter of its prescription of a strong monarch ruling in the interests of his people. Yet there were influences other than the Enlightenment on the Swedish King. He was, in the first place, a firm admirer of Bolingbroke's idea of a 'Patriot King' who stood above the party strife characteristic of Sweden's 'Age of Liberty' (1719–72) which his *coup d'état* in August 1772 had summarily ended. He greatly admired his uncle, Frederick the Great of Prussia, who provided a practical model of kingship. Gustav III also sought to emulate his famous predecessors on the Swedish throne and particularly the great warrior-king Gustav II Adolf (1611–32), the real creator of Sweden's seventeenth-century greatness. His aim after 1772 was not a restored absolutism of the kind created by Karl XI after 1680 and swept away after Karl XII's death in 1718, but a constitutional monarchy. Gustav III aimed to restore the kind of balance, above all in relations between crown and aristocracy, which he believed had existed before 1680. Yet this constitutional balance was to favour the monarchy. Gustav III's relations with the Swedish nobility were always tense and in the second half of his reign they broke down completely. By the later 1780s, a formidable opposition had emerged and the King was to be assassinated by a group of noblemen in March 1792.

Enlightened reform was only one strand in Gustav III's policies, but it was an important dimension particularly during the first decade of the reign. The 1770s saw a series of initiatives inspired by the Enlightenment. The King had read and admired Beccaria, and he promoted some limited but important legal reforms. Torture was abolished shortly after the August 1772 *coup*; other punishments were made more humane; while the King tried, not altogether successfully, to reduce the number of crimes subject to the death penalty. Measures were also taken to speed up the administration of justice and to make it more even-handed. Gustav III's reign saw the beginnings of religious toleration in Sweden. In 1781 a very limited measure of toleration was given to Catholic and non-Lutheran foreigners, and in the next year, Jews were permitted to live in certain specified towns and were granted freedom of worship.

The ideas of the physiocrats and, to a lesser extent, of Adam Smith were apparent in the economic policies pursued by the King and his advisers. The grain trade was freed in two stages (1775, 1780); barriers to internal trade were removed and import duties lowered; an effort was made to curb the power of the guilds; and a free port was established at Marstrand on Sweden's North Sea coast in 1775, in order to encourage trade with Western and Southern Europe. A new silver-based currency was introduced in 1777 and this restored much-needed monetary stability. Measures were taken to improve the position of the peasants, by encouraging freehold tenure in the belief that this would also benefit agricultural production. The status of illegitimate children was raised and measures were introduced to improve Stockholm's sanitation and to create the beginnings of a public health service. Gustav III's policies closely resembled those of other reforming monarchs, though his enlightened professions and his self-conscious adoption of the mantle of 'philosopher-king' were more obvious than his actual achievements. In one important respect his *régime* was less liberal than the one it supplanted: the freedom of the Swedish press was actually curtailed shortly after the *coup* in August 1772. Many of his policies were less radical than those pursued in other countries, and the initiatives were largely confined to his first decade on the throne. By the later 1770s, the King's enthusiasm for the *philosophes* was waning, and in the 1780s he was pre-occupied with the deepening social, economic and political crisis and with the more adventurous foreign policy which his domestic problems encouraged him to adopt.

The case of France is considerably more complex. An established theme of eighteenth-century French history is not the success of reform but its failure. The conventional view is that this was primarily responsible for the downfall of France's *ancien régime* and even for the Revolution which began in 1789. The failure of enlightened absolutism in France is, at first sight, surprising. France was the homeland of the *philosophes*, the physiocrats and the *Encyclopédie*, and a country which played a central role in the Enlightenment. Yet French central government during the second half of the eighteenth century does not seem to have been much influenced by enlightened ideas, while efforts at reform were usually short-lived and mostly unsuccessful. Some ministers admired the *philosophes*, but none sought to admit them to power. In any case, the leading figures of the French Enlightenment were often hostile to the political and social *status quo* and this reduced their impact on government policy.

There were several more substantial reasons for the limited impact of enlightened ideas. The kind of decisive royal leadership and support for reforming initiatives which was elsewhere crucial for success simply did not exist in eighteenth-century France. Louis XV (1715–74) was weak and lazy, and he played little part in internal government, while his successor Louis XVI (1774–93) was pious and well-meaning, but similarly timid. To the extent that enlightened absolutism was a monarchical phenomenon, it was not experienced in France. Neither king was capable of playing the kind of central, directing role in the implementation of reforms assumed by other continental rulers. The French monarchs were also unwilling to support ministerial advocates of reforming policies when they were attacked at court or by the *parlements*. Louis XV acquiesced – and, indeed, contributed to – the dismissal in December 1770 of the able and energetic Choiseul, leading minister since 1758; while six years later, Louis XVI did nothing to save Turgot – the most plausible representative of enlightened absolutism in France and a minister whose own links with the physiocrats were particularly close. No reforming minister received the kind of support essential if his measures were to be implemented and the resulting opposition overcome. The extent of that opposition was more formidable than elsewhere. In every continental country the authority of central government was imperfect, and resisted by traditional forces within the state, but this was especially true in France. The Catholic Church, town guilds, corporations (such as universities), municipalities, above all the provincial Estates and the *parlements* in Paris and in the provinces: all these bodies possessed powers which it was difficult to override and which were a significant obstacle to reform.

This situation was exacerbated by the French Monarchy's financial plight, especially after the middle of the eighteenth century. This was caused by frequent and expensive wars, together with a tax system which exempted many members of the privileged orders from personal, direct taxation and instead imposed heavy burdens on the peasantry. By the end of the Seven Years War in 1763, the French state was effectively bankrupt, and a desperate situation was made worse by the costly intervention in the American War after 1778. This provided considerable incentive for reform and especially for a modernis-ation of the fiscal system, but it also ensured that any innovations would be resisted by the corporate bodies, determined to defend their individual and collective privileges. In the struggle which ensued, the Monarchy and the central government were usually the losers.

Eighteenth-century France was not completely untouched by enlightened government. In three different respects successful efforts were made to introduce reforms inspired by the ideas of the *philosophes*, though the overall impact was significantly less than in most European states. At certain points, individual ministers were able to carry through some reforms. During the

1760s, the Choiseul ministry implemented a radical reform of the grain trade, freeing it from government regulation in exactly the way advocated by the physiocrats. This remarkable measure was undermined by poor harvests and consequent famine and rising prices, and it was abandoned after 1769, though a further unsuccessful attempt was made to promote free trade in grain in the mid-1770s. Choiseul himself was clearly influenced by the Enlightenment and the thoroughgoing re-organisation of France's colonial empire which he carried out was also directly inspired by physiocratic ideas. In 1767, a measure of free trade was introduced, where previously the strict mercantilist principle that colonial commerce should be reserved to the ships and merchants of France had been upheld. After Choiseul's fall in 1770, the 'Triumvirate' which ruled during the final three years of Louis XV's reign (1771–74) and then Turgot, briefly Controller-General in 1774–76, sought to introduce enlightened measures. But neither was able to secure the kind of consistent royal support needed for enduring changes to be imposed, and this was even more true of the later attempts of Necker and Calonne to carry through reforms.

The failure of reform at the centre should not obscure two successful examples of innovative policies in the localities. The waning of the authority of the crown and central government, especially after the 1750s, gave the provincial intendants considerable autonomy, and some were sufficiently strong-minded to exploit their opportunity. Many of these men were admirers of the *philosophes* and particularly of the economic teachings of the physiocrats. Individual intendants introduced enlightened policies, which aimed to encourage agriculture and manufacturing, to improve communications and to better the condition of the poorest Frenchmen by a variety of humanitarian and social welfare policies. The exemplar of the reforming local administrator was Turgot as Intendant at Limoges between 1761 and 1774. Such enlightened administration was also apparent in the two new territories acquired by France during the 1760s: the duchy of Lorraine, which reverted to the French Crown in 1766, and the island of Corsica, purchased from the Republic of Genoa two years later. The French case here corresponds exactly with the general European pattern. Governments enjoyed a freedom of action in newly-acquired provinces which they never possessed within their core territories, where more account had to be taken of established élites and traditional privileges. The Austrian

Habsburgs, for example, had considerably more scope to introduce radical policies in the former Polish province of Galicia (annexed in 1772) than elsewhere in the Monarchy.[56] In a similar way, both Lorraine and Corsica experienced enlightened government from their new French rulers.

The island of Corsica is a particularly illuminating example.[57] As soon as Louis XV's troops had defeated the Corsicans, the French officials sent to the islands enthusiastically introduced wide-ranging reforms. These administrators improved communications, encouraged trade and industry, introduced humanitarian social policies, and pioneered increased educational provision. They even considered, though never attempted, the physiocratic prescription of a single tax on land, and they sought to increase fiscal and judicial equality. Here, as in Lorraine and in certain French provinces, administrators successfully implemented reforming policies. Overall, however, eighteenth-century France experienced a relatively small dose of enlightened government, and its history was dominated not by the success of reform – as in most other European countries – but by its failure. In this, as in the Revolution which began in 1789, France offers a sharp contrast to the other continental states. During the second half of the eighteenth century, the rest of Europe experienced an age of reform which is primarily to be explained by the ideas of enlightened absolutism.

1. Social Forces and Enlightened Policies

DEREK BEALES

ENLIGHTENED ideas and attitudes, and the policies influenced by them, are commonly explained by twentieth-century historians as in some sense the product of social forces – perhaps as the result of pressure exerted from below by a class or classes or groups, perhaps as the outcome of less easily identified but more fundamental economic and social tendencies. To take a distinguished example, Albert Soboul, one of the most notable Marxist historians of the French Revolution, which he saw as the triumph of the bourgeoisie, was equally convinced that the *Encyclopédie*, the central text of the French Enlightenment, published between 1751 and 1765, should be regarded as a manifesto of the bourgeois spirit.[1] He conceived 'enlightened absolutism' as the characteristic product of those eastern European societies which experienced in the early modern period 'the second enserfment', that is, a strengthening of landlord domination over the peasantry.[2] Other historians have explained enlightened absolutism as an effect or a concomitant of industrialisation – or of 'proto-industrialisation'.[3]

Eighteenth-century writers themselves, however, seldom discussed such social forces, and particularly rarely – perhaps never – when explaining the spread of Enlightenment. They did not, that is, see Enlightened ideas and the policies influenced by them as the transient products of contemporary social structure or of particular social and economic – or even political – pressures. On the contrary, these authors imagined – as for example d'Alembert maintained in the *Discours préliminaire* of the *Encyclopédie* – that Enlightened views, strengthened by scientific and other discoveries, had been spreading remorselessly through European society at least since the time of the Renaissance, essentially by their own force as ideas.[4] Voltaire in his innovative history of the world, the *Essai sur les Moeurs* (1756), declared that,

37

apart from war, *opinion* had proved the most powerful force in modern history; and by *opinion* he clearly meant the prevailing attitudes of the educated rather than any form of active pressure from below.[5] The chief contributory factor mentioned, as for example by Beccaria in his *Crimes and Punishments* of 1764, was the invention and agency of printing.[6] It was Enlightened writers who originated the notion of a seventeenth-century scientific revolution, they saw themselves as its heirs, and they conceived the spread of Enlightenment to be essentially its consequence.[7] For some authors it was merely or largely a question of the triumph of reason or of self-evidence – *évidence* as the French economic theorists known as the physiocrats called it[8] – or of an unquestionable good. It was supposed that when rulers and their advisers adopted an Enlightened programme their personal intellectual convictions had much to do with it. The more optimistic writers believed that, once governments embarked on such policies, and especially wherever a measure of religious toleration and some freedom of expression existed, the spread of Enlightenment would gather momentum, and so promote further cognate legislation.[9]

Even those writers in the tradition of Montesquieu who related many of men's varying attitudes and customs to the nature of the particular society they lived in, and condoned practices like polygamy if they were characteristic of a civilisation remote in time or place, did not treat the central ideas and policies of the Enlightenment in the same way, but saw them as absolutely and universally valid. Montesquieu himself was obviously torn between description and prescription. Here are some of his remarks in *De l'esprit des lois* (1748) on the subject of punishments:

> The severity of punishments is fitter for despotic governments, whose principle is terror, than for a monarchy or a republic. . . .
> It would be an easy matter to prove that in all, or almost all, the governments of Europe, penalties have increased or diminished in proportion as those governments favoured or encouraged liberty. . . . I was going to say that [torture] might suit despotic states . . . – but nature cries out aloud and asserts her rights.[10]

Adam Ferguson put forward in his historical writings the strikingly modern view that each age should be judged on its own terms, but he still did not make connexions in the twentieth-century manner between Enlightened attitudes and contemporary social forces.[11]

Contemporary critics of Enlightenment also missed this trick. Rousseau himself, in all his denunciations of civilisation, inequality, cosmopolitanism, rationalism and Enlightenment, did not suggest that they were the product of social forces as modern historians use the term.[12] A still more virulent critic of the *philosophes*, Linguet, particularly criticised their readiness to attack the institution of slavery, claiming that civilisation, and Enlightenment itself, rested on the subjection of the majority of men. In other words, he was exceptionally aware of social forces and constraints, so much so that he ranks as an inspirer and precursor of Marx.[13] Yet he believed that Enlightenment, far from arising out of the existing state of society and furthering the self-interest of the Enlightened, challenged the former and contradicted the latter. Even for him, therefore, ideas had a life of their own, and purely intellectual developments might have powerful, if often baleful, effects in practice. Herder came nearest to dismissing the Enlightenment as time-bound, but he too saw it as a movement of thought divorced from social forces and deserving of criticism precisely for that reason.[14]

So far as there was awareness that social forces were related to Enlightened policies, it was the hostile rather than the friendly forces that were noticed. Frederick and Voltaire talked obsessively of their battle against the *infâme*, that is, 'superstition' and 'fanaticism', the Church, theology and religion, especially in their Roman Catholic versions. Monks and nuns in general, and Jesuits in particular, were very commonly seen as obstructing the progress of Enlightenment.[15] So were more secular vested interests like the *parlements* in France and the Estates in other continental countries, or guilds and Universities.[16] With regard to serfdom, three of the more radical men of the Enlightenment, Frederick II of Prussia and Joseph II of Austria for their own dominions, and Rousseau for Poland, were all agreed in the 1770s, despite their widely differing standpoints, that, however desirable its abolition might be in principle, no rapid or radical reform could safely be attempted, given the strength of aristocratic resistance and the fact that the institution had to be seen as part of the constitution.[17]

As for the people at large, all thinkers admitted them to be in general hostile to Enlightenment.[18] True, there were occasions when pressure for certain Enlightened policies was seen to come from below. As Joseph II travelled round his territories, he made himself available to all his subjects of whatever class, and thus collected tens of thousands of petitions. Of these, though many were protests against change, some advocated measures of an Enlightened character, such as the abolition of tax-farming – though generally on grounds of personal hardship rather than intellectual persuasion.[19] Some of the demands made in the great peasant revolts of the 1770s in Bohemia and Russia may be classed as Enlightened.[20] More specific cases of pressure for Enlightened reform include the petition of the medical faculty of the University of Vienna in 1773 against too severe torture, and the request from a group of Moravian Protestants in 1777 for religious toleration.[21] But it was notorious that, even in the relatively tolerant intellectual climate of Britain, popular outcry in the shape of the Gordon riots prevented mitigation of the anti-Catholic code in 1780; and the same had happened with the relaxation of laws against the Jews in 1753.[22] Petitioning was a weapon that could be used by opponents as well as supporters of Enlightenment, as the Duke of Parma showed when he deployed it in his campaign to get rid of his Enlightened minister, Du Tillot, in 1771.[23] Within months of freeing the Danish press, Struensee saw it join in hounding him from office.[24] The rebellions of the period could be wildly reactionary: the Belgian insurgents of 1787 asked for a return to the position of two hundred years before, and the Hungarians in 1790 insisted on the restoration of Latin as the language of administration.[25] Even the majority of the French Third Estate's *cahiers* of 1789 have been described as profoundly conservative.[26]

Further, it was always a question whether a government was right to give in to pressure, and especially to rebellion, whatever the nature of the opposition's demands.[27] Joseph's accessibility to his subjects was deplored by an Enlightened minister like count Firmian in Milan, and even by an advanced ruler like Leopold of Tuscany who claimed to favour the introduction of representative government everywhere.[28] Joseph's aim in soliciting petitions was not to find out what his subjects wanted and then do it. The hostility he found to the tax-farm in Milan only strengthened his hand to impose on a reluctant bureaucracy a policy he was already advocating. Though in the 1770s he had bowed to aristocratic pressure on the question of abolishing serfdom, in the 1780s he tried after all to undermine the institution. And he told the recalcitrant Estates of Brabant in

1789: 'I do not require your permission for doing good.'[29]

Contemporaries, then, had some conception of the relation between Enlightened policies and such social forces as opinion groups and even classes. But their awareness of general social and economic tendencies was minimal, and often mistaken. To take a classic case, population was rising in most places most of the time in the eighteenth century. Yet it was well into the century before observers realised it. They were simply ill-informed; they could not know better before the days of full and regular censuses.[30] The physiocrats and other early economic theorists were convinced that only agriculture among economic activities could create wealth. Here such experts had much justification, given the state of many contemporary economies. The first classical economists, Adam Smith included, not merely failed to predict an industrial revolution, they thought such a development impossible. Again, their mistake is understandable. Among other things, they could not be expected to foresee the massive exploitation of mineral fuel and power resources. Moreover, recent writing about 'the first Industrial Revolution', the British Industrial Revolution, has shown how little impact industrialisation actually made, even in the country most affected, until far into the nineteenth century.[31] But the track-record of eighteenth-century thinkers in dealing with economic developments is not impressive. So, although contemporary discussion must always be taken seriously, and its apparent shortcomings explained rather than condemned, historians must plainly go beyond it in their attempts to analyse the relationship between social forces and Enlightened policies.

In this chapter I shall try to throw light on the problem by considering certain examples of Enlightened policies on the basis both of contemporary and modern analysis.

Voltaire, the arch-priest of the French Enlightenment, agreed with Kant, the greatest figure of the German *Aufklärung*, that at the heart of Enlightenment lay questions of religion; and the great rulers who are known as Enlightened shared their opinion.[32] To enlarge religious toleration, to reduce the influence of the clergy and churches generally, to exclude it altogether from a growing range of affairs now conceived to be purely secular, to control the study of theology, to attack what was seen as 'superstition' and 'fanaticism' – these were always and everywhere aspects of Enlightened statesmanship. Since this is almost the only generalisation that can validly be made about

Enlightened policies across the whole of Europe, this seems to be the best area to take for discussion. How far, then, did social forces contribute to the adoption and implementation of such policies? I shall take certain crucial instances which happen to have been well studied: the Calas case in France, the progress of toleration in Hamburg, Joseph II's toleration edicts of 1781–82, and the position in Britain's North American colonies.

Jean Calas, a Protestant citizen of Toulouse, was executed there in 1762 by the barbarous method of being broken on the wheel, having been convicted of the murder of his son. The motive for his alleged crime was said to have been that his son was on the point of converting to Roman Catholicism. Voltaire took up the case, made it into a great public issue in France and elsewhere, and was largely responsible for the quashing of the verdict in 1765.[33]

The case is best approached through Voltaire's writings. In the *Traité sur la tolérance* of 1763 he castigates the bigotry of the Roman Catholics of Toulouse, which he declares responsible for Calas's unjust condemnation and cruel punishment. But, whatever had happened in the time of persecution under Louis XIV before and after he had revoked the toleration Edict of Nantes in 1685, during the decades immediately before the Calas case relations between the Protestants and Catholics of the city were generally easy, even cordial; the penalties of the laws against Protestants were seldom invoked; Catholic 'fanaticism' appeared to be diminishing; and although religious prejudice played a part in the judgement of the Court against Calas, it has to be recognised that legal procedure was weighted against a defendant of any faith, especially if (as Calas did) he changed his story from one implausible version to another; and the barbarity of the punishment had nothing to do with the victim's creed. Such religious hysteria as was aroused by the case in Toulouse seems explicable largely as a transient product of the hardships and defeats experienced in the war that France happened to be waging against Protestant Powers. As a witch-hunt, the short-lived and localised campaign against the Protestants around Toulouse paled into insignificance beside the exactly contemporary crusade against the Jesuits, who were expelled from France on the initiative of the *parlements* with scant regard for truth and justice, to the applause of the *philosophes*.[34] Within a decade after the reversal of the verdict against Calas by the *parlement* of Paris, the *parlement* of Toulouse itself was leading the way in bending the law in the interests of Protestants. Prejudice such as Calas encountered seemed to be becoming rarer. Over the whole of France in the eighteenth century, down to the Revolution, with occasional setbacks, religious tolerance appeared to be growing, as part of a general tendency towards greater rationality and secularism and away from 'superstition' and 'fanaticism'.[35]

Some of the chapters in Voltaire's *Traité* appear to argue for virtually complete toleration of all religious or irreligious opinions that are not themselves actively intolerant and politically subversive – including those of Buddhists, Deists and Jews. But, when it comes to specific demands, he does not venture to ask for civil equality even for French Protestants. He proposes merely the introduction of a measure resembling the restricted toleration extended by the Act of 1689 to some Nonconformists in England, that is, the right to worship and teach and practise some professions, but not the right to any public office. He does not even go so far as to advocate the introduction of a French version of the Indemnity Acts which had further eased the position of English Dissenters, the first of which was passed when Voltaire was in England in 1727.[36] This tension within his argument is revealing. He evidently realised that his own view, that a man's private and peaceable opinions and religious practices were a matter of indifference to the state, had no prospect of commanding general support in France. The Protestants themselves hesitated to put forward demands on their own account, for fear of provoking a reaction and making their actual position worse. Public opinion as a whole had certainly not adopted the views of radical figures of the early Enlightenment like Bayle and Locke, or of Voltaire himself. Nonetheless, opinin had moved significantly since the days of Louis XIV's *dragonnades*. It was a question of the slow defusing of religious issues, of a growing acknowledgment, found among clergy as well as laity, that some matters once thought to be matters of earthly as well as spiritual life and death need not be taken quite so seriously, of a slow and very incomplete approach to secularisation. Ultimately, in 1787, encouraged by Joseph II's measures, Louis XVI issued an edict of toleration that gave some legal recognition to this limited but momentous change of spirit and behaviour.[37] During the Revolution, however, it soon became clear that to take toleration further, to give Protestants civil equality, was too strong a step for many Catholics to accept.[38]

In the Lutheran imperial city of Hamburg, as compared with

France, a greater, though still very small, degree of legal toleration existed even at the beginning of the eighteenth century, as a result of the struggles of the Reformation, the Thirty Years War and the terms of the Peace of Westphalia of 1648. Some non-Lutherans were allowed to worship publicly and to enjoy certain rights of citizenship. The city became the centre of German-language Enlightenment in the second half of the eighteenth century. A movement to extend toleration started in the 1760s – at almost the same moment as the Calas case – involving a Patriotic Society and pamphlet and pulpit warfare. Eventually a small concession was made to the Calvinists in 1785. In all the debates on the subject no one seems so much as to have mentioned Calas or Voltaire. Equally, the long-standing and wide-ranging toleration notoriously practised by Frederick II seems to have exerted no influence. The admiration felt for Frederick by most German Protestants was tempered in the case of Hamburg by the bullying attitude he adopted towards the city for political and commercial reasons. It was of some importance that an adjoining territory, the port of Altona, controlled by Denmark, enjoyed a much wider measure of toleration, designed to attract trade and inhabitants away from Hamburg. The economic disruption caused by the Seven Years War may have promoted change. Unlike in France, some of the non-tolerated churches themselves dared to campaign for relief. But the most effective forces seemed to be, first, a slow decay of religious ardour and strife, associated with a diminution in the perceived role of religion and the clergy – as in Toulouse and, more widely, in France – and secondly the impact of the toleration edicts of Joseph II, who as Holy Roman Emperor was Hamburg's nominal sovereign.[39]

Joseph had very little power as emperor, though what influence he could exert over the largely independent German states was used to extend religious toleration.[40] However, from 1780 to 1790 he exercised more or less absolute power as ruler of an enormous collection of inherited lands including all of modern Austria and Hungary, most of what are now Belgium, Luxemburg and Czechoslovakia, and substantial portions of present-day Poland, Romania, Yugoslavia and Italy. His mother and predecessor, Maria Theresa, though she gave up trying to convert Protestants and others to Roman Catholicism by force, continued to exclude them from office in nearly all her territories, allowed non-Catholic worship only where treaties required her to do so, and, when heretics were discovered in her relatively prosperous central provinces, had them transported as colonists to inhospitable parts of Transylvania and the Banat of Temesvar. Joseph told her that her policy was 'unjust, impious, impossible, harmful, ridiculous'.[41] His own edicts varied a little from province to province, they were not in all cases fully publicised, and the resounding principles of their preambles were belied by the limitations of their detailed provisions: in most of his territories, among Christians other than those in communion with Rome, only the Greek Orthodox, Lutherans and Calvinists were affected; and they were allowed merely to build meeting-houses that did not too closely resemble churches, in certain circumstances to run their own schools, to engage in hitherto restricted occupations, and to hold office. But these provisions, restricted though they seem, went much further than any other legislation then operative in Catholic states, and further than most Protestant countries. Joseph's treatment of the Jews, permitting them to enter a wider range of employments and to set up their own schools, although it was directed towards assimilating them into the German-speaking population at large, was far more sympathetic than any other ruler's. The introduction of civil marriage in 1784 made the legal position of non-Catholics better still.[42]

The emperor himself sometimes expressed views on toleration that sounded as radical as Voltaire's, though they derived as much from German and Italian writers like Thomasius and Muratori as from French. He maintained that, within certain limits, an individual's religious beliefs and observances should not concern the state. At other times, however, he talked of toleration as a means to recover dissenters for Catholicism. He was much influenced by the desire to encourage immigration into the remoter and sparsely inhabited parts of the Monarchy, and to discourage emigration to more tolerant states, partly with a view to enlarging his army, and partly from the conviction, instilled into him by his tutor, Beck, that states which persecuted and drove into exile worthy and industrious subjects, as France had done in 1685 and Salzburg in 1732, were bound to suffer economically.[43]

Some religious groups had ventured to put pressure on him and on the government, particularly the Greek Orthodox of Transylvania and the Protestants of Moravia. When some of the latter requested toleration at the end of Maria Theresa's reign, Kaunitz, her chief minister, told her that it was 'quite impossible in the present state of affairs for her not to grant it to the dissenters of Moravia, as [will be the case] elsewhere when sooner or later the same state of affairs comes to pass'. She rejected this advice, and she did not live long enough for its validity to be tested.[44] 'Fanaticism' had somewhat diminished in the Habsburg lands: for example, executions for witchcraft at last died out even in Hungary in the 1750s, when 'the belated sea-change of an enlightened mentality among the educated anticipated radical legal reforms under Maria Theresa'.[45] But in several parts of the Monarchy, such as the Netherlands,[46] it is plain that any pressure from Protestants in favour of toleration was more than balanced by hostility to it on the part of Catholics. Joseph's toleration edicts clearly went far beyond what public opinion in his lands would have demanded. He had the support, it is true, of a small party among the nobility and clergy and of many government officials, some of whom had studied in Protestant Universities and most of whom had been influenced by the writings of the 'cameralists', theorists whose rationalistic manuals on statecraft were widely used in the training of the army of bureaucrats required to run the more than three hundred German states.[47] His radical relaxation of the Viennese censorship in 1781 led to the *Broschürenflut* ('flood of pamphlets') of the next few years, and most of the pamphleteers supported him on religious issues.[48] Since his toleration edicts survived the *débâcle* of the rest of his policies at the end of his reign, it would appear that the pamphleteers had by then persuaded opinion to accept them. One of the writers claimed as much: 'This whole Reformation, the introduction of toleration, would never have been brought to pass so easily if the pamphleteers hadn't bombarded the people's minds with so many booklets, and said ... exactly what the people needed to know.'[49] But the initiative of the emperor himself was fundamental. The Moravian Protestants only put forward their demands because of his known sympathy with them.[50] If the social conditions of the big city made possible the *Broschürenflut* once the censorship had been relaxed, the history of Austria before and after the reign of Joseph II makes it clear that the forces in favour of liberalisation could be contained with ease if the government so wished. Indeed, Joseph himself learned from the Belgian revolt and Hungarian opposition that in the end he did need – and would be unlikely to receive – his subjects' 'permission to do good'.

It is almost as though social forces were operating at one

remove. In certain countries – Britain, Holland and parts of Germany – social forces in the sense of religious groups and associated political factions, with *some* relation, though an equivocal one, to social classes and economic development, had played a part in promoting what came to be regarded as Enlightened policies, or in creating a situation favourable to toleration – but this had happened in the sixteenth and seventeenth centuries, before the days of so-called Enlightenment. Then, later, self-consciously Enlightened rulers in other countries – without much encouragement from indigenous social forces, because religious dissidents were weak and economic and social development was less advanced – carried such policies further than they had been taken in their original homes.[51] In some cases, paradoxically or ironically, these initiatives in turn exerted influence back on more developed and yet apparently more conservative countries, as Joseph's edicts did in Belgium, France and Hamburg.

At least that seems to be the pattern until one looks across the Atlantic. Several of the colonies that became part of the United States had been founded by religious groups not tolerated in the mother country. Though there were established churches in some colonies, they did not compare in power, status and wealth with those of Europe. By the end of the seventeenth century some states were already exceptionally tolerant, and the Quaker foundation, Pennsylvania, accorded full religious liberty. Hostility to church establishments in all their aspects was naturally strengthened in the struggle against Britain, especially since most of the religious conservatives were also loyalists and many of them emigrated during and after the Revolution. It proved easy to make the new United States, by their constitution of 1787, officially secular; and something approaching complete religious freedom was achieved. Here social forces – among the whites, that is – can be said to have made possible the realisation in the New World of a programme of Enlightenment devised in the old. But the forces concerned included the obscurantist demands of rigid Puritans; it was a question of sect and religion rather than of class and *philosophie*; and it owed nothing whatever to industrialisation.[52]

The story of toleration has something in common with the story of numerous other issues that were seen as related to religion. Many of the great men of the Enlightenment attacked – to take some examples – the judicial use of torture, the indiscriminate application of the death penalty, the penalties imposed on the corpses and families of suicides; and belief in the authenticity and efficacy of religious relics, in witchcraft, diabolic possession, exorcism and so on and so forth. In many states these practices and attitudes were decaying slowly, though unevenly, without the direct influence of abstract thinkers being easily discernible, and without much legislative encouragement. This happened with the treatment of suicide and suicides, and with the exaction of the death penalty, in both France and England. In England, indeed, the strange situation prevailed that Parliament, if less bloodthirsty than has sometimes been claimed, was nonetheless continually making more and more offences capital, supported by at least some elements of public opinion, while fewer and fewer executions were actually taking place, presumably in response to growing humanitarianism.[53] The root-and-branch proposals for criminal law reform made by Beccaria can be seen as extreme extensions of tendencies already operating, for example in Frederick's Prussia, and his ideas owed much to the progressive circle in which he moved. But they cannot be said to have had very rapid effect in his own country: in the 1770s Maria Theresa and Joseph abolished torture in their central lands, but could not induce the legal authorities in Milan to do the same; and Beccaria himself was never given much scope as a government servant.[54] In Russia, on the other hand, Catherine the Great tried, and with a degree of success, to impose some of Beccaria's recommendations from above, thus creating or much extending a body of sympathetic opinion. The main social force on her side was a section of the aristocracy, seeking recognition of their own civil rights.[55] So, in many parts of Europe, a growing 'social force' favouring Enlightened policies was visible, a movement towards rationality, secularisation, humanitarianism and sentimentality – a movement supported by, but wider than, such institutions as reading societies and Freemasons' lodges. Among its sympathisers were Enlightened clergy, aristocrats and bourgeois, but it was opposed by many, probably most, members of the same classes.[56] It sometimes gave Enlightened rulers and writers a fair wind. They in their turn, by proposing and enacting stronger measures, could cause it to blow more strongly. But if they went only a little too far, they would suddenly find themselves caught up in a reactionary hurricane.

Toleration is unusual among Enlightened policies in that war, diplomacy and international contacts played a large part in its history. The degree of toleration found in the model states of

England, Holland and the United States was to a large extent the outcome of armed struggle. Many German states were only as tolerant as the Peace of Westphalia required. It was the terms under which the Habsburgs acquired Hungary and Transylvania that ensured their grudging toleration of Protestantism in both provinces before 1781, even extending to Unitarianism in the case of the latter. In late eighteenth-century Poland, the land of 'second enserfment' *par excellence*, Roman Catholicism was gaining ground, and the limited toleration previously established was threatened. Its survival largely depended on the intervention of Russia and on the dismemberment of the country in successive partitions after 1772, a fact which Voltaire adduced in lending his support to Catherine's Polish annexations.[57] Suppression of the Jesuits was another policy whose success owed much to considerations of power-politics: it was adopted by Maria Theresa chiefly to please her Bourbon allies, and without her acquiescence the pope would not have dissolved the Society.[58]

In most fields, though, war and diplomacy mattered little, but there remained an important element at least of international imitation. The south and east of Europe commonly lagged behind the west and north, especially in the early stages, though later Prussia and Russia – and then America – were set up as examples to more conservative regimes. Study of each policy, of course, reveals a somewhat different pattern. An exceptionally interesting instance is inoculation against smallpox, the only major advance of eighteenth-century medicine. The method adopted was 'variolation', infection of the subject with smallpox itself. This practice appears to have begun in Europe within the purview of folk medicine, that is among the superstitions that the Enlightenment derided, though it had also been known in the more respected ambiences of China and the Near East. The first European country in which it became acceptable among the upper classes was the England of the 1720s, and Voltaire backed it after seeing its operation there. The medical profession generally considered it dangerous and of doubtful benefit, and an otherwise progressive doctor like the elder Van Swieten in Austria stubbornly opposed it. Many Jesuits were in favour, while some radicals were against. The English example continued to be important at later stages of its diffusion. Probably nothing did more to commend it in Austria – apart from the frightful death-rate from the disease – than the patronage of English royalty. Maria Theresa, after nearly dying of it herself and losing several of her close relatives from it, in 1768 accepted from George III the loan of the services of Dr John Ingenhousz, had her family inoculated and arranged an inoculation feast at Schönbrunn for a group of sixty-five children. This ceremony – highly utilitarian compared with the usual Habsburg round of pilgrimages, retreats, crib-blessings and foot-washings – owed little either to the great Enlightened thinkers or to any social force other than the broad movement of opinion already discussed. In the same year Catherine II had herself inoculated by another Englishman, Dr Thomas Dimsdale.[59]

Outside the range of the battle for some measure of religious toleration and against the more obvious superstitions, most issues divided the Enlightened. The disunity of the *philosophes* and the *Aufklärer* is particularly obvious in the realm of economic policy. If there could be broad agreement on the abolition of guilds and of impediments to internal trade, support for international commercial freedom, a policy passionately advocated by the physiocrats, was not unanimous. The numerous attempts to impose free trade in grain, made by those who believed in its manifest rightness, were generally frustrated in the end by opposition and riot. In so far as this policy was recommended as a result of economic and social pressure rather than as the fruit of new economic theory, it favoured and was favoured by the large capitalist, probably noble, agriculturalist rather than by the peasant, industrialist, merchant or consumer. The classic case is the French: in 1763 free trade in grain was enacted as part of a radical new departure in economic policy made acceptable by the indebtedness of the country at the end of the Seven Years War. After a brief honeymoon period, the shortage of corn in a succession of bad years aroused the fury of the consumers and awoke the scepticism of the Enlightened. The measure had to be repealed in 1770.[60]

Attempts have been made to establish a close linkage between 'proto-industrialisation' and certain Enlightened policies in the Habsburg Monarchy.[61] Both Maria Theresa and her son, in common with other Catholic rulers, succeeded in reducing the very large number of saints' days which workmen were allowed to treat as holidays. The two rulers also introduced in and after 1774 a scheme to establish enough schools to make possible universal primary education in their dominions. It has been argued that both these reforms arose from the necessity to discipline the growing number of industrial workers, mainly engaged in manufacturing textiles, especially in Bohemia. So far as the reduction of saints' days is concerned, it is certainly true that the principal motive of the government was to extract more work from its subjects. But the measure was not designed with special reference to industrial workers; the problem was conceived to be universal, which meant that in the Monarchy it primarily concerned agriculture. As for education, the motives were much more complex. The ignorance of the mass of the population alarmed the rulers for many reasons, because it handicapped the development of industry no doubt, but much more because it hampered agricultural and commercial development, more still because it made the vast conscript army less efficient and narrowed the state's choice of servants, and perhaps most of all because it implied a lack of knowledge of Christianity, and particularly of Catholic doctrine, among the ordinary people of the Monarchy.[62]

In the context of this discussion, it is a striking fact that most *philosophes* and *Aufklärer* were against universal education precisely because they thought it hopeless to expect Enlightenment to permeate a whole nation.[63] Some writers, it is true, believed that universal education could create an enlightened nation. One such, probably Baron Holbach, proclaimed this belief in his *Essai sur les préjugés* of 1769: truth is made for Man and should always be uttered; the multitude can and should be rescued from superstition; society can and should become more equal. Frederick the Great was stung to publish a reply: 'prejudices', he declares, 'are the reason of the people'; it is unsafe to tell the people the whole truth; the clergy will always have more power than the *philosophes*; the first essential is to maintain respect for the ruler and the aristocracy. The view that Enlightenment must necessarily be confined to the elite was widely shared, if seldom so brutally stated. Diderot denounced the king's assault on the *Essai*, but not in print.[64]

There is, though, a sense in which Enlightened policies were actually embodied in what could reasonably be regarded as a social force. I am thinking of the bureaucracy, especially in what Marc Raeff calls 'the well-ordered police state', 'police' in this context being an attempt to translate *Polizei* or *police* as used in the eighteenth century, when it almost meant 'civilisation' or 'ordered society'. The states of Germany and in the German tradition, with Prussia at their head, developed relatively large bureaucracies which spawned and tried to enforce vast codes of regulation, often restrictive, sometimes ridiculous, not always effective, but at least in part justified on theoretical grounds by cameralist principles taught in state-controlled Universities, and tending to order, rationalise and secularise society in the name of a ruler who had identified himself with an impersonal state. They developed a 'bureaucratic ethic', sometimes obstructive and tradition-bound, but at its best or most creative uncorrupt, selfless, benevolent, progressive and meritocratic.[65] The notion and practice of 'the career open to the talents' have been too little studied. How does it come about that Joseph II could write in 1763: 'Everything to personal merit – if this rule is inviolably observed, what geniuses will not emerge who are now hidden through slackness or noble oppression!'[66] Such sentiments were appearing in English radical circles at the same time, but they can hardly be imagined yet in the mouths of England's or France's monarchs – or even in Frederick the Great's, despite his introduction of competitive examinations for the civil service in 1770.[67] The article *Génie* in the *Encyclopédie* contains nothing so subversive. It is difficult to see how Joseph's ideas can be products of 'the second enserfment', or of bourgeois influence, or of proto-industrialisation.

Historians could not get very far in discussing the relationship between social forces and Enlightened policies if they acquiesced in the more extreme criticisms sometimes made of any historical analysis that uses modern concepts. It is only too evident that contemporaries failed to grasp fully what was really going on.[68] Not only were they at sea about economic developments. I am not sure that any eighteenth-century discussion of the suppression of the Jesuits allowed for such a basic factor as the self-interested hostility of other religious orders. I should be surprised to learn that anti-clericalism was an eighteenth-century concept, yet we can hardly doubt that it was an eighteenth-century force. But it is possible to go too far from contemporary explanations. The Enlightenment was by definition an intellectual movement, and intellectual history possesses a degree of autonomy. Sometimes Enlightened policies were imposed from above that were more advanced than were being demanded by, or were appropriate to, the balance of social forces in the society concerned; and this happened essentially because of intellectual development at the top of society, perhaps among the elite as a whole, more commonly among a limited group, and in some cases largely because of the ruler's own attitude. Below that level, but not among the uneducated, a much more dilute and generalised Enlightenment was slowly and unevenly seeping through society and affecting policy, and that too represented a partly autonomous change of attitudes. That both movements – which were of course interrelated – were limited in scope, at least in Europe, was shown by the numerous bouts of popular hostility aroused when Enlightened measures were hastily or inopportunely introduced. The Enlightenment seems actually to have widened the intellectual gap between educated and uneducated, rulers and people.[69]

Education does not always lead to Enlightenment, but without education there can be no Enlightenment. The fact that the mass of the people so often resisted Enlightened measures reflected the limitations of eighteenth-century popular education. In all European countries it was necessarily in the hands of the clergy, Catholic or Protestant. When an Austrian minister, count Pergen, fought to exclude monks at least from secondary teaching, Maria Theresa, Joseph and Kaunitz all agreed that the thing was impossible.[70] When the Society of Jesus was dissolved in 1773, the Catholic states lost their best teaching instrument. Afterwards, individual ex-Jesuits remained indispensable to their educational systems, though the rulers of Prussia and Russia enticed as many of them away as they could.[71] The alternative to monks and nuns as teachers was not lay men or women, but non-monastic clergy. The pessimism of so many *philosophes* about the possibility of mass Enlightenment becomes more comprehensible against this background. It was virtually impossible in the eighteenth century to imagine a society capable of establishing, let alone affording, any kind of secularised system of universal education. Yet, with education in the hands of clergy, there seemed no hope of enlightening the masses.

2. The Italian Reformers

M. S. ANDERSON

ANY brief discussion of efforts at enlightened reform in eighteenth-century Italy is difficult because of the deep and long-standing divisions within the peninsula, divisions which make satisfactory generalisation almost impossible. The most obvious aspect of these divisions was the splitting of Italy into a considerable number of separate states, many of them with long and proud histories of their own. All these states were now weak, most of them very weak, by the standards of the major European powers: many of them during the eighteenth century fell under the rule or dominant influence of foreigners. Thus the kingdom of the Two Sicilies, territorially much the largest, was a Habsburg possession from the Spanish Succession war until 1734, when Spanish control was reasserted to last for another generation. The duchy of Milan (Lombardy) was in Austrian hands from early in the century until the French conquest in 1796–97; and from 1737, with the extinction of the native Medici dynasty, the grand duchy of Tuscany was also a Habsburg possession, though never formally annexed to the main block of central European territories. Of the smaller states Parma was assigned to a Spanish Bourbon prince in 1732, annexed by the Habsburgs in 1738 and given independence once more under another Spanish Bourbon in 1748, while in Modena the Este family survived as rulers largely through Austrian support.

The fragmentation of eighteenth-century Italy, however, was much more than merely political. There were deep lines of division, economic, social and administrative, within the states as well as between them. Different parts of the same state might well have very different economic resources and be quite differently affected by economic change or crisis. This is seen clearly in the impact of the food shortages, sometimes outright famines, which struck central and southern Italy in the mid-1760s. Though in the south Calabria and the Abruzzi suffered very badly in the famine of 1763–64, Apulia was in a far better

2.1   The Italian Peninsula in the later eighteenth century

position and the province of Trani actually gained population.
In the Papal States the situation was very serious in Umbria and
the Marches, while the Romagna and the area around Ferrara
were hardly touched at all. A greater obstacle to any sort of
enlightened and rationalising reform, however, was the tenacious
survival everywhere of a multitude of administrative and judicial
bodies, deeply rooted in tradition and buttressed by legal and
historic rights, whose intersecting and overlapping powers made
much of the peninsula a bewildering administrative kaleidoscope.
This situation was impossible to reconcile with the demand for
uniformity, simplicity and rationality, to be imposed by a

powerful central government, which was the essence of the Enlightenment as an active political phenomenon. Thus in the grand duchy of Tuscany, which had been built up by the gradual extension of the power of Florence, the areas which fell under the city's control had normally been allowed to retain their own forms of government; and even in Florence itself the Medici had permitted a good many republican institutions to survive, merely subjecting them to new ones of their own. The result was that there were in the city almost fifty judicial tribunals of different kinds, so that one of the newly-appointed regents for its Habsburg rulers reported to Vienna in 1737 that 'The government of this place is a chaos almost impossible to penetrate; a mixture of aristocracy, democracy and monarchy ... the only means of untying this Gordian knot is to cut it and start a new system'.[1]

In the south things were even worse. The legal system (or lack of system) of the kingdom of Naples was an amalgam of more than a dozen different and separate ones, while in both Naples and Sicily a great many barons had very wide powers of jurisdiction in their own fiefs. Endless local variations meant endless inequity and unfairness. An enquiry of 1733 among eighty communes in different parts of Lombardy disclosed that some of them paid, as a proportion of the capital value of their land, ten times as much in taxation as others. Underlying this often chaotic irrationality was the widespread assumption that administration meant no more than the enforcement of traditional rights and the following of legal precedents. All these old regime systems of government were permeated by legalistic preconceptions; very often they were operated and upheld largely by tribes of lawyers, conservative and tradition-bound, whose numbers and influence became as the century progressed a target of many would-be reformers.

Yet even before mid-century significant forces making for change were beginning to show themselves. More and more there can be seen among an educated élite a feeling that Italy, however glorious her past, had now sunk into a position of humiliating inferiority vis-à-vis the advanced states of western Europe – France, Great Britain and the Dutch Republic. From this she could, it was increasingly argued, escape only by becoming more productive and better educated, by sweeping away unnecessary restrictions on trade, by reducing drastically the deadweight of unproductive clerics and lawyers with which society was everywhere to varying extents encumbered, by attacking feudal rights which crippled agriculture and helped to keep the peasant poor. Ideas from abroad and translations of foreign books played some part in producing such attitudes. The Neapolitan economist and political writer Antonio Genovesi, for example, their most important exponent in the middle decades of the century, had read very widely in the works of French economists and travellers, and as he grew older more and more took England as his favourite example of economic and political achievement. (He was also, perhaps rather contradictorily, a great admirer of the reforming activities of Peter the Great in Russia.) But the intellectual impulses towards change which can be seen so clearly in several parts of Italy were far from being merely derivative. In Naples, where they were most vocal, Genovesi and other figures such as Bartolommei Intieri and Ferdinando Galiani were doing more than merely echo what had already been said in Paris or London. A regaining of intellectual vitality can also be seen in a marked revival of some of the Italian universities during the first half of the century. That of Pisa, notably, was to produce a considerable number of the administrators who later attempted, with varying degrees of seriousness and success, to reform the government of several of the Italian states. Bernardo Tanucci, who for a generation before his fall from power in 1776 was the most important figure in the government of the Two Sicilies, was only the most notable of these graduates.

Side by side with these growing intellectual forces which favoured change and reform went another which was in practice a good deal more powerful. This was the increasingly widespread hostility in ruling circles to the privileges, wealth and influence of the Catholic church, the feeling of many ministers and most intellectuals that these were now grossly excessive and damaging and must be reduced. The number of clerics of all kinds who swarmed everywhere was one of the things which struck foreign travellers in Italy most forcibly: Montesquieu on his visit in 1729 called the country 'a monk's paradise'. In the south (without Sicily) in the 1740s, for example, monks and priests made up well over $1\frac{1}{2}$ per cent of the population, a very high ratio by the standards of Europe in general, while twenty years later as much as 5 per cent of the inhabitants of Milan were ecclesiastics of some kind. Moreover numbers meant riches. A third of all the wealth of the kingdom of Naples was alleged in the 1740s, probably correctly, to be in clerical hands, and in the 1760s one reformer claimed that the more than 180 religious establishments in the city of Milan enjoyed an income which would have maintained 18,000 active working citizens. More and more, both officials and writers, even those completely untouched by the theories and ideals of the Enlightenment, felt that the accumulation of so much wealth in clerical hands and its removal to a large extent from ordinary productive uses must be resisted and reversed. The fact that the church was everywhere privileged in regard to taxation, often grossly so, did much to strengthen these feelings. So did the belief that for large numbers of men and women to live idly in monasteries and convents was a serious drag on production and economic progress (the fact that Protestantism made this impossible in Britain and the Dutch Republic was often put forward as a major reason for their economic success). So again did the workings of the right of asylum, which allowed criminals to find refuge from the law not merely in churches but often in a wide variety of other buildings belonging to religious bodies. Hostility to clerical wealth and privilege was the factor which most consistently underlay almost every official effort at reform, whether inspired by the ideals of the Enlightenment or by purely empirical and practical considerations, in eighteenth-century Italy. Except perhaps in the Papal States all such efforts contained, to a greater or lesser degree, such an element. Even within the church itself there were significant figures who saw clearly that too much wealth and power brought with them serious dangers and who wished to strengthen it spiritually by restricting its authority in secular affairs. L. A. Muratori, a priest as well as the greatest of Italian historians, advocated in his *Della regolata divozione dei Cristiani* (On the Manner of a Well-Ordered Devotion) (1747), a book which became 'the classical statement of Catholic reforming ideals in the eighteenth century',[2] a form of religion which laid heavy stress on morality and sincere belief and which was relatively simple in its ceremonies. In the middle years of the century, during the pontificate of the moderate and conciliatory Benedict XIV (1740–58), it seemed indeed for a time that some compromise between the church and the pressures for change which were now gathering strength might well be possible.

The attack on clerical power and privilege changed its character as the century progressed. Until the 1760s it expressed itself in largely legal terms, in 'jurisdictionalism', efforts to limit clerical and especially papal powers by legal and historical arguments. This found its most striking intellectual expression in the *Istoria civile del regno di Napoli* of Pietro Giannone (1723), one of the relatively few Italian books of this period to have a serious impact in other parts of Europe. Later, as Jansenist[3] influences, mainly from France, became more influential in Italy, and as hostility to the Jesuits as the main prop of papal power and religious conservatism in general became stronger, feelings intensified. By the 1760s and 1770s, when efforts at enlightened reform were approaching their height in several states, much of the peninsula was affected by genuine anticlericalism. The most extreme statement of such attitudes, the *Di una riforma d'Italia* of Carlantonio Pilati (1767–69), even demanded the abolition of canon law and in effect separation of church and state, as steps towards returning religious life to a state of primitive purity. Very few reformers were prepared to go to anything like these lengths: all of them continued to think of themselves as good Catholics. Throughout the century, however, it was hostility to the wealth and influence of the church more than any other factor which gave efforts at reform much of what unity of thought and feeling they possessed. The final suppression in 1773 by Pope Clement XIII of the Society of Jesus, so feared, disliked and envied throughout much of the Catholic world, was itself one of the most striking exercises of arbitrary power in Europe during this period; but it won the unanimous support of all the Italian reformers.

Reforms inspired at least in part by the ideals of the Enlightenment hardly began in Italy until after the mid-century. In Tuscany and Milan, the states in which they were most active and successful, they reached their climax between the 1760s and the 1780s. In Naples, where the gulf between advanced theories and intractable and depressing reality was widest of all, efforts at reform continued even later. It is important to remember, however, that all this had been preceded in some parts of Italy by more than a generation of attempts, though often hesistant and unsystematic ones, at constructive change. Enlightened reform did not emerge from a vacuum. It followed and to some extent sprang from a period of essentially empirical reform, of efforts at change which were practical and *ad hoc* and which had relatively little of the intellectual and theoretical backing which sometimes became important from the 1750s onwards.

The earlier decades of the century saw in the first place a series of attempts to assemble the information which was often essential for successful change. The most important aspect of this was an effort in many states to compile an accurate register of land-ownership, a *catasto*, without which effective reform of the tax-system in particular was hardly possible. In the duchy of Milan work on such a register began in 1718, though it was broken off in 1733 and not taken up again until the 1750s. In Savoy-Piedmont one was begun in 1728 and another, though a very unsatisfactory one, was completed in Naples in 1742. The same quest for essential information can be seen at an even more basic level in occasional efforts to produce detailed local or regional maps, something hitherto totally lacking in many areas, especially in the south. The Austrian administration in the 1720s made the first attempt at such a map of Sicily (there was to be no other until the British occupation of the island from 1806 onwards) and in 1750 a pamphlet by a Neapolitan grandee demanded the drawing of an accurate map of the city of Naples of which, he wrote, contemporaries had an idea 'no clearer than they have of the cities of Japan or Tartary'. Efforts of this kind, humdrum as they may seem, had their own importance and were genuine anticipations of later developments. More strikingly, the same decades saw a number of efforts to limit clerical rights and assert more effective state control over the church and its wealth. Concordats with the papacy made attempts in this direction in Savoy-Piedmont in 1741 and 1742, while another signed by the government of Naples in 1741 provided that in future church property should at least pay taxes at half the rate levied on that of the laity; it also attempted to limit the right of sanctuary and the exemption of clerics from customs duties. In Tuscany there was a series of similar efforts in the 1740s, though they were not formalised in a concordat: in 1743 a new press regulation attacked the hitherto very strict clerical control there of printing and publishing, while the effective workings of the Inquisition in the grand duchy came virtually to an end. In other words, many 'enlightened' policies and attitudes towards the church can already be seen in action in different parts of Italy, though often with only limited effect, long before any ruler or government had been seriously influenced by the ideas and ideals of the Enlightenment.

The first half of the century also saw attempts to come to grips with another problem, in many ways a more intractable one, which was later to preoccupy many would-be reformers at the height of the Enlightenment – that of feudal privileges and their social and economic effects. These attempts were most far-reaching and effective in Piedmont, where King Victor Amadeus II, from the end of the Spanish Succession struggle onwards, made a sustained effort to cut down the hitherto great powers of the landed nobility and to rule through an effective centralised bureaucracy – policies which were later to be fundamental to administrative reform as practised by most of the enlightened absolutists. In 1719 he ordered that all noble fiefs illegally acquired should be confiscated and incorporated in the royal domain: the edict was given force by the setting-up of a special court to apply it, and eight hundred fiefs were in fact confiscated in this way. In 1729 a new law-code attacked feudal rights of jurisdiction and the seigneurial courts which enforced them and were the most flagrant of all the challenges which noble privilege offered to the claims of central government. These measures were the most effective taken by any Italian ruler during the century. Elsewhere attempts to cope with the problem of feudalism were less whole-hearted and their achievements correspondingly smaller. Nowhere except in Sicily were the retarding effects of feudal survivals more apparent than in the kingdom of Naples; and there the first years of Bourbon rule after 1734, when reforming energies were running relatively high, also saw legislation which sought to weaken the position of the barons. In 1738 in particular a new law attempted to remove serious criminal cases from the scope of feudal justice and to reserve the right to try them to the state alone. But the forces of conservatism were strong and the obstacles to effective central control in an area so backward and undeveloped were immense. The result was that some years later, in 1744, as the reforming impulse lost its momentum, this legislation was repealed. In Tuscany a few years later efforts, though more cautious ones, to undermine the position of the nobility and strengthen that of the central government can again be seen. In 1747 there was an attempt to restrict the entailing of estates and the resulting consolidation of the position of the important landowning families. Two years later there was further legislation aimed at preventing the abuse of feudal jurisdictions; and in 1750 the government demanded that nobles should prove their right to titles and noble status which they had acquired in past centuries.

In a few cases Italian regimes in the first half of the century did not merely attack the relatively specific (though very large) problems of the church and the nobility but also attempted reform on a broader front by improving their whole administrative

machinery, notably their systems of taxation. Here again Piedmont took the lead. New central institutions, a Council of State
and General Council of Finance, were established in 1717; and
there was a serious effort at tax reform in 1731–32. Central
control over the provinces was strengthened by a system of
intendants modelled on that which had evolved in seventeenth-
century France. There was also a movement, though a slow and
cautious one, towards a system of local government which would
be uniform throughout Savoy-Piedmont and would override the
regional and communal variations which increasingly irked tidy-
minded ministers in many parts of Europe. In the duchy of Milan
the Austrian regime set up a commission in 1718 to investigate
the tax system; and in 1730–33 this produced far-reaching plans
for a unified structure of direct taxes, though the outbreak of the
Polish Succession war meant that these remained merely a paper
exercise. Later significant administrative reform was carried out
there, notably under an innovating governor, Count Pallavicini,
in the middle years of the century. In 1749 an essential step
towards a modern bureaucracy was taken when sales of
government offices were forbidden; and in the same year the
catasto begun back in 1718 was at last completed and the way
thus opened to fairer and more effective taxation of property. In
1749–50 all the main direct taxes and some indirect ones were
united in a single tax-farm. Though the government still delegated
their collection to a syndicate of financiers who made a profit
from the arrangement, at least the system was simplified and
made rather more efficient. In 1755 efforts began to reform the
municipal corporations which, as in so many parts of Europe,
were the most vocal of all supporters of a selfish and class-bound
conservatism. In Naples also the 1740s and 1750s saw a series
of proposals meant to make the tax system more fair and effective;
but, all too predictably, they had little practical effect: a historian
has described them as 'great in conception, but inadequate,
incomplete, ineffective in action'.[4]

All this adds up to a picture of serious and sometimes successful
efforts at reform in several of the most important Italian states.
But it was not 'enlightened' reform. Even as late as the 1750s
these efforts at change were hardly influenced by the intellectual
currents which were beginning to flow so strongly in Paris and
which were already arousing sympathy and interest in some parts
of Italy, notably in Naples. They were based rather on expediency
and purely empirical considerations. They sprang from the desire

of rulers for bigger revenues and more effective control of their territories, even from the instinctive dislike of officials for the jigsaw puzzle of competing jurisdictions and vested interests which they struggled to administer. The most effective and constructive changes carried out anywhere in Italy before the 1760s were in Savoy-Piedmont; but no Italian ruler of the century was less accessible to any sort of enlightened idealism, any idea of political freedom, humanitarianism or change by consensus, than Victor Amadeus II. The mainspring of his reforms was the need to extract more men and money from his territories and thus be better able to defend and if possible expand them. His new law-code of 1729 was notably backward-looking in its rejection of any idea of equality before the law and in its retention of severe traditional punishments and the liberal use of torture. The practical reformers of the mid-century – in Naples Tanucci (essentially a lawyer and to the end of his life ignorant of many aspects of the Enlightenment or actively hostile to it); in Milan Pompeo Neri (a Tuscan who made his reputation by the successful completion of the *catasto*); in Savoy-Piedmont Giovan Battista Bogino (who until his fall in 1773 did much to modernise the administration of Sardinia, the most backward part of the entire Italian world) – were all in varying degrees conventional and authoritarian in their attitudes. All thought in terms of power, a power which expressed itself primarily through legal mechanisms. All were indifferent to the liberal, tolerant and humanitarian elements which bulked so large in the Enlightenment.

Even their successors from the 1760s onwards, often much more open to new ideas and sincerely anxious to put them into effect, were almost always officials and administrators first and radical intellectuals very much second. In Italy proposals for far-reaching change came predominantly from government servants, men deeply involved in practical problems and the tasks of day-to-day administration. Thus in the duchy of Milan the strengthened reforming impetus of the 1760s owed much to the brothers Pietro and Alessandro Verri, both of whom served for much of their lives in the Austrian administration there. So did Cesare Beccaria, whose *Dei delitti e delle pene* (*On crimes and punishments*) (1764) became almost from the moment of its publication the most influential intellectual product of the entire Italian Enlightenment. At the other end of the peninsula most of the leading advocates of radical and enlightened reform in Naples – Giuseppe Palmieri, Ferdinando Galiani, Gaetano

Filangieri and others – were government officials. In Italy more than in many other parts of Europe the proposals and initiatives of the later decades of the century cannot be seen simply in intellectual terms, in those of an Enlightenment at least partly distinct from the real world. They were the work of men who knew that world and its difficulties, not of theorists or pure intellectuals.

Enlightened reform was to be seen in its most radical, sustained and self-conscious form in Tuscany under the rule of the Grand Duke Peter Leopold, the younger brother of the Habsburg Emperor Joseph II, from 1765 onwards. Over the next quarter of a century, until he left Florence for Vienna to succeed his brother in 1790, Leopold undertook in his territories a series of changes which make him, in many respects, the most interesting and impressive of all the 'Enlightened Absolutists'. Tuscany was reasonably well suited to be the scene of an experiment of this kind. Though its internal administration was still traditional and riddled with inefficiencies, its territory was not unmanageably large. From the beginning of his reign Leopold was supported by a number of very able administrators, mostly natives of the duchy – Pompeo Neri fresh from his successes in Milan, Alberto Tavanti, Francesco Maria Gianni and others. Partly because of this the measures taken in Tuscany to cope with the threat of famine in the mid-1760s were much more energetic than in either Naples or the papal territories, the other Italian states mainly affected. But the decisive factor was the character and outlook of the grand duke himself. Leopold was the only Italian ruler to be deeply and personally influenced by the ideas and writings of the Enlightenment. His whole reign was a sustained effort to give practical effect to some of the most radical thinking of the age on government and on the relationship which ought to exist between ruler and subjects. By the early 1770s he was being strongly influenced by the economic and social theories of the physiocrats in France: the elder Mirabeau, one of the most prominent figures among them, dedicated one of his books to him and spoke of him as 'the Solomon of our century'.[5] Moreover many of his most influential advisers in the first years of his rule were also strongly influenced by the physiocratic belief in a natural and therefore harmonious economic and social order which would assert itself when the man-made obstacles which prevented its realisation had been removed. Though this largely physiocratic cast of thought weakened with time Leopold's regime remained to a remarkable degree one founded upon theories and ideals of good government. But side by side with a lively interest in the most advanced political and social theory of the age Leopold also felt a deep sense of responsibility for and to his subjects. It was this which underlay the most remarkable of all his initiatives, the draft constitution of 1782 (see pp. 67–8 below) and it is entirely in character that before he left for Vienna in 1790 his last significant action was to arrange for the publication of a detailed account of his handling of the state finances during his reign. By doing this he was fulfilling what he saw as a fundamental political obligation and at the same time giving a final demonstration of his belief that effective reform was possible only if his subjects, or at least the more educated element in them, were involved in the process.

Every aspect of life in Tuscany now felt the wind of change more strongly than ever before. From 1767 onwards the trade in grain was freed by a series of enactments from the restrictions, both within the duchy and as regarded exports, from which it had so long suffered. Tax-farming was ended in 1768. The Florentine guilds, here as elsewhere a restrictive legacy of the past, were abolished in 1779. For almost two decades from 1766 onwards there were rather unsuccessful efforts to improve the position of at least some of the peasants by giving them long leases of lands belonging to the grand ducal patrimony or to religious bodies now falling under state control. To reform and rationalise the chaotic traditional administrative structure, on which the changes of the 1740s and 1750s had had only very limited effect, was an even more difficult task than to develop the economy. However in a piecemeal and empirical way a good deal was achieved. In one province after another – in Arezzo and Volterra in 1772, in the Pisa area in 1776, in that around Siena in 1777, in 1782 in Florence itself – new magistracies were set up, each backed by a General Council with considerable powers over the collection of the land tax, the supervision of communal property and the appointment of communal officials. Similarly piecemeal judicial reforms in the later 1760s and early 1770s paved the way for a more general and ambitious one in 1786. The new criminal code of that year abolished judicial torture and the death penalty.

Any systematic effort to reform the administration and the judicial system tended to mean also some reduction in the powers and prerogatives of the church; and in Tuscany efforts in this direction were marked by the strength there by the later 1770s of Jansenist influences. Nowhere in Europe, not even in the Habsburg dominions under Joseph II, were official attitudes to the church so deeply influenced by Jansenism and its anti-papal and often puritanical tendencies. From 1771 onwards there was a sustained effort to restrict the powers of church courts to purely religious matters, while from 1773 a considerable number of monasteries which the government regarded as superfluous were suppressed. In 1782 the tribunal of the Inquisition ceased to exist, while in 1786 it was made clear that in future papal nuncios were to be merely ambassadors with no power over the church in Tuscany. In the same year the bishop of Pistoia, Scipione di Ricci, in a diocesan synod, attempted with the backing of Leopold to enforce a series of essentially Jansenist reforms – to make church ceremonies simpler and less expensive, to improve the education of parish priests and to bring religious orders under stricter episcopal control. This was the high-water mark of anti-papal feeling and zeal for church reform in eighteenth-century Italy. The synod was meant to pave the way for one with members drawn from the entire grand duchy, perhaps chaired by Leopold himself. However it soon became clear that the Pistoia decisions were opposed by both the majority of the Tuscan bishops and by a great weight of popular and deeply conservative religious feeling. Leopold, who was cautious as well as idealistic, made therefore no real attempt to enforce them.

The grand duke was not a totally admirable character. He shared his elder brother's taste for petty puritanical interference with the lives of his subjects (in such things, for example, as his prohibition of card-playing in public). Like Joseph II, he came to rely heavily on a system of secret policing and surveillance in the carrying-out of his policies. Yet there was in him a wide streak of genuine political idealism of a kind hardly paralleled by any other ruler of the period. This is seen most strikingly in the remarkable project for a Tuscan constitution which he ordered Gianni to draw up in 1779 and which took its final form in 1782. This was in fact less radical than Leopold's own original version; nevertheless it was a remarkable document. It would have created an elaborate structure of local and regional assemblies, capped by a general one for the entire grand duchy. This was to have powers which drastically limited those of the ruler. He could not on his own initiative alter the boundaries of Tuscany, declare war or make any foreign alliance. Nor could he raise troops, build fortifications, alter the judicial system or interfere in any judicial proceedings. The elected deputies were also given positive powers. They were to oversee state expenditure and the judicial system and suggest new legislation. The constitution therefore explicitly involved a great transfer of authority from the ruler to at least the better-off and politically conscious part of his subjects; it spoke of Leopold's desire to return to them 'their full natural liberty'. This transfer, moreover, was to be made purely in pursuance of an ideal. There was no popular demand for it, and almost all the grand duke's advisers were opposed to such a wholesale surrender of his prerogatives. Leopold believed firmly that the constitution was needed as a matter of principle, not merely of expediency. He claimed that 'every government must have a constitution' and sincerely believed that strictly limited monarchy was the best of all constitutional forms. This political idealism was shaped by several writers of the Enlightenment (notably Turgot) and also influenced by contemporary example; Leopold refers at different times to habeas corpus in England and to the choice of popular representatives in the Swiss cantons, while he owned and studied a copy of the constitution of Pennsylvania.

The Tuscan constitution never came into force, largely because of the opposition of Joseph II, who hoped to make the grand duchy an integral part of the Habsburg dominions and not a mere *sekundogenitur* ruled by a younger member of the family. But it remains the most impressive single document produced by the movement for enlightened reform in Italy and the most forward-looking political proposal generated by the entire European Enlightenment. It clearly anticipated many of the later constitutions of the nineteenth and twentieth centuries.

The duchy of Milan, the other important Italian state to experience in the generation before the French Revolution a systematic reforming impulse, produced no such far-reaching document. But the efforts there to improve the economy and the administration were comparable in scope to those in Tuscany. The reorganisation of the land tax (based on the new *catasto*) had been completed before the end of the 1750s; and in 1765 this, together with the guilds and the grain trade, was placed under the control of a new supreme economic council whose head until 1780, Gianrinaldo Carli, was a strong reforming influence. Farming of indirect taxes was abolished in 1770, while government revenues formerly alienated to landowners and the Milanese patricians began to be regained. These were constructive changes. The new tax system did something to stimulate agricultural development: it helps to explain the fairly rapid recovery of the duchy from the devastation and losses of the Austrian Succession war in 1741–48. But how far enlightenment in any high-flown intellectual sense of the term, as opposed to practical need and felt experience, had much to do with these changes is open to question. The Verri brothers and Beccaria, the leading figures in the Milanese Enlightenment, had little interest in either agriculture or industry and knew little about them. The administration of the duchy, in which they were interested, was extensively overhauled in 1770–71; in particular the state's financial accounts were unified for the first time and a separate treasury created to supervise them. This new system, however, was swept away with characteristic thoroughness by Joseph II in 1786. In that year he abolished all the main existing organs of central government and replaced them by a new council subdivided into specialised departments. At the same time the duchy was split into eight districts each administered by an intendant reporting directly to the new council, while a new centralised judicial system meant the sweeping away of all existing jurisdictions. This was a rational and highly centralised system of government, the first such that Milan had ever had. But it involved so much sudden change and flouted so many interests and traditions that it aroused at once fierce opposition. Even Pietro Verri, hitherto a great admirer of the emperor, now turned against him. Joseph's attitude to the church helped to intensify this feeling. The relatively moderate efforts which had been made from the end of the 1760s to limit the accumulation of property in clerical hands and the church's powers of censorship were notably intensified after his assumption of power in 1780. The setting up in the following year of an ecclesiastical commission to supervise church affairs, the abolition in 1783 of ecclesiastical exemptions from taxation, accelerated closures of monasteries and convents, led on to the establishment in 1786 at Pavia (the most important intellectual centre of Jansenism in Italy) of a general seminary to train parish priests for the entire duchy. But all this, like the administrative changes, was too sudden, too sweeping and too little understood by those it was meant to benefit. In his Italian province, as in so many other parts of his empire, Joseph's reign thus ended in an atmosphere of bitter discontent.

In both Tuscany and Milan, efforts at reform, though far from completely successful, had real achievements to their credit. The kingdom of the Two Sicilies, the third major Italian state where such efforts were seen in the later decades of the century, presents a different picture. In Naples there was certainly no lack of discussion of the need for change or of projects for bringing it about. During the 1780s writers such as Filangieri, Palmieri and Francesco Maria Pagano displayed a radicalism deeper than any Tuscan intellectual had dared to show under the Grand Duke Leopold. But to translate all this into action was a very different thing. The kingdom (Sicily even more than the mainland) was backward and undeveloped, the eighteenth-century European equivalent of a Third World country. More than any other Italian state it was dominated by a feudal nobility; it was claimed in 1789 that of over 2700 cities and villages in the kingdom of Naples only 200 were not under some form of feudal control. The mass of the population distrusted all secular authority and found it hard even to conceive of radical change. 'The people', wrote Palmieri, 'is made up of grown-up children who do not know enough to recognise what is good for them and have to be led by the hand to it.'[6] The vested interests – the nobility, the church, the very numerous lawyers and other groups – which opposed all real change were numerous and powerful. It was they who made up what a recent writer has called 'the complex and paralysing mess of Neapolitan political life'.[7] King Ferdinand IV, who assumed effective authority in 1767, a vulgar, ignorant and frivolous man, was quite incapable of giving the reformers the backing they needed. Minor and peripheral changes might be attempted; but from any whole-hearted attack on deep-seated abuses the court and the ruler always drew back, especially since real reform very often meant some loss of revenue at least in the short run. Tanucci tried in 1759 and again in 1773 to attack the feudal jurisdictions which undermined so seriously the authority of the central government; but these efforts were neither systematic nor sustained. He launched in the early 1770s a considerable attack on clerical privileges. He was even able to ensure that in Sicily half or more of the land confiscated from the Jesuits when their order was suppressed and its property seized by the state in 1767 actually went to small peasant proprietors, while those on the mainland probably also benefited. But this hardly touched the structure of baronial power which was the greatest obstacle to real change. Though hardworking and honest, Tanucci became increasingly pessimistic about what could be achieved; and throughout his long period of office his attention was concentrated largely on a rather narrow range of legal and judicial issues.

His successors were more adventurous. The 1780s saw in Naples an attack on guild privileges, while criticism of those of the barons became more outright and vocal than ever before. In 1782 the setting up of a new supreme council of finance, of which both Filangieri and Palmieri were members, was an effort to bring some order into the chaotic tax system. Two years later there was even a proposal for the introduction of a system of general primary education on the Austrian model, though in such an environment this could be no more than a fantasy. In Sicily an actively reforming viceroy, the Marchese Domenico Carraciolo, struggled in 1781–86 to modernise one of the European societies most resistant to change. He too attacked the jurisdictional rights of the nobility, closed a number of monasteries, struggled to reduce the number of religious feast-days and in 1782 abolished the Inquisition in the island (only three old women were found in its prisons). But again all this had merely superficial effect. Carraciolo failed in his efforts to carry out a survey of land ownership, a *catasto* on the model of those in Piedmont, Tuscany and Milan; and though his successor, the Prince of Caramanico, was able in 1789 to abolish the last remnants of serfdom in Sicily and Palmieri made a similar effort on the mainland two years later, it was beyond their powers to do anything effective to improve the position of the peasant. The impulse towards reform from above lasted longer in the Two Sicilies than in other parts of Italy. Not until 1794 did it die away amidst the growing fears aroused by events in France. But it was a failure; and its failure shows unmistakably that for successful enlightened reform much more was needed than merely the demands or schemes of intellectuals. Naples in the second half of the century had a more stimulating intellectual life than any other Italian state and produced a greater volume of challenging political speculation. Yet in practical terms there was little change. Savoy-Piedmont two generations earlier under Victor Amadeus II had had little active intellectual life. But there, without the benefit of 'enlightenment', there was real change with permanent and constructive effects. In eighteenth-century Italy, as in Europe generally, some societies were simply more reformable than others and some rulers more willing than others to embark on serious reform. No amount of writing or talking, however high its intellectual level, could alter this fact.

Striving for constructive change was not confined to the larger Italian states. In some of the smaller ones, declining entities such as Genoa and Venice, there was little sign of it. In others, however, it was sometimes quite marked. In Modena during the later 1760s and 1770s there were proposals for serious economic reform along essentially physiocratic lines. In Parma, under a French chief minister, Du Tillot, an outwardly impressive structure of enlightened government was built up until his fall in 1771: it was Parma which, in 1758, was the first Italian state to expel the Jesuits from its territory. In Modena in the 1780s under a reforming minister, Ludovico Ricci, there were serious efforts at economic development; here again a new *catasto* compiled in 1788–91 was meant to serve as the foundation of an improved system of taxation. Even in the Papal States, which one observer in 1784 called 'the home of sluggishness and irresolution',[8] the government tried in 1786 to create some kind of economic unity through the workings of a new tariff system. But none of this differed essentially from much of what was being proposed or attempted in Tuscany, Milan or the Two Sicilies; and though in principle reform perhaps ought to have been easier in the smaller and more manageable political units this was not always the case. Local rivalries and jealousies, in particular, were often a serious obstacle in the smaller states. Change in the duchy of Parma was hampered by the ingrained rivalry between the cities of Parma and Piacenza, and that in Modena by the similar bad feeling between Modena and Reggio, while the tariff reform in the Papal States was impeded by Bologna's dislike of government from Rome.

Enlightened reform in Italy was always at bottom weak even in the states where it seemed to have effect. Active support for it from society in general was everywhere very limited. Its typical proponent was a well-educated, public-spirited and forward-looking official, usually of good and often of noble birth. Its foothold in society was therefore a narrow one, while the forces which opposed it were numerous and powerful. Often the most intractable opposition of all came from the mass of the population, from ordinary men who distrusted all novelties, who feared that free trade in grain might mean higher prices for their staple food, who deeply resented any interference with traditional religious ceremonies and observances. Many of the reformers were quite aware of this fundamental difficulty but hoped vaguely that education and experience of the practical advantages of change might overcome it. 'I have conceived the hope', wrote one Neapolitan intellectual in 1751, 'that the golden age which in the past was only the unsound (*corotta*) fantasy of poets may in the future become reality when knowledge has been spread more widely and has penetrated more deeply in the countryside, among peasants, shepherds and people of that kind who today are a brutish, low and savage rabble.'[9] This was mere wishful thinking, however. In the 1770s and 1780s, as the movement for enlightened reform took more radical shape in the writings of Filangieri and others in Naples or in the constitutional proposals of Leopold in Tuscany, it seems likely that superstition and a taste for the miraculous and the occult were becoming even stronger and more widespread in many parts of the peninsula. The gulf between self-consciously progressive forces and popular feeling was widening.

The particularly intense resentment which could easily be aroused by any effort to alter traditional religious observances can be seen most clearly in Tuscany. There the departure of Leopold for Vienna early in March 1790 was followed almost at once by widespread rioting in Prato, Pistoia, Leghorn and finally Florence itself, against the efforts of Ricci and others to simplify forms of worship and weaken what they regarded as popular superstitions. In Leghorn, for example, the rioters demanded the reopening of churches which had been closed, the re-establishment of banned religious confraternities and the exposing for public veneration of a particularly cherished image of the Virgin. These demands were very soon merged with an equally active hostility to the freeing of the grain trade and with pressure for the return of the traditional *annona* system of public granaries, official price-fixing and government control. The house in Florence of Gianni, who was regarded as the main author of the hated freedom of trade, was attacked and he himself forced to flee to Bologna. On both counts the regents whom Leopold had left in control were forced to give way. Many of the religious reforms were withdrawn, while the export of grain and olive oil was forbidden and official prices for them fixed. Such an end to the greatest programme of reform undertaken in any Italian state was the most forcible of all demonstrations of the difficulties faced by any effort at real change. This contrast between the enlightened idealism of the few and the unthinking conservatism of the many had disastrous potentialities. These were to be seen clearly a few years later in the crushing of the Neapolitan Jacobins in 1799, one of the great tragedies of modern Italian history. A really effective breaking of the eighteenth-century mould could be achieved only by more powerful and more ruthless forces which the peninsula itself could not provide. These were soon to be supplied by the armies of the French Revolution and Napoleon.

3. Pombal: the Paradox of Enlightenment and Despotism

KENNETH MAXWELL

'Small powers need much more than great powers to take carefully considered action, because the first do not have the resources to repair the errors they make, whereas the latter always have the means to recuperate.' (Instructions of D. Luis da Cunha to M. A. de Azeredo Coutinho [1738]).

THE Anglo-American historical tradition tends to assume that the eighteenth century is synonymous with the Enlightenment and the Enlightenment with liberty: especially those liberties enshrined in the United States Constitution and the Bill of Rights, rights which have as their fundamental objective the protection of the individual from the state. But in Central, Eastern and Southern Europe, the Enlightenment was more often married to Absolutism than to Constitutionalism. Here it was not the individual seeking protection from the state, but the state seeking protection from overpowerful individuals. Marc Raeff has gone so far as to characterise the legal formulations of eighteenth-century reformism in the German territories and Russia as being aimed at the creation of 'a well-ordered police state'. And indeed state is the operative word; certainly in Portugal also.

We also assume when discussing the eighteenth century that liberty was the handmaiden of progress. This is not the view from the European periphery. Here the more common union was of 'Order and Progress'; not the happy union of 'Order and Freedom' Macaulay claimed as the great attribute of eighteenth-century England.

Portugal unfortunately is rarely considered in these debates. To be sure it represents a complicated and contradictory case, and some of its peculiarities are worth considering in a comparative context. Like the better known examples of the reception and uses of the Enlightenment in Eastern and Central Europe, the Portuguese case unsettles our rather optimistic and linear views of the role of the *philosophes* in the making of the modern world, but it does have the merit of helping to unravel some of the paradoxes which lie behind the coexistence of the reformist and authoritarian traditions in Southern Europe.

I

If the eighteenth-century rationalist writer needed a stereotype of superstition and backwardness, Portugal almost invariably was that chosen. Voltaire summed up the attitude well. Writing about the gold-rich Portuguese monarch D. João V (1707–50), he observed: 'When he wanted a festival, he ordered a religious parade. When he wanted a new building, he built a convent, when he wanted a mistress, he took a nun.'[1] There was of course a kernel of truth to these prejudices. D. João V did indeed spend vast sums of Brazilian gold to build his great palace monastery at Mafra. In 1750 Portugal, with a population of less than three million people, had a veritable army of clergy, numbering 200,000 according to some estimates. And the Portuguese did burn people at the stake in public ceremonies as late as 1761. The torture and public destruction in 1759 of members of two of Portugal's most distinguished families, accused of attempting to assassinate King D. José I, particularly shocked foreigners. Although their consternation, it should be noted, was caused less by torture, and the breaking of limbs and burning, than by the fact that the victims were aristocrats, not commoners nor, as had been more traditionally the case in the Iberian peninsula, the Jews and heretics for whom such treatment had been commonplace for centuries.

The contrast between the views of foreigners and the image of the eighteenth century within Portugal, however, is striking. The period, especially after the 1750s, is seen as being the very embodiment of the Enlightenment. Among the developments singled out is the legislative activity which left few aspects of Portuguese life untouched. This included the establishment of the first system of public, state-supported education, the root and branch reform of the University of Coimbra, the reduction of the power of the Inquisition, the abolition of slavery in Portugal, and the modernisation of the army. A Royal Treasury with centralised accounting systems and uniform fiscal powers was established and its first head was designated, following British practice, the King's chief minister. Above all, the reconstruction of Lisbon after the devastating earthquake of 1755 is held up as a model of Enlightenment town planning. In the colonies, Brazil most especially, the reform of the whole structure of administration can be claimed; the setting-up of joint stock companies, the outlawing of discrimination against Amerindians in Portuguese America and Asians in Portuguese India; the ending of the distinction between so-called 'old' and 'new' Christians (that is the descendants of Portuguese Jews compelled to embrace Christianity in 1497). Portuguese historians will agree that eighteenth-century Portugal was governed by an authoritarian and absolutist regime. It was, however, a regime inspired by an absolutism of reason, and essential to the process of re-establishing national control over the economy and revitalising the state.

There are several special reasons for the contrast of views from inside and outside, and it is as well to enumerate some of them at the outset because it is important to be aware of what they are, and how they originated.

First, eighteenth-century Portugal is almost inseparable from the dominating figure of the Marquês de Pombal. Sebastião José de Carvalho e Melo (1699–1782), born in Lisbon, came from a family of modest gentry who had served as soldiers, priests and state functionaries within Portugal and occasionally overseas. His father Manuel de Carvalho e Ataide (1668–1720) had served in the navy and the army. His paternal grandfather had been a Lisbon gentleman of some means holding two entailed estates, and his uncle, Paulo de Carvalho e Ataide, was an archpriest of the Lisbon Patriarchy, as well as the proprietor of an entailed estate in Oeiras, a small town close to Lisbon. Pombal inherited the Oeiras property and later constructed his elaborate country house there. The powerful minister's ancestry was, therefore, neither as grand as he claimed at times, nor so modest as his enemies implied. Pombal's background was much like that of many of the ministers the absolutist monarchs chose to strengthen their power and enhance that of the state. His title had been awarded late in life. He became Marquês de Pombal in 1770 when he was 71, having been made Count of Oeiras in 1759. Since he is known to history as Pombal this practice is followed here; but it is important to remember that this noble status was not that confirmed by hereditary aristocracy but obtained by service to the monarch and to the state.

Pombal was the eldest of twelve children, four of whom died young. One of his brothers died in Portuguese India and his sisters took religious orders. Two of his brothers, Paulo de Carvalho e Mendonça (1702–70) and Francisco Xavier de Mendonça Furtado (1700–69) became very close collaborators in his administration. Paulo de Carvalho, a priest, elevated to cardinal by Pope Clement XIV, was Inquisitor General and president of the municipal council of Lisbon (the latter position was filled by Pombal's son following the death of his brother). Mendonça Furtado served as governor-general of the Brazilian province of Grão Pará and Maranhão (essentially at the time the Amazon river valley), and later in Lisbon as secretary of state for the overseas dominions. The family was a very close one, and since neither of these brothers married, they pooled their financial resources and property in Pombal's interest. There is a remarkable portrait of the three with linked arms at the Oeiras mansion with the title 'Concordia Fratrum.'

When a young man Pombal had suffered rejection at court. Following the death of his father and with the family facing severe financial difficulties and dependent on his uncle the archpriest, he had left the capital city to administer a rural property belonging to the family in Gramela, north of the town of Pombal in central Portugal. During this period (1723) in the face of opposition he married a widowed niece of the count of Arcos, Teresa de Noronha, an arrangement which related him to the high nobility. It was a childless marriage and following his wife's death he later married again while Portuguese envoy in Vienna. His second wife was Eleonore Ernstein Daun, niece of the Marschall Heinrich Richard Graf von Daun, and the marriage (1745) brought the personal blessing of the Empress Maria Theresa. The Portuguese envoy in Rome sourly observed that this was a marriage which guaranteed Pombal a position as secretary of state in Lisbon. Maria Theresa in fact took a more than usual interest in the connection, telling Pombal's new wife in private correspondence that she 'owed the preservation of the monarchy' to the Daun family.[2] It was Maria Anna of Austria, the Queen Regent of Portugal, who recalled Pombal from Vienna in 1749 to join the ministry in Lisbon. Pombal was 50 at the time. His preeminence and power were to coincide exactly with the reign of D. José I (1750–77), a monarch who preferred opera and hunting to government, and who placed virtually complete authority in his minister's hands.

Pombal himself has a very curious historical image. To some he is a great figure of enlightened absolutism, comparable to Catherine in Russia, Frederick the Great in Prussia and Joseph II in the Habsburg lands; to others he is no more than a half-baked philosopher and full-blown tyrant. The controversial nature of Pombal's rule and the parochial nature of discussion about it was to a large degree of his own making. Pombal in many ways wrote his own history in a series of self-justifications (*apologias*) after his fall in 1777, following his royal patron's death. Almost a year and a half after his removal from office, the complaints against him led to a famous judicial case in which Pombal was subject to grave accusations of abuse of power, corruption and other types of fraud. The Marquês employed all his energy to combat these attacks and organised his defence in a very audacious manner, claiming he had never acted without royal authority. During the entire winter from October 1779 to January 1780, the 80-year-old Pombal was interrogated. He always maintained that the King was responsible, and that he (Pombal) was merely following orders from his master 'of sacred memory'. A committee of five judges examined the evidence, but was divided on how to proceed. The new Queen, Maria I (1777–99), cut the process short in 1781 by issuing an edict declaring Pombal deserving of exemplary punishment but instituting no proceedings against his person because of his age and feeble condition.[3] These extraordinary proceedings and Pombal's *apologias*, or parts of them, fascinated Europe. Extracts from his writings, and commentary about his activities, were by the mid-1780s circulating in multiple editions and several languages.

A second reason for the partisan nature of discussion about enlightened absolutism in Portugal was the result of Pombal's formidable promotion of state propaganda. The most striking example of this activity was the concerted assault on the Jesuits. The Pombaline administration stimulated and subsidised throughout Europe a virulent campaign against the order. Pombal himself was intimately involved in the writing and formulation of the remarkable piece of propaganda known as the *Deducção chronólogica e analítica* (the *Chronological and Analytical Deduction*). This divided the history of Portugal between the useful and the disastrous linked essentially to the influence of the Jesuits.

It upheld a rigorous regalist view concerning the Church in Portugal. Professor Samuel Miller described the work not unjustifiably as 'a monotonous repetition of all the accusations ever leveled at the Jesuits by anyone at any time'.[4] The history of the assault by the Portuguese and Spanish crowns on the Jesuit missions along the Uruguay river in South America during the late 1750s was also encapsulated, and was for many defined, by another piece of state supported and financed propaganda, the *relação abbreviada*. Published in Amsterdam in Portuguese, Italian, French, German and English, the *relação abbreviada* was an account of the joint Portuguese and Spanish military campaign against the Jesuit missions in what is now the southern borderlands of Brazil. Some 20,000 copies are estimated to have been distributed. It was a major weapon in the Europe-wide battle which led to the suppression of the Jesuits by Pope Clement XIV in 1773. The success of the Portuguese propaganda offensive had much to do with the sympathetic reception his ideas found among other Catholic reformers in Southern Europe. Bernardo Tanucci, for example, the powerful minister of Charles of Bourbon, King of Naples (later Charles III of Spain), regretted the ferocity and pure *raison d'état* of Pombal, but approved his objectives. The *relação abbreviada* and *Chronological and Analytical Deduction*, therefore, represented an official historiography, which the Jesuits were and remain dedicated to the task of refuting. If anyone should think that this struggle for historical memory is over, they need only read the comments in the Jesuit magazine *Brotéria* (Lisbon) published to mark the bicentennial of Pombal's death in 1982. 'Pombal's methods owed nothing to our own contemporaries,' the leading article asserted. 'His methods were in effect an anticipation and mixture of the methods of Goebbels and Stalin.'[5]

A third area of public controversy, also involving surreptitious use of subsidised propaganda to influence public opinion, grew from the disputes between the Portuguese and the British over trade, the port wine sector and the Portuguese government's attempts to stimulate manufacturing in Portugal. Both sides had recourse to pamphleteering, industrial espionage and even industrial sabotage, and this had a large impact on the image of Portugal in England, and by extension has influenced English historiography. A further cause of the fractured image of eighteenth-century Portugal was undoubtedly the impact of the Lisbon earthquake of 1755, the literary and public reaction to which T. C. Kendrick devoted a whole volume some years ago.[6] The great earthquake of All Saints Day 1755 reduced one of the richest and most opulent cities of the epoch to ashes, and provoked an extraordinary philosophical debate about optimism, God and natural phenomena.

The British Consul Edward Hay, writing to London two weeks after the earthquake, provided a concise eye-witness account of the event:

> The first shock began about a quarter before 10 o'clock in the morning, and as far as I could judge, lasted six or seven minutes, so that in a quarter of an hour, this great city was laid in ruins. Soon after, several fires broke out, which burned for five or six days. The force of the earthquake seemed to be immediately under the city. . . . It is thought to have vented itself at the quay which runs from the Customs House towards the King's palace, which is entirely carried away, and has totally disappeared. At the time of the earthquake, the waters of the river rose twenty or thirty feet.[7]

About one third of the city was totally destroyed. Estimates of casualties ranged from 10,000 to 40,000. At the time the latter figure was widely believed though the true figure was probably closer to 15,000 – no small number for a city of 250,000. The royal family, who were at Belém, outside the city, escaped what would have been certain death in their collapsed palace in Lisbon. The bewildered and frightened king placed full authority in the hands of the only minister who showed any capacity to deal with the catastrophe, Pombal.

The scope of the destruction was colossal. The Royal Opera House, only completed a month before, was in ruins. Of Lisbon's 40 parish churches, 35 had collapsed, many onto parishioners who had been at mass when the earthquake struck, crushing them to death. Only 3,000 of Lisbon's 20,000 houses were habitable. The palace of the Inquisition on the Rossio had crumbled and many townhouses and palaces of the aristocracy were destroyed. At one mansion alone two hundred paintings were lost, including a Titian and Rubens and a library of 18,000 books and a thousand manuscripts. In the king's library 70,000 books perished.

It was the earthquake which propelled Pombal to virtual absolute power which he was to retain for another 22 years until the King's death in 1777. He took quick, effective and ruthless action to stabilise the situation. Looters were unceremoniously hanged, bodies of the earthquake victims were quickly gathered and, with the permission of the Lisbon church authorities, taken out to sea, weighted and thrown into the ocean. Rents and food prices, and the cost of building materials were fixed at pre-earthquake levels. No temporary rebuilding was permitted until the land was cleared and plans for new construction drawn up. Military engineers and surveyors were charged with making inventories of property rights and claims, and implementing the myriad of practical decisions to ensure that sanitary and levelling operations were carried out safely. They were also charged with drawing up plans for the new city. It was these practical-minded engineer officers who, under the closest scrutiny from Pombal, developed the economical Pombaline architecture and grid of streets, and the focal point of the scheme, the great water-front square, which make Lisbon to this day a classic example of eighteenth-century town-planning. It was also highly significant that the new square, placed on the old royal plaza, was called, as it remains, the Praça do Comércio – the place of commerce. The new Lisbon was thus intended to be a pre-eminently mercantile and administrative centre. As the rest of Europe debated the meaning of the earthquake for the philosophy of optimism, the reaction in Portugal was more prosaic. Pombal's architectural and city planning was intended to celebrate national economic independence and a modern, well-regulated and utilitarian state.

The pragmatism of Lisbon's reconstruction is something of a paradigm for Pombal's activities in government and represented a good example of the role he wished to see the state perform. It was a role deeply rooted in a pragmatic assessment of options, a mixture of eclectic borrowing and innovation, and the selective intervention by the state in society to promote what was conceived to be the national interest.

II

Two distinct economic phases marked eighteenth-century Portugal. In 1750, gold remittances to Lisbon from Brazil, Portugal's vast South American colony, were at their height. By 1777 remittances from Brazil were less than half what they had been. Brazilian gold was extremely important to Portuguese commercial relations with Britain, financing the vast trade imbalance in Britain's favour. The fall in gold production severely affected Anglo-Portuguese commerce. Between the mid-1750s and the mid-1760s, British woollen exports to Portugal, which composed 70 per cent of the British trade, declined by well over 50 per cent. Yet, if the first half of the eighteenth century was marked by the rise of gold production and the second half by its fall, both cycles took place within a broader system, the parameters of which changed surprisingly little between the 1660s and 1807, the year of the Napoleonic invasion of Portugal. To use a Braudelian shorthand, the chronological framework might be called the long eighteenth century and the geographical framework, the Atlantic.

Briefly stated, the economic characteristics of this long eighteenth century are, first, the pre-eminence of colonial, mainly Brazilian staples (sugar, tobacco, cacao, hides, cotton) and specie (gold and silver), the latter obtained by contraband with Spanish South America. Second, the period is marked domestically by the rise of viticulture and port wine production, and by the capture by port of the British wine market. The British Factory in Oporto had initiated the wine trade in 1678 as a substitute for the re-export trade in Brazilian sugar and tobacco which they had lost to British West Indies competition. As a result of parliamentary pressure within England the British merchants in Portugal obtained a tariff beneficial to them in 1697, and under the terms of the Methuen Treaty of 1703, Portuguese wines paid a rate one third below that on French wines. By the end of the eighteenth century, nine-tenths of all port wine exported went to England where Portuguese wine captured three-quarters of the English wine market. Third, Portugal remained a chronic grain importer, from Northern Europe at the beginning of the century, from North America – especially Virginia and the Carolinas – towards the end, a factor which helped develop North American taste for Portuguese wines. Fourth, the growth and decline of manufacturing industry in Portugal was inversely proportional to the rise and fall of Brazilian gold production. Hence, Portuguese domestic manufacturing thrived prior to 1700 and again after 1770, but languished during the golden age.

Social, institutional and strategic questions fit similar chronological and geographical frameworks. Gold from Brazil allowed the Portuguese monarchs the luxury of avoiding recourse to the nation's ancient representative (and tax granting) institution. It was no accident, therefore, that the last Cortes met in 1698, and was not to meet again until 1820. The eighteenth century, therefore, saw the apogee of the absolutist state in Portugal.

The gold boom also had a profound impact on the entrepreneurial structure of Portuguese society. The prosperity gold brought had encouraged many small merchants and speculators to enter the colonial market. Likewise, the favourable conditions Portuguese wine enjoyed in the English market led to a speculative boom in wine production. The colonial boom adversely affected the larger merchants and the expansion of vineyards adversely affected the aristocratic producers of the Douro valley in northern Portugal.

The whole period from the 1660s through 1807 is also marked by the dominant, the Portuguese felt domineering, presence of influential foreign merchant communities in Lisbon and Oporto, protected by treaties, enjoying special legal status and their own courts, appointing their own chaplains and enjoying religious toleration. Through their entrepreneurial skills and access to capital, foreign merchants penetrated the whole fabric of the metropolitan and colonial economy, directly (as in the case of the port wine merchants) or indirectly (as in the case of the merchants with interests in the commerce with Portuguese America). The British merchants in Lisbon and Oporto (the so-called factories), their privileges guaranteed by the Cromwellian treaty of 1657, reinforced by the Methuen Treaty of 1703, were not the only foreign merchants in Portugal to be sure, but they were the most prominent by far. 'A great body of His Majesty's subjects reside in Lisbon, rich, opulent, and everyday increasing their fortunes and enlarging their dealings' was how Lord Tyrawley described the British community to London in 1752.[8] The traveller, William Costigan, some time before Noël Coward made a similar observation, wrote: 'That excepting of the lowest conditions of life, you shall not meet anyone on foot [in Lisbon] some hours of the violent heat every day but dogs and Englishmen.'[9]

Brazilian gold was not the only link between the English and the South Atlantic colonial complex. 'The foreign merchant houses by means of their great capital had made themselves absolute mistress of the metropolitan and colonial commerce', commented a Portuguese contemporary:

Few or rare were the Portuguese merchants in a condition to

do business with their own funds, none with goods that were not foreign. All the commerce of Brazil was made on credit and the greater part by salesmen of the foreign houses and by itinerant traders (*commissários volantes*) who took manufactures from Portugal to America and did business on the account of the foreigner receiving a commission for their work and a bonus for extra service.[10]

The *commissários volantes*, who bought goods in the metropolis, sold them personally in America, and returned with the proceeds, were one of the essential elements in the trans-Atlantic commercial connection. These itinerant traders often travelled under false pretences and carried merchandise in their shipboard accommodation, avoiding outlays for commissions, freight charges and warehousing.

From the point of view of the British merchants this commission system and the long-term investment it represented (at least a year, usually two or three), was inherent in the colonial trade, and the great advantage the British merchants enjoyed over their Portuguese rivals was the ability to sustain this extension of credit based on their own capital resources. The profits, however, were worth the risks. The Lisbon Factory in 1769 estimated an average profit of sales to the Portuguese traders between 12 and 15 per cent. The return of goods disposed of within Lisbon was relatively low, from 10 per cent. The profits of the British merchants dealing with agents outside Lisbon was even higher, 15 to 17 per cent. However, in the Brazil trade the return was from 25 to 30 per cent.[11]

Finally, the eighteenth century is characterised by the ongoing Anglo-French struggle which increasingly compromised Portugal. Lisbon tried to accommodate both, but by its very Atlantic nature, Portugal was tied inextricably to Britain, so that although it always sought the best of all possible worlds by remaining neutral, in the event of Anglo-French hostilities it was very rarely able to maintain neutrality long. Such then in very broad outline is the context with which the Portuguese state had to deal in the eighteenth century. It was a context which offered options, as well as imposed limitations on what might be done.

III

There were two distinct but inter-related aspects to the intellectual environment in eighteenth-century Portugal which influenced the way governments thought about the problems
confronting them. First, there was the immediate background of
intense debate over fundamental questions concerning philosophy
and education. Secondly, there was a considerable body of
thought about various aspects of Portugal's political economy.

The philosophical debate brought into question the role of the
Jesuits since they held a near-monopoly of higher education and
were, in the view of their opponents, the upholders of a dead
and sterile scholastic tradition, ill-suited to the age of reason.
The Jesuits held the exclusive right to teach Latin and philosophy
at the college of arts, the obligatory preparatory school for
entrance into the faculties of theology, canon law, civil law and
medicine at the University of Coimbra. The only other university
in Portugal, at Evora, was a Jesuit institution.

The stimulus in Portugal as elsewhere in Europe was the
intellectual achievements of Descartes, Newton and Locke, who
during the seventeenth century had promoted a bold break with
the tradition of authority – be it biblical or Aristotelian – and
promoted the merits of reason, experience and utility. The most
important works to emerge from this school in Portugal included
those of Martinho de Mendonça de Pina e Proença, who
attempted to adapt to Portugal some of Locke's theories; the
writings of the New Christian Dr Jacob de Castro Sarmento,
who introduced Newtonian ideas into Portugal; and the works
of António Nunes Ribeiro Sanches, another New Christian who
had left Portugal in 1726, working thereafter in England, Holland,
Russia and finally in France (from 1747 until his death in 1783)
where he was a collaborator of the Encyclopedists, and wrote on
medicine, pedagogy and economics. Most influential of all was
the Oratorian, Luís António Verney (1713–92), the author of *O
Verdadeiro Metodo de Estudar* (*The True Method of Education*) (1746).
Luís António Verney lived most of his adult life in Italy, where
he was a friend of the leading Italian thinker, Ludovico António
Muratori. He served for a time as secretary to the Portuguese
envoy at the Vatican, a relative of Pombal's first wife, Francisco
de Almada e Mendonça. Paraphrasing Newton, Verney wrote
that 'Philosophy is to know things by their causes, or to know
the true cause of things'.[12]

As elsewhere in Europe, however, much substantive discussion took place in private debating or philosophical societies. One major circle of critics of the *status quo* in Portugal had, since the turn of the century, centered around the house of the Ericeiras, made famous by the third count, D. Luís de Meneses (1632–90), a proponent of mercantilist development and Colbertian economic policies in Portugal during the late seventeenth century. Several short-lived conclaves of individuals, organised to discuss scientific and philosophical questions, had developed under the Ericeiras' protection. One of them, the *Academia dos Illustrados*, met during 1717 at the Lisbon house of Pombal's uncle. The fourth count of Ericeira (1674–1743), one of the most distinguished men of the *Academia dos Illustrados*, was director of the *Academia Real de Historia Portuguesa* founded in 1720. He had sponsored Pombal's election to the Academy on 24 October 1733. Pombal was the author of a eulogy of the fifth count of Ericeira, which appears to have been first published in London.

Priests also played an important role in the introduction of new ideas into Portugal. Some of the most outspoken advocates (as well as practitioners) of educational reform came from within the religious establishment. The activity of the Oratorians was notable. The congregation of the Oratoria de S. Felipe de Nery, a society of secular priests, had taken the lead in Portugal, as elsewhere in Catholic Europe, in the introduction of scientific experimentation and in the conflict with the Jesuits over pedagogical models. The Oratorians were strong promoters of the natural sciences, and also stressed the importance of the Portuguese language, grammar and orthography, which they believed should be studied directly and not through Latin.

In addition to this philosophical debate, part of a Europe-wide phenomenon, there existed an important current of thinking more specific to Portugal. This was a body of ideas and discussion about governance, economy and diplomacy, which had emerged during the first half of the eighteenth century among a small but influential group of Portugal's overseas representatives and government ministers. Sometimes members of this group were called pejoratively the 'foreigners' [*estrangeira-dos*] because of their supposed infatuation with foreign models. Yet their preoccupations were in fact intimately a product of a Portuguese milieu. D. Luís da Cunha, Portuguese ambassador successively to the Dutch Republic and to France, was the most formidable of these thinkers, and author of a comprehensive analysis of Portugal's weaknesses and the means to remedy them.

These discussions, unlike the disputations of the philosophers and pedagogues, in the main took place in private, and built on the longer tradition of Portuguese economic and diplomatic thinking which had emerged from the experience in the decades following the re-establishment of Portugal's independence from Spain in 1640. Less concerned with the specific impact of the golden age on Portugal, this debate focused on the broader parameters of Portugal's location within the international system, and confronted directly both the constraints and the options with which a small country like Portugal, part of Iberia but independent of Spain, had to live. Central to these discussions was the problem of retaining and exploiting the considerable overseas assets Portugal controlled in Asia, Africa and in South America; and developing a mechanism to challenge English economic domination without weakening the political and military alliance with Britain needed to contain Spain.

D. Luís da Cunha in particular had placed Portugal's problems in the context of its relation with Spain, its dependence on, and economic exploitation by, England, and what he believed were Portugal's self-inflicted weaknesses in terms of lack of population and spirit of enterprise. This sad mental and economic condition he attributed to the excessive number of priests, the activity of the Inquisition, and the expulsion and persecution of the Jews. The Methuen Treaty had, in his view, only benefitted England. He praised the short-lived attempt by Ericeira to introduce manufacturing industry in Portugal during the late seventeenth century. He proposed the creation of monopolistic commercial companies on the Dutch and English model. As to the purpose and impact of these commercial enterprises he had no illusions. 'There is no doubt', he wrote, 'that such companies are at base no more than monopolies defended by the state, because they take from the people the liberty to engage in certain commercial activities. But states should take such action when they see that although such intervention prejudices some subjects, in other areas it produces a greater utility.'[13]

Pombal of course also believed that the state should intervene in such a manner, and he also drew models from his interpretation of the experience of other European countries. From 1739 until 1744 he had represented the Portuguese king in London. These were critical years, the era of the War of Jenkins' Ear and

Vernon's attack on Cartagena, the great bastion of Spanish strategic control of the trade routes from Spanish South America. The period was crucial for the crystallisation of expansionist and imperial ideas and mythology in Britain. For Pombal the British threat to Portugal's vast and rich dominions in America became a major preoccupation. It was essential he believed to understand the origins of Britain's commercial and military superiority, and Portugal's economic and political weakness and military dependency. Pombal's sponsors within Portugal during the reign of D. João V had been especially involved in discussions over theories of government and strategies for economic development. One of them, Cardinal da Mota had supported the establishment of the royal silk factory of the Rato in Lisbon in the 1730s and in doing so had identified in some detail one of the central problems facing governments in eighteenth-century Europe, the choice between the state's long-term interest in mercantilist development and its short-term interest as tax gatherer. D. Luis da Cunha's highly complex and sophisticated critique of Portugal's international relations and social and mental condition was contained both in his 'instructions' for Marco Antonio de Azeredo Coutinho (1738) and in his testament (1748). Here the link to Pombal's thinking was direct. Azeredo Coutinho, who was secretary of state for foreign affairs, had previously been ambassador to London (1735–38) and before that ambassador to France (1721–28). Pombal and Azeredo Coutinho were cousins, although Pombal referred always to his distinguished relative as 'uncle', and it was at his 'uncle's' instigation that Pombal embarked on his diplomatic career.

In London Pombal, who had become a member of the Royal Society in 1740, set out to investigate the causes, techniques and mechanisms of British commercial and naval power. His remarkable library reflected his interests and was in fact a veritable treasure house of mercantilist classics. He used his extensive reading to formulate his famous critical account of the unfair advantages the British enjoyed in Lisbon and Oporto, advantages for which Portuguese merchants in Britain had gained no reciprocal privileges.

From London Pombal had moved to Vienna. Here he was no less observant, and in terms of contacts, he was very well placed, since he became the very intimate friend of Manuel Teles da Silva, a Portuguese *émigré* of aristocratic lineage who had risen high within the Austrian state and had been created Duke Silva

Tarouca by the Emperor Charles VI in 1732. We possess an extraordinarily frank and intimate correspondence between the two men. In 1750 Silva Tarouca, excited by Pombal's rise to power, wrote to congratulate his friend and to remind him of their conversations and hopes for the future: 'We are not slaves of fashion and foreign practices,' Silva Tarouca told Pombal, 'we conserve unalterably the names and external practices and national establishments; but still less are we slaves of ancient habits and preoccupations. If there is puerility in fashions there is folly in the obstinacy of old ways. When great new dispositions are necessary, they should always be put forward by ancient names and in ancient clothing.'[14] The Duke's formulation in many ways is a succinct description of the methods Pombal was to make his own.

IV

The new administration first concentrated on the colonial situation. One of the earliest initiatives reformed the methods of taxing Brazilian gold production. Vigorous measures were introduced to control contraband and provide incentives for those who co-operated with the authorities. The new government then moved to bring protection to the commerce and producers of the two most important export crops of Brazil, sugar and tobacco. On 1 April 1751, inspection houses (casas de inspeção) were established in Bahia, Rio de Janeiro, Pernambuco and Pará, to guarantee 'the good and just prices of these two most important articles'.[15] The inspectors included representatives of the merchant community and of the sugar and tobacco producers, elected through their respective municipal councils.

Strategic and security problems also focused attention on South America. The Treaty of Madrid, signed in January 1750, upheld the Portuguese claim to the Amazon basin. The new Lisbon administration had to implement the agreements which called for the evacuation of the Jesuits and their Indian neophytes from the Uruguayan missions, and envisaged a survey of the line of demarcation between Spanish and Portuguese America by two joint commissions.

The 'very secret' letter to one of the Portuguese commissioners revealed the full extent of Pombal's aims and hopes for Portuguese America. 'As the power and wealth of all countries consists principally in the number and multiplication of the people that inhabit it', he wrote, 'this number and multiplication of people is most indispensable now on the frontiers of Brazil for their defence. . . .' As it was not 'humanly possible' to provide the necessary people from the metropolis and adjacent islands without converting them 'entirely into deserts', it was essential to abolish 'all differences between Indians and Portuguese', to attract the Indians from the Uruguay missions, and to encourage their marriage with Europeans.[16] The instructions to the other commissioner, Pombal's brother Mendonça Furtado, recommended that he secure the liberation of the Indians, introduce married couples from the Azores, and encourage trade in African slaves. In practice this meant that the Indians should be removed from the tutelage of the Jesuits.

Pombal's brother, soon after his arrival in Brazil, recommended that a commercial company be established to facilitate the supply of African slaves to the Amazon region, hence releasing the pressure from the colonists to enslave and mistreat the native Indian population. He also wanted to see investment in the Amazonian economy to help develop exports. In 1755 in response to this advice Pombal established a monopolistic company for Grão Pará and Maranhão. The Company was given the exclusive right to all commerce and navigation for a period of 20 years. Simultaneously, Pombal drew up a decree ordering the expulsion from Brazil of those itinerant traders (commissários volantes) who had flooded the colony during the golden age as commission agents of foreign, mainly British merchant houses established in Lisbon.

The two measures were linked. The hidden objective of the Brazilian monopoly company was in fact broader than its regional focus might indicate. Pombal hoped that by granting special privileges and protection, national merchant houses would be able to accumulate sufficient capital to compete more effectively with British merchants in the colonial trade and by extension in Portugal proper. And by striking at the itinerant traders he had removed one key linkage between the foreign merchants in Portugal and the Brazilian producers. Pombal's aim in establishing the Company of Grão Pará and Maranhão was 'to restore the merchant places of Portugal and Brazil the commissions of which they were deprived, and which are the principal substance of commerce, and the means by which there could be established the great merchant houses which had been lacking in Portugal . . .'.[17]

Pombal was giving close attention simultaneously to the port wine sector, which by 1755 had been facing four years of unsteady markets and declining prices after three decades of unbroken prosperity. This produced recriminations between large and small producers, the former blaming the latter for saturating the market, as well as between Portuguese producers and English shippers.

In the Upper Douro Valley, the traditional source of the wine exported from Oporto, the major producers were large landowners – some secular, some religious orders. Substantial capital was needed to engage in the wine production, since it was necessary for the landowners to build and own lodges, presses, casks and vessels for transport down river. By the early 1750s, however, the traditional producers were facing a severe economic and entrepreneurial challenge with small producers selling from two to eight times more wine to the English merchants than the big vineyard owners. Wine cultivation, moreover, had spread into many other regions of Portugal, and even Douro producers bought wine elsewhere and passed it off as port. In 1755, the principal vineyard owners petitioned Lisbon for relief and assistance. Pombal responded by establishing the General Company for the Agriculture of the Vineyards of the Upper Douro (*Companhia General da Agricultura das Vinhas do Alto Douro*). The company statutes had been largely based on a plan put forward by the large noble vineyard owners of the Douro, that an officially designated zone for port wine production be established, any wine produced outside this area being prohibited from the export trade. The Douro producers reacted enthusiastically when Pombal announced his planned company in 1756, though the minister encouraged a mix of investment in the company's first capital fund from both producers and Oporto businessmen. To protect the larger estates, Pombal also abolished all entails yielding less than 100 milreis per annum in the north of Portugal, and 200 milreis in the south. He also issued legislation instituting strict primogeniture for entailed estates, prohibiting daughters from inheriting more than 4,000 cruzados unless there were no male heirs.

The Upper Douro Company aimed essentially to protect the Upper Douro Vineyard owners from the vast expansion of vine cultivation by smaller producers which had occurred over the previous decades. The title of the Company is invariably mistranslated (indeed it is very often mistranscribed even in

Portuguese), transforming *vinhas* (vineyards) into *vinho* (wine), thus misrepresenting entirely the major objective of the company. The company established a restricted production zone and exclusive name (a *'nom d'appellation'*) almost a century before the French.

Pombal's intervention in northern wine production was not unlike that in the colonial entrepreneurial nexus. Using state power to protect the large export producers, all opposition was ruthlessly suppressed, on a day-to-day basis, by the stringent enforcement of the Douro Company's monopoly rights, and more generally by the vigorous exercise of military and judicial authority. This policy, as with the protection afforded the producers of cash crops in Brazil, aimed at stabilising prices and market conditions. But as John Croft, a British wine merchant, observed in 1788, the exclusive zone in the Douro included 'the vineyards only of the principal Gentry and Religious Houses, excluding those of the menial Vintagers and Farmers'.[18]

The Company's purpose was not to seize the port wine trade from the British, though a monopoly to supply Brazil was granted to it. The privileged position port wine enjoyed in English markets was, after all, a result of *English* tariff manipulation (in favour of Portuguese wine), and was exactly the sort of reciprocal and mutually beneficial trading of which Pombal approved and had no intention of disrupting. On the surface, therefore, neither the Brazilian nor the Douro monopoly companies posed any direct threat to British economic power in Portugal, and neither violated the treaties which circumscribed Anglo-Portuguese economic relations.

Pombal's monopoly companies nevertheless met objections on several levels – not all of them made explicit. The fundamental objective in the colonial trades was certainly to diminish British influence, but the methods employed were subtle, pragmatic and enveloped in subterfuge. The problem that could not be avoided was that the British relationship was circumscribed by treaties which it was not desirable to do away with for political and security reasons. This, in fact, was the conundrum that had faced all Portuguese nationalists since the mid-seventeenth century, when the circumstances which led to those treaties had much to do with guaranteeing Portugal's independence from Spain. One way of taking action against British influence, however, while avoiding open confrontation over the actual terms of the treaties, was to use a variety of techniques in Portugal and within the colonial setting to shift concessionary economic advantages away from foreigners to national merchant groups. At the same time, the state intervened to protect agricultural producers – especially those of colonial cash crops in Brazil and the export wine producers in Portugal itself – in each case by seeking to control production and thereby assure stable markets and prices. Father Mansilla who was Pombal's liaison with the General Agricultural Company of the vineyards of the Upper Douro, put the policy with respect to the English succinctly: 'The objective', he wrote, 'was to hurt them in such a way they cannot scream'.[19]

V

Pombal's intervention within the entrepreneurial structure of the Portuguese economy did not go unopposed. The promulgation of the Company of Grão Pará and Maranhão's monopoly privileges and Indian emancipation from religious tutelage had provoked an immediate response from the dispossessed traders and Jesuits. Both found an organ for agitation in the *mesa do bem commum*, a rudimentary commercial association established in the late 1720s. In the face of these protests, Pombal acted swiftly. The commercial fraternity of Espirito Santo was dissolved, and the offending deputies were banished. The confiscated papers of the *mesa* revealed the extent of Jesuit involvement, and Pombal interpreted and dealt with the protest as if it were a conspiratorial uprising against royal power. The *mesa do bem commum* was abolished in September 1755. In its place Pombal established, in December 1756, a *junta do commercio* or board of trade, charged with the regulation of 'all affairs connected with commerce'.[20] Headed by a *provedor*, it consisted of a secretary, advocate and six deputies (four from Lisbon and two from Oporto), who were Portuguese born or naturalised subjects. The members of the junta were bound to a strict secrecy in their deliberation.

Meanwhile, a serious popular reaction to the Douro Company was developing in the north of Portugal, especially within the city of Oporto. Here were concentrated a variety of different occupational groups adversely affected by the establishment of the Company. The Coopers for example, whose guild had exclusive rights for cask production, feared the Company's power to requisition their services. The city tavern keepers were also adamantly opposed to the Company's charter which reduced their number to 95 (from an estimate 1,000 taverns in 1755). The Company had been granted a monopoly of supply and employed inspectors to control the quality of the wine sold at retail, and to verify if the wine sold was indeed *company* wine. So intense was working people's hostility to the Company that Belleza (a leading vineyard owner) advised Pombal in October 1756 that 'only fear prevented them from revolting'.[21]

Four months later, Belleza's fears proved justified. On 23 February 1757, according to the account of the municipal government, some 5,000 rioters besieged the house of the Judge conservator of the Douro Company, Bernardo Duarte de Figuei-redo. They forced him to concede the freedom to buy and sell wine as had been the case before the Company's establishment. The mob next stormed Belleza's mansion, ransacked the company's archives along with the house. Pombal reacted to the uprising – a mob action which dissipated as quickly as it had arisen – with ferocity, treating the event as an act of *lèse majesté* and giving full power to João Pacheco Pereira de Vasconcelles to head an official investigation. Pacheco Pereira was a director of the Douro Company, Judge of Oporto Customs House and a rich landowner of the Douro Valley.

The tribunal sat from April to October 1757, and tried 478 people; only 36 were absolved. The court condemned 375 men, 50 women and young boys: a total of 442 people. Some escaped and were hanged in effigy, but on 14 October 1757, 13 men and one woman were hanged, quartered and their limbs placed on spikes for 15 days. Ten women and 49 men were exiled to Africa and Portuguese India, and the remaining prisoners were flogged, sent to the galleys or imprisoned. Most had all their goods confiscated. Oporto itself was placed under military occupation, people could not hold meetings after dark, wear capes, carry arms or loiter, and to enforce these restrictions a further 2,000 troops were billeted in the city, adding to the existing force of 2,400 troops.

The creation of the Company of Pará and Maranhão also had unintended consequences. First, it linked the attempts to assert national control over sections of the colonial trade to wider geostrategic questions. Secondly, it brought Pombal, and no less importantly his brother, Mendonça Furtado, into headlong conflict with the Jesuits because Pará and Maranhão happened to be a stronghold of the Society's missionary activities, and the scene of bitter disputes between them and the colonists. The

Jesuit Indian villages in the South took up arms to defend themselves and oppose the implementation of the Treaty of Madrid, provoking a joint Spanish–Portuguese military campaign against them. The image of militarised Indians under Jesuit control opposing unilaterally the Iberian monarchs had a significant effect on the European mind. Voltaire, in *Candide*, portrays a sword-wielding Jesuit riding on horseback. More significantly, the events surrounding the attempted implementation of the Treaty of Madrid fortified Pombal's conviction that the presence of the Jesuits in Portuguese territories was an obstacle to his plans.

As opposition grew, Pombal moved to consolidate his own power. The socio-economic situation in Portugal had strictly limited the group from which he could choose his collaborators. The foreign dominance of commercial activity had restricted the Portuguese almost exclusively to internal and colonial trade. There was a handful of Portuguese houses in Lisbon which had the experience of exchange business, book-keeping methods and general commercial expertise to engage in business with foreign markets, among them the most notable were Bandeira and Bacigalupo, Born and Ferreira, and Emeretz and Brito, and even in these houses the Portuguese were in partnership with foreigners.

It was here, however, that Pombal found three of his most active associates. José Rodrigues Bandeira became the first *provedor* (superintendent) of the new *junta do comércio* and director of the Pernambuco Company. António Caetano Ferreira and Luís José de Brito were both to play significant roles in the formulation and execution of economic policy. A second potent group of entrepreneurs came from the Cruz family, brought into the minister's favour by the activity of the Oratorian, António José da Cruz. José Francisco da Cruz, a merchant with interests in Bahia and the tobacco trade, was closely involved in the formulation of the statutes of the Company of Pará and Maranhão, and became *provedor* and deputy of the company, administrator of the customs house in Lisbon, and was a close adviser to Pombal on various financial matters. His brother Joaquim Ignácio, who had made a most profitable marriage to an immensely rich Brazilian heiress, succeeded him in all his posts. The fourth brother, Anselmo José, succeeded to the Cruz fortune and became contractor to the tobacco monopoly. His daughter married Geraldo Wenceslão Braancamp, director of the Pernambuco Company and deputy to the *junta do comércio*, later to become Anselmo José da Cruz's heir.

The careful farming of the royal contracts was an important part of the state aid to those individuals Pombal hoped would found the 'great merchant house' that he wished to see established in Portugal. Tobacco was one of the major and most lucrative of the royal monopolies farmed out to the private parties. The contracts at Pombal's disposal were not confined to the metropolis or to the royal monopolies. Ignácio Pedro Quintela, himself linked to the tobacco interest and a member of both Brazil Companies, held the contract of tithes (*dizimos*) in Bahia. The right of collection of the *dizimos* in Brazil had been given up by the Church in return for fixed salaries paid by the state. The collection of the tax was farmed in the overseas council to private individuals, usually on a three year basis, just like any other metropolitan or colonial contract. Quintela also held the rights during 1754 and 1755 to the collection of duties on all non-fleet shipping entering Rio de Janeiro. In a similar way, José Rodrigues Esteves, another director of the Pernambuco Company, held the right to the duty paid on slaves entering Bahia.

The establishment of the Brazil Companies was closely linked to initiatives concerning manufacturing industries. In 1757 the *junta do comércio* took over the bankrupt silk factory in the Lisbon suburb of Rato. The factory had been founded during the 1730s by Cardinal da Mota and two French entrepreneurs. Capital had been raised, a company formed and a large building constructed. The early years had been difficult and during the 1740s the Company's deficits considerable. The royal takeover placed the *junta do comércio* in control, and determined that the directors of the factory should be chosen equally from among the deputies of the junta and from the directors of the Company of Grão Pará and Maranhão. The statutes were drawn up under the influence of José Rodrigues Bandeira, and the factory's product's were granted exemption from duties at the custom house.

The type of manufacturing agglomeration envisaged was based partly upon the industrial configuration of the period. Pombal also drew on his own analysis of the success of manufacturing enterprises in England. Instead of large capital expenditure on plant and equipment such as undertaken at the silk factory at Rato, he had seen industrial concentrations in England grow from individual units where only a small initial outlay had been necessary and returns were immediate. The English, he commented, 'only study the methods to make simple or cheap the means to establish them'.[22] The concept of manufactory as a co-ordinating centre resting firmly on the household producer was systematically applied to the Royal Silk Factory in 1766, and of the small units of production taken under its control only part were involved in silk production *per se*. By 1776, the Rato factory consisted of the central building with manufacturing, accounting, retailing functions, and an associated network of dependent individual workshops in other parts of the city. At the factory itself at least 91 looms were in operation and perhaps over 200 looms dispersed in a large number of small units of production. These independent producers were integrated with the factory for marketing and dependent on it for their supply of raw material.

It was the intimate connection between the factory and the monopolistic company which was of most significance. The same powerful directors controlled both enterprises, created a close and profitable relationship, and allowed for a fluidity of funds and mutually beneficial aid. The Company of Pará and Maranhão paid no dividends until 1759, and it is probable that capital from this source was secretly used to encourage the manufacturing enterprise. The Company's monopoly also provided an assured and protected market, particularly after the establishment of a second company in the important and populated market of Pernambuco and Paraíba in 1759. As the British factory complained, the 'companies have the sole permission and privilege to supply the Brazilians [within their monopoly area] and some of the directors openly declared, that their views and design are to prefer the exportation of the commodities of their own country's produce, which consequently must find a sale, when no other goods are in competition with them either in quality or price'.[23]

Both Brazil companies took special care to favour Portuguese manufactured goods, and well they might for it was the same group of investors, directors and merchants who benefitted at each stage of the process. The growth in export of raw cotton stimulated by the Brazil companies, and the flexible interpretation of the function of the silk factory, saw the creation of a lucrative triangle which brought profits at all points: as interest on the loans that stimulated the production of commodities that were to be shipped to the metropolis in company ships, and manufactured in company supported workshops, to be shipped back again on company ships to the colonial consumer, who bought them with company credit. The lucrative triangle was not the only part of the system (raw cotton after the Peace of Paris in 1763 provided a profitable re-export business) but it was the vital core. During the early years it powerfully boosted both the commercial and manufacturing activities, not to mention the capital accumulation, of the new Pombaline merchant oligarchy.

Pombal's attempt to introduce new business practices was very apparent in his attitude toward the upper Douro landowners. He wished to have four or five noblemen on the Company Board of Directors. The reason as he explained it was to encourage cohabitation: 'Where the nobility serves with men of business promiscuously and indiscriminately,' he wrote, 'it destroys the irrational and very prejudicial preoccupation that commerce is mechanical'. The election of aristocrats to the board was desirable, he continued, 'so that every two years, four or five nobles will leave trained in this very important science through the practice gained in the job'.[24] One of the most significant initiatives of the *Junta do Comércio* was to establish a school of commerce (*aula de comércio*) whose statutes were promulgated in April 1759. This school was to teach Italian and double book-keeping methods, and to give preference to sons of Portuguese businessmen in its three-year course of study.[25]

These attitudes produced adverse reactions from among the self-styled 'puritanos' among the Portuguese nobility. 'Puritanism' in Portugal referred to the old concept of 'purity of blood', that is the absence of Jewish or Moorish ancestry, a condition which since 1496 was required for office holding and state honours. The minister also sought to raise taxes 'without differences and without privileges whatsoever'.[26] The statutes of all Pombal's commercial companies used the allurement of ennoblement as an incentive to invest and not only offered to non-noble investors certain exemptions and privileges which were the prerogative of the nobility and the magistracy, but also admitted them to membership of the military orders. As to the nobles who invested, involvement in commercial matters was held not to prejudice their status but actively to aid its advancement. Incentive was also held out to the class of magistrates by permitting them to become shareholders in the companies, and making that involvement entirely compatible with the exercise of their administrative or legal functions.

The *mesa do bem commum* affair, the attack on contraband and the regulation of colonial commerce had already brought an identity of interest between the dispossessed itinerant interlopers, their English creditors and the Jesuits. The favours bestowed on Pombal's collaborators produced an identity of interest with the discontented nobles, for the group opposed by the interlopers and supported by Pombal also represented a potent challenge within Portuguese society to aristocratic privilege. The reaction was not slow in developing.

The crisis came to a head with the attempted regicide in 1758. D. José was returning to the palace after an evening visit to his mistress, the wife of Marquês Luís Bernardo of Tavora, when his carriage was fired upon. The King was wounded sufficiently seriously for the Queen to assume the regency (7 September 1758) during his recuperation. There was official silence on the incident until early December, when in a large dragnet operation a substantial number of people were arrested, including a group of leading aristocrats. The most prominent prisoners were leading members of the Tavora family and the Duke of Aveiro. The residences of the Jesuits were simultaneously placed under guard.

The King established a *Supreme Junta de Inconfidência* (9 December 1758) presided over by the three Secretaries of State and seven judges, but in fact dominated by Pombal. The tribunal acted with dispatch. On 12 January, the prisoners were sentenced to their several grotesque fates. The Duke of Aveiro was to be broken alive, his limbs and arms crushed, exposed on a wheel for all to see, burnt alive, his ashes thrown into the sea. The Marquês of Tavora Velho was to suffer the same fate. The Marquesa of Tavora was to be beheaded. The limbs of the rest of the family were to be broken on the wheel, but they were to be strangled first, unlike the Marquês and the Duke, whose limbs were to be broken while they were still alive. The sentence was carried out the next day in Belém.

The treatment of the conspirators was not out of keeping with eighteenth-century European practice. What was unusual was the status of the victims. The Duke of Aveiro was the second most powerful aristocrat in Portugal, head of the court nobles and president of the supreme court. The Marquês of Tavora Velho was a General, Director-General of the Cavalry, and had served as viceroy of India.

The day before this spectacular punishment, eight Jesuits were arrested for complicity, among them Father Gabriel Malagrida, a Jesuit missionary and mystic. Malagrida, who had been born in Italy, had gone to Brazil in 1721. After a brief sojourn in Lisbon between 1749 and 1751, he had returned to Brazil where he ran foul of Pombal's brother. Malagrida had also published a pamphlet attributing the Lisbon earthquake to Divine Wrath. Pombal had gone to great efforts to explain the disaster as a natural phenomenon. Pombal personally denounced Malagrida to the Inquisition, at the head of which he had installed his brother, Paulo de Carvalho. On 21 September 1761, after an *Auto da Fe*, Malagrida was garrotted, burnt, and his ashes thrown to the wind.

The execution (judicial murder might be a more appropriate phrase) of the half-mad old Jesuit in 1761 was the final straw to many Europeans who saw only Portugal's backwardness and Pombal's tyranny. It was a curious historical irony, however, that the last individual burnt at the stake by the Portuguese authorities at the instigation of the Inquisition should have been a priest and a member of the order which had been the very spearhead of the Counter-Reformation. As if to emphasise the point, the Chevalier d'Oliveira, an adventurer who had recommended that D. José establish a Portuguese Church on the model of the Church of England and later became a Lutheran, was burned in effigy alongside Malagrida. Clearly Pombal used the attack on King José I as a means to crush both aristocratic opposition and the Jesuits in Portugal. He also used the occasion to strike at the small traders whom he accused of plotting with the Jesuits against his plans to abolish their guilds and hence their representation.

VI

The first decade of Pombal's pre-eminence had seen important initiatives in several areas of state policy. Some of these had resulted from planning, others from unforeseen developments. In terms of economic and social policy, Pombal embarked on an ambitious plan to re-establish some measure of national control over the riches flowing into Lisbon from Portugal's overseas dominions, Brazil most especially, and to do this he adapted many of the techniques he had seen elsewhere in Europe to the Portuguese situation. He had also been faced with implementing the Madrid treaty involving a major effort to delineate and survey Brazil's vast frontiers. In both cases the Jesuits provided a major obstacle to his plans. On the southern border a military campaign had been needed to defeat the Jesuit missions. In the north of Brazil the missions ran into a headlong conflict with Pombal's brother. The disenchantment of the old nobility – the *puritanos* – upset at their exclusion from office and by the favours bestowed on the merchants and businessmen, as well as the escalating dispute with the Jesuits, together with the resentment of small merchants and tavern keepers excluded by the new monopolies, combined to produce a series of violent reactions, riots and assassination attempts. Pombal reacted ferociously, not only against the popular classes but also against the high nobility and the Jesuit order.

One immediate consequence of Pombal's drastic measures was to clear the way to action by the government on several fronts. The 1760s thus marked a period of consolidation and amplification of the reforms initiated during the previous decade. Pombal's administration moved aggressively in several areas to increase the power of the state. This involved the erection of a new system of public education to replace that of the fallen Jesuits; the assertion of national authority in religious affairs and church administration; the stimulation of manufacturing enterprise and entrepreneurial activity; a strengthening of the state's tax gathering power, its military capabilities and its security apparatus. In each case the necessary legislation was encapsulated within a reformed, codified and systematised system of public law where the reason for these measures was clearly outlined, justified and explained.

Central to the efforts of rationalisation and centralisation was the creation in 1760 of a royal treasury (*Erário Regio*) which received and recorded all the crown's income. Pombal appointed himself Inspector General of the Treasury, a position designed to be closest to the Monarch and by implication that of chief minister. The crime of *lèse majesté* was expanded to include attacks against the king's ministers, while Pombal obtained for himself a personal *corps* of bodyguards: something not seen in Europe since the pre-eminence of Richelieu in France.

Just as the Pombaline state engaged in propaganda to enhance its image and influence opinion elsewhere in Europe, so too did its legislation outline in sometimes tiresome detail for domestic audiences the objectives and antecedents of the policy changes, as well as the substance of the measure itself. In this respect the corpus of legislation establishing secular authority over the areas which had previously fallen under papal or ecclesiastical jurisdiction required special argumentation. In no country had the Counter-Reformation been so thoroughly embedded, or the Order that so exemplified the utramontane claims of papal supremacy – the Jesuits – been so warmly received, or Jesuit control so strongly established over the education of the élite. The struggle with the papacy was an inevitable result of the expulsion of the Jesuits. The Vatican, for its part, was horrified by developments in Portugal. The Papal Nuncio in Madrid was instructed in 1760 to warn the Spanish monarch, Charles III, that 'in that Kingdom [Portugal] . . . occult Hebrews and obvious heretics . . . benefit in every way from the greatest favour of the minister [Pombal]'.[27]

The occasion for the break with Rome, as so often in such cases of Regalist and Ultramontane conflict, was a dispute over a papal dispensation for the marriage of Maria, princess of Brazil, the heir apparent, to her uncle, the King's brother, Dom Pedro. In the face of foot-dragging by Rome the Papal Nuncio was expelled from Portugal on 15 June 1760; and on 2 July the Portuguese Envoy and all the Portuguese in Rome were expelled, including Pombal's eldest son, Henrique. The breach lasted for nine years, an important period during which Pombal moved to create a secular state fortified by a systematic rejection of papal claims of jurisdiction. Again Pombal turned to precedent. *Placet*, the right to exclude ecclesiastical documents; *exequatur*, the power to approve the delivery of papal documents to their Portuguese recipients; and the overall claim of *recursua ad principen*, the power of royal courts to hear appeals from ecclesiastical courts, had all been claimed by Portuguese monarchs since the thirteenth century, and were the substance of perennial disputes with the Papacy. Pombal used all these justifications to place the Church firmly under State control.

Equally important was the secularisation of the Inquisition, where Pombal installed his brother, Paulo de Carvalho. Book censorship, previously the Inquisition's responsibility, was usurped by a newly-created royal censorship board. The police powers of the Inquisition were taken over by a new intendant-general of police. The justification for secular claims over Church matters and the state's seizure of jurisdiction was provided by the Oratorians, especially in the writings of António Pereira de

Figueiredo and João Pereira Ramos de Azeredo Coutinho. Azervedo Coutinho's task was to justify the installation of bishops without recourse to Rome. The French experience proved especially attractive to the Portuguese ecclesiastical reformers. It was not that they became Jansenists or Gallicans as Rome darkly hinted, but that they took and adapted from others what suited their purposes which were in essence regalist and Catholic: that is which accepted the state's supremacy but did not wish to see Catholicism itself overthrown. They wanted papal authority circumscribed, the fraternal orders and regular clergy purified and their numbers limited, and they wished to see the power of the bishops expanded and made more effective. Again these moves did not go unopposed. The bishop of Coimbra specifically condemned the regalist writings the royal censorship board approved. Like so many of Pombal's opponents, he was swept into jail, and replaced by one of the minister's closest advisers and apologists, Francisco de Lemos (1732–1814), the reforming rector of the University of Coimbra. Lemos was the brother of Azervedo Coutinho, the regalist apologist. Both men were Brazilians.

The expulsion of the Jesuits had left Portugal virtually bereft of teachers at both secondary and university level. Not surprisingly the establishment of a state-sponsored system of secondary education and the reform of the University of Coimbra drew directly on the recommendation of the Jesuits' old enemies, the Oratorians and Luis Antonio Verney, the latter by now a paid consultant to the Portuguese government. The subtitle of Verney's famous work, the *Verdadeira metodo de estudar*, in fact, summed up the radicalism as well as the limitations of his educational philosophy. It was a method 'intended to be useful to the Republic, and the Church, commensurate to the style, and necessity of Portugal'.

To implement the educational reforms, Pombal had first established a position of Director of Studies to oversee the creation of a national system of secondary education. Later a Junta for the Provision of Learning was formed to prepare for the reform of higher education. A leading figure was Fr. Manuel de Cenáculo Vilas Boas (1724–1814), a brilliant scholar who had been educated at Coimbra and Rome and was an expert in Greek, Syriac and Arabic. The son of a candlemaker, Cenáculo became the reforming provincial of the third order of St Francis, president of the royal censorship board, confessor and preceptor to the heir apparent Prince Dom José and first bishop of Beja. He was a major influence in the reform of the University of Coimbra.

The Pombaline educational reforms had a highly utilitarian purpose: to provide a new élite to staff the state and church bureaucracy. It was here that the reforms would find their perpetuators and defenders. It is no accident that an almost classic statement of this process should have come from the pen of D. Francisco de Lemos, installed by Pombal as the reformer of the University of Coimbra:

> One should not look on the university as an isolated body, concerned only with its own affairs as is ordinarily the case, but as a body at the heart of the state, which through its scholars creates and defuses the enlightenment of wisdom to all parts of the monarchy, to animate, and revitalize all branches of the public administration, and to promote the happiness of man. The more one analyses this idea, the more relationships one discovers between the university and the state: the more one sees the mutual dependency of these two bodies one on the other, and that science cannot flourish in the university without at the same time the state flourishing, improving and perfecting itself. This understanding arrived very late in Portugal, but at last it has arrived, and we have established without doubt the most perfect and complete example in Europe today.[28]

In 1762, however, external events again intervened. The Diplomatic Revolution of the mid-1750s and the subsequent Seven Years War (1756–63) at first did not involve Portugal directly. Pombal believed, in fact, that he could avoid participation and sustain traditional Portuguese neutrality. Yet relations with Austria were obviously adversely effected, and with Charles III's accession in Spain in 1759, the conditions were created for the Third Family Compact between France and Spain (August 1761). French objectives involved shutting British commerce out of the continent, including Portugal, thereby forcing Lisbon from its neutrality into the broader struggle. In 1762 Spanish forces invaded Portugal on two fronts and Lisbon was obliged to seek British assistance. But Pombal's request for on the basis of 'the common cause' did not permit the British the opportunity to hold Portugal to closer observance of what the British regarded as obligations in the commercial sphere, where they now recognised Pombal was intent on challenging their dominant position. The British nevertheless sent troops and a commanding officer of considerable reputation, Wilhelm Graf von Schaumburg-Lippe-Stemberg, who arrived in July. He was the (illegitimate) grandson of George I, and was accompanied to Portugal by the Queen's brother, Prince Charles of Mecklenburg. The war itself was a dilatory affair with each side's weaknesses often proving more significant than its strengths. Both sides relied extensively on foreign troops and officers, though Portuguese popular opposition to the Spaniards proved decisive in places, especially in the North. Peace in any case was made before the end of the year (November).

The consequences of the Seven Years War were twofold. It provoked Pombal to begin a major overall of the Portuguese military under Graf Lippe's direction. And it compelled him to institute a comprehensive tax on all income and production by reviving the *décima*, a tax granted in the mid-seventeenth century to the Portuguese monarch by the Cortes in time of national emergency. The recourse to British assistance also forced much greater discrimination in dealings with Britain and the urgent extension to Brazil of many of the reform measures being implemented in Portugal itself, largely because Pombal viewed British successes against Spain's overseas empire as a potential threat to Portuguese America as well. These colonial measures included fiscal and military reforms, the diversification of cash crop production (coffee, cotton, cochineal, indigo) as well as the encouragement of colonial manufactures (cloth, iron). Several of the officers who participated in the military campaign of Graf Lippe, for example, his adjutant Johann Heinrich Böhm, and two of the more effective Portuguese aristocrats, the Marquês of Lavradio, and Luis de Sousa 'Morgado of Mateus', were sent to Brazil as military chief of staff, viceroy and governor of São Paulo respectively. The colonial garrison was later reinforced by two of the best Portuguese regiments.

As with the limited entrepreneurial capacity of Portugal, Pombal's reforms in the educational and administrative spheres always confronted the obstacle of the lack of personnel to implement them. For this reason he was compelled to make use of foreigners and to rotate from institution to institution the few modernisers he had at his disposal. The professors hired in Italy for the new college of nobles, for example, were subsequently moved to the University of Coimbra. The creation of human capital was a slower process than the accumulation of wealth through the manipulation of tariffs, or the concession of lucrative monopolies. Pombal's efforts to create an enlightened generation of bureaucrats and officials was to benefit his successors; he himself meanwhile relied on a very small group of collaborators, including a handful of enlightened gentry and aristocrats such as the Morgado de Mateus and the Marquês of Lavradio, and reform-minded clergy of modest origins or colonial backgrounds such as Cenáculo and Francisco de Lemos. Many of these men accumulted several posts just as his business associates accumulated positions in the management of fiscal and commercial affairs. Above all, throughout his government he used members of his family, especially his two brothers. With his powerful will and ruthlessness, Pombal was able to mobilise these scarce human resources and succeed in putting into practice a series of extraordinary measures. But his narrow base of operations was always a fundamental threat to the long-term success of his reforms.

There were also limits to what could be accomplished by legislation. Antonio Ribeiro Sanches, reviewing a copy of the law prohibiting discrimination against those of Jewish origin like himself, wrote in his diary, 'but can this law extinguish from the minds of people, ideas and thoughts they have acquired from their earliest years?'[29] Sanches, of course, highlighted the key weakness. The legal formulations of the Pombaline state were justified as an application of natural law, a secularised system which was a logical construct where reason (*boa razão*, as it was expressed) rather than faith or custom defined justice or injustice. In practice, however, the explicit constructs of the state were underpinned by the unstated networks of personal relationship, clienteles and self-interest. Such self-interest was seen in fact by Pombal as a means to fortify the state's objectives in economic policy, as well as in government. Yet to work, this required a vision which set the national interest above private interests. While Pombal ruled, this overall objective prevailed, though at the cost of continuous personal intervention and much repression. And as Pombal grew older, and as his brothers died, he became more and more repressive, suspicious of even his closest collaborators should they show too much independence or oppose his desires. Among the victims of his ire even the Oratorians were included.

The importance and the influence of Pombal's Austrian connections in his reforms have not been sufficiently appreciated. The impact was clearest in the relationship with Silva Tarouca. Pombal received a very wide array of visitors while Portuguese envoy to Maria Theresa's court, including Gerhard van Swieten, who was his personal physician as well as being that of the Empress-Queen. The reforming Dutch doctor was a principal agent of Maria Theresa's educational reforms (see below, p. 173).

In Vienna, the principle mechanism of educational reform had been the Censorship Commission under van Swieten. In Portugal the *real mesa censoria* (the royal censorship board) established in 1768, was intended to secularise the longstanding religious control and prohibitions which had controlled the introduction of new ideas into the country. Thus the *real mesa censoria* superseded the Inquisition and became the judge of what was 'good or bad' for the Portuguese reading public. The censorship of the state in this instance was paradoxically intended to provide a means of stimulating enlightenment. The *mesa* often released books which had been condemned previously by the Inquisition – among them Voltaires' *Oeuvres*, Richardson's *Pamela*, Montesquieu's *De l'Esprit des Lois*, and Locke's *Essay on Human Understanding*. But the limitations placed on readership are also illuminating. Works which were deemed harmful to religion remained excluded, and the principal censors were drawn from the reforming wing of the church, including the erudite Fr. Cenáculo. As in other areas Pombal took from the example of others what suited him.

Despite claims and fears in Rome and more traditionalist circles, the activities of the royal censorship board were exemplary. Dominated by reform-minded churchmen, its members carefully analysed the literary production of the high Enlightenment (and some works of a less elevated nature) and equally carefully removed from the Portuguese editions whatever they deemed detrimental to Catholic dogma; or, as was sometimes the case, they restricted circulation to those they believed should be aware of the offending works in order to be better able to refute their message. If measures of Pombaline organs such as the censorship board seem cautious and contradictory to purists, they seemed eminently dangerous, even sacrilegious, to traditionalists. Verney, for example, was viciously attacked by the Jesuits and their apologists. The Portuguese reformers were not free thinkers; they were seeking to promote what they believed would be useful to the state. In the context of eighteenth-century Portugal this was a major innovation.

VII

Pombal's final decade was overshadowed by a changed economic environment, and this had significant results for Portugal's political economy and indirectly for the well-being of his economic collaborators. The 1770s brought a major contraction in Brazilian gold production which produced a prolonged decline in colonial re-exports and consequently in the capacity to import, especially from Britain. The economic crisis was – in effect – a balance of payments problem caused by the diminution of gold production from Brazil, combined with a fall in Brazilian sugar prices as a result of the re-establishment of peace in 1763 and the increasing production in both the British and French West Indies. These economic changes, however, created an economic environment in Portugal conducive to the growth of manufacturing industries. As the Portuguese historian Jorge Borges de Macedo has demonstrated, the recession predated and accompanied Pombal's celebrated 'industrial' development. Of the manufacturing establishments set up by the *junta do comércio* during Pombal's regime, no less than 80 per cent were authorised after 1770.

The favourable organisational and entrepreneurial framework created during the 1750s greatly facilitated the new manufactories. Many of the establishments founded after 1770 produced luxury goods – silks, hats, chinaware, tapestry, decorative jewellery, ribbons and buttons. Often they received monopoly privileges, exemption from taxation, and the special protection of their supplies of raw materials. In addition a comparatively large number of cotton textile enterprises were set up, which were in part a response to the growing exports of cotton from Brazil. Sixty per cent of the workshops, established with the direct participation of the *junta do comércio* were in Lisbon and Oporto. The cotton manufacturers were almost exclusively located in coastal areas close to their seaborne raw materials.

In addition to the reorganisation and the establishment of royal factories on the model of the royal silk factory agglomeration, the state encouraged private manufacturing enterprise with exclusive or monpolistic protection. There was also fluidity of funds between the royal and private establishments. Joaquim Ignácio da Cruz, for example, used funds from the royal silk factory to encourage his silk stocking and paper box factories in Tomar, as well as the cotton textile workshops there. The labour cost of both royal and private privileged establishments was kept low by the imposition, sometimes violently, of a system of apprenticeship. A general protective framework for both types of enterprise was provided by a flexible tariff policy of exemptions and prohibitions in favour of Portuguese production.

A substantial number of entrepreneurs aided by the junta continued to be foreigners, and twenty-seven of fifty-two royal decrees issued for the foundation of new workshops went to them, and a third of these to Frenchmen. Typical of the more important recipients was Jacques Ratton, who had been born in Dauphiné in 1736, shortly before his parents emigrated to Portugal to establish a business in Lisbon. Ratton had been one of the first to develop the re-export trade with France in Brazilian cotton after the Peace of Paris and was to engage in a whole range of manufacturing enterprises in the production of calico, hats, paper, and cotton textiles. At Marinha, an Englishman, William Stephens, contracted by Pombal personally, and with whom Pombal was to develop a close friendship, set up a glass manufactory with junta aid and became a highly successful industrialist.

The Pombaline manufactories, despite the traditional bases on which they were constructed, and although surrounded by an ancient system of privileges, monopolies and apprenticeships, were essentially a new industry, fostered by varied techniques within a protective tariff regime, and made possible by the changed economic environment. The intimate relationship between the government's means of encouragement and its recognition of the economic and technical base available to it, both in the resilient household industry of the interior, and among the skilled artisans of the coastal cities, contributed to the general success of the establishments set up after 1770. Also important was the mobility of funds and of directing personnel between the monopolistic companies, royal silk factory, *junta do comércio*, royal treasury and the new manufactories. Mining output decline in Brazil had served to create an economic environment in Portugal favourable to the development of import substituting manufacturing enterprises, and this growth was also fostered by the skilful state intervention.

The special relationship and interconnected interests which

had been established during Pombal's first decade – particularly between the Brazil companies and the royal silk factory – had also placed his collaborators in an excellent position to take advantage of the changed economic situation. The 1770s in fact found the new merchant oligarchy firmly and powerfully entrenched, and a remarkably small, compact and inter-related group of men in positions of great power and influence both in government and in the world of commerce.

The state also continued to reward with noble titles the new great merchant dynasties it had so carefully stimulated and aided since 1750. The attack on *Puritanismo*, that is, the caste-like exclusivity of the hereditary aristocracy, was part of the process which had seen the ennoblement of Pombal's collaborators among the businessmen and the participants in his state-supported economic enterprises. In 1768, 'puritanism' was formally outlawed by royal decree. Joaquim Ignácio da Cruz obtained the entailed estates and title of Sobral, an ex-Jesuit possession, and the title passed to his brother Anselmo on his death. José Francisco da Cruz obtained the *morgado*, or entailed estates of Alagoa, a title which became that of his sons, aspiring pupils at the college of nobles. When in 1775 the inauguration of the equestrian statue of D. José I 'the Magnanimous', and officially, designated Portuguese 'Caesar Augustus', took place in the great new commercial place on the water front of Lisbon, it was Anselmo José da Cruz who stood at the right hand of the beplumed and ceremonious Marquis de Pombal. By 1777 the great merchant houses Pombal envisioned twenty years before had come to fruition. By state intervention and economic circumstance the Pombaline oligarchy had been created.

Yet the problem of perpetuating these changes remained. This concern was at the bottom of the educational reforms. Pombal believed that his most important innovation was the reform of the University of Coimbra. He had also placed one of his closest collaborators in this area, Father Cenáculo, as tutor to the heir apparent, Prince José. In order to diminish, and it was strongly suspected to pre-empt, the succession of the King's pious daughter Maria, Pombal arranged for the marriage of the young prince to his aunt, D. Maria Benedita, one of Pombal's strongest supporters in the royal circle. This marriage occurred when it was obvious that the King, Pombal's sponsor, had little time to live, and that the minister was unlikely to survive once the King was gone. The young Prince José, who was a fervent admirer of the

Habsburg Joseph II, was to survive for only a decade after his grandfather's death, however, succumbing to smallpox in 1788.

Pombal exercised vast powers, but his authority had always depended on royal support. This was both his strength and his weakness. His own position depended on the King's survival, a link the conspirators against the King's life had seen in the late 1750s. When the King died in 1777, Pombal was soon forced out of office. Many, though not all, of his associates followed. Cenáculo was ordered to resign as the prince's tutor, and required to take up residence at his bishopric of Beja. The reformer Francisco de Lemos of the University of Coimbra was replaced. Pombal's sister, the abbess of the convent of Santa Juana, was deposed. The Jesuits incarcerated since 1760 and the surviving aristocratic conspirators were released, as was the Bishop of Coimbra, who had opposed the regalist writers.

Pombal's fall was rapid. The situation became sufficiently threatening for him to withdraw first to Oeiras, and then to move north to his properties near Pombal. He had to travel incognito, but his empty coach was stoned, and troops were called out to prevent his Lisbon house from being burnt; the crowd had to be satisfied with burning his effigy instead. There was an explosion of denunciation and satires. Abandoned by many of his allies (though not all) Pombal prepared to face his enemies both in the juridical proceeding instituted, and also through a systematic written defence of his policies and actions.

Abroad, the fallen minister received credit for positive achievements. 'No one can deny him original talents for achieving views', commented the *Gazzetta Universale*, 'By the means of commerce, of agriculture, of population, he has laid the foundations of Portuguese independence, viewed with an envious eye by the greedy rivalry of Great Britain.'[30] It was one of the paradoxes of Pombal's image that although the catastrophic decline in British commerce with Portugal had been caused by a profound change in the economic system, following the decline of gold production in Brazil, the memory of the bitter disputes of the 1750s associated and credited the measures Pombal had taken during that period with the consequent achievement of a more balanced trade with Britain. As a result it was for the supposed results of the creation of privileged Brazil companies that his regime was simultaneously praised and condemned, depending on the personal interest or nationality of the observer. Defence of the measures of the 1750s became tantamount to a defence of Pombal himself.

In fact, the new regime was faced immediately with the need for a decision over the future of one of the most notorious of Pombal's creations, the Company of Grão Pará and Maranhão, and the question of the prorogation of the monopoly soon became the subject of a propaganda battle. On one side stood the Company's directors, and on the other those interests suppressed since the late 1750s and now vociferous in their hostility to all things Pombaline, backed by some of the Company's debtors in Brazil who saw in the change of regime an opportunity to escape their obligations. Strong pressure was brought to bear on the new ministry to suppress the monopoly and open the trade of Pará and Maranhão to all. The Company's directors stressed its national objectives and its success in reducing the dependency on Britain and claimed the capital invested in America had stimulated cotton and rice production. The directors of the Pernambuco Company, also faced with the threat of extinction, stressed its regulatory functions and the capital investment employed to reestablish sugar and tobacco production. It was precisely the Companies' investment, however, which had been the origin of the huge debt of the colonists, and the regulation of the supply of metropolitan merchandise to colonial production was blamed by the colonists for having produced high prices, shortages and exploitation, as well as resulting in the exclusion of non-Company merchants.

The Council of State voted to abolish the Company of Grão Pará and Maranhão, and the failure to prorogue the Pernambuco Company's monopoly was a logical consequence. The new era, claimed one of the anti-monopoly memorialists, would bring the 'liberty of commerce and the competition of businessmen, and mark the end to private privileges, half understood taxes, and a thousand vexations'.[31]

The extinction of the Companies was a visible triumph for the free traders and the old system, as well as for the Companies' debtors in Brazil. Yet the achievement was more apparent than real. The situation of 1777 was not that of the 1750s, despite the reappearance of the old debate. During the intervening years the economic system had been transformed. Despite the popular hysteria that accompanied the change of regime, Pombal's collaborators were far too deeply embedded in the social structure, and too closely associated with the collection of revenue and the government's fiscal agencies, to disappear by the mere abolition of the monopoly privileges. The deeper socio-economic factors which underlay their position in society made the attack on the monopoly Companies, in as far as it was an attack on them as a privileged group, at best a matter of form. Moreover, the results of the vast investment in Brazil, particularly in cotton, and the close interconnection between the Company of Grão Pará and Maranhão and the local fiscal and administrative structure, could not be obliterated by the stroke of a pen in Lisbon. In fact the 'extinct Company' remained a very real force, retaining administrators in Brazil, and actually continued trading during the 1780s. In the metropolis the removal of a central and unchallenged focus within the government was by no means a disadvantage to the interests of the opulent merchant houses which had arisen during the Pombaline era.

The most immediate consequence of Pombal's fall was that the vital directing influence of the centralised administrative structure, already overburdened and backlogged, faltered. An old aristocrat, the Marquês of Angeja, who lacked administrative experience and was ignorant of economic matters, became the president of the royal treasury. Under his dilatory care the treasury's role weakened, and its previously closely-supervised administrative machinery itself became negligent and infinitely more susceptible to corruption. The debilitation of this vital central government agency, together with the lack of a clear-cut focus of power within the new regime, created a situation where the privileged interests which for so long had been encouraged, protected and used by the state to further its nationalistic and imperial pretensions, found themselves in a position to manipulate the state for their own advantage. Their key roles in the royal treasury and the customs' administration, their directing influence in the royal manufactories, and their personal wealth and influence as contract holders and opulent merchants placed them in an unassailable position which the weakness of the state served to exaggerate. Thus while Pombal's enemies tilted at the windmills of the privileged companies, Pombal's collaborators increased and strengthened their wealth and influence.

As the royal treasury's supervisory role was undermined, the other great administrative agent of Pombaline government was weakened. By the Queen's failure to reappoint deputies, the *junta do comércio* membership was reduced to three. The administration of the manufactories, disassociated from the junta in 1778, was placed under the care of a new body, the *junta das fábricas*, which was concerned exclusively with metropolitan establishments, and the activities of the old junta in encouraging colonial manufacturing enterprise lapsed. The new regime also witnessed the state's retreat from the direct administration of the royal manufacturing enterprises. Again, 'liberalisation' was claimed to be the objective, but this did not mean the removal of privileges, the special protection of raw materials, or easy access to colonial markets. The royal establishments were alienated into the hands of the private capitalists closely involved in their establishment. Meanwhile the expansion of the manufacturing industry continued with added momentum. Between 1777 and 1778, 263 new workshops were established. By comparison the previous reign had seen a mere 96.[32] The new *junta das fábricas* provided a new focus for metropolitan manufacturing interests.

Thus while the reputation of Pombal, 'that great man, known as such to the middle and thinking class of his nation', as Ratton wrote much later, went into eclipse, the group he had favoured remained and prospered.[33] Liberalisation, far from undermining their power and influence, served to provide a cover for the manipulation of the state in their own interests, and their takeover of most of the enterprises established by the state. The merchant-industrial oligarchy retained the lucrative soap and tobacco monopolies from which they acquired gigantic profits. Ratton calculated that the contractor of the tobacco monopoly gained in one year more than the treasury had received from the contracts disposal in forty years. Extension of contract periods at set annual prices also benefitted the contractors to the detriment of royal revenues. The opulence of these Portuguese noble businessmen of the last quarter of the eighteenth century was praised by poets and pamphleteers and impressed visiting *literati*. It was 'the large and magnificent houses' of the Quintellas, Braancamps and Bandeiras that the English poet Robert Southey observed at the turn of the century, and whose 'clearing display of false taste and ill-judged magnificence' William Beckford noted in the 1780s.[34] 'Is there any one who does not do business?' asked Bernardo de Jesus Maria in his *Arte e Dicionario do Comércio e Economia Portuguesa*, published in Lisbon during 1783. 'Good customs and much money', ran the contemporary jingle, 'make any kind of knave a gentleman'.[35]

VIII

The peculiarity of Portugal in the eighteenth century, and of Spain too for that matter, is in part attributable to the coincidence of Enlightenment with the struggle of old powers to be great again, by adapting self-consciously the techniques they believed their competitors had used to surpass them.

The role of intellectual construction is therefore important in understanding eighteenth-century Portugal. Many years ago Fritz Hartung distinguished clearly between absolutism, that is, a form of government which is not hampered by parliamentary institutions, but which voluntarily submits to laws and acknowledges the rights of subjects, from despotism, which is unchecked tyranny. Pombal's Portugal was a hybrid, part absolutist, part despotic. The attempt to increase state power by improving the efficiency of the administration and the army and by stimulating the economy, that is, by a policy of mercantalism, was not uniquely or especially a characteristic of enlightened absolutism, and certainly Pombal did all these things. Yet Pombal also acquired and established for the Portuguese state four key monopolies of power – over coercion, over taxation, over administration and over law-making, which were indeed the enlightened absolutist's task and aim.

Pombal's particular combination of methods, however, reflected Portugal's unique position. There was a counterpoint between opportunity and necessity in all his activities. Many of his most important interventions were both reactive and creative at the same time. The army reforms followed the Spanish invasion of 1762. His overhaul of the educational system was the inevitable result of the expulsion of the Jesuits. The resulting break with the Papacy forced a re-evaluation of state–church relations. The increased emphasis on manufacturers followed the creation of an economic environment favourable to import substitution. But simply to enumerate opportunities does not imply they are inevitably exploited. Often opportunities are wasted, lost or unseen. In fact it in no way diminished Pombal to recognise the intimate relation between opportunity and response. Indeed it was his skilful manipulation of circumstances which made possible his successes in the economic and social sphere. Essentially, the all-powerful minister placed the power of the state decisively on one side of the conflict that had developed between Portuguese entrepreneurs, as a consequence of the gold boom.

He chose the large established merchants over their smaller competitors because he saw the small merchants as mere creatures or commission agents of the foreigners whom he hoped the large Portuguese merchants, with the state's assistance, would be able to challenge. Pombal had similarly thrown the state's support behind the large producers of the Upper Douro, protecting them from competition and, in the process, stabilising prices and quality. He had also given the Douro Company a retail monopoly, alienating the independent tavern keepers as well as the small producers. When they revolted, he crushed them ruthlessly. Likewise, in the long struggle over pedagogy and education, by adopting the recommendations of the Oratorians and of Verney he chose one side in an existing quarrel. The dispute with the Papacy was part and parcel of the reform movement in Catholic Europe which sought to diminish ultramontane pretensions and in effect nationalise the church. The reformation and codification of laws served to impose duties and obligations on subjects, not to grant citizens individual rights.

Pombal was a pragmatic and subtle adapter, one who almost always pressed against the limits of what was possible within the constraints with which he had to work – those of the long Portuguese eighteenth century and the Atlantic system. Within these parameters Pombal could draw on the considerable body of past Portuguese thinking as well as his own observations in London and Vienna. Above all he did not hesitate to act. Indeed in his action is his monument – for better or for worse. And whether it was better or worse depended mainly on who you were. To the great merchant houses he helped create, he was a hero. To the small traders he suppressed, he was a tyrant. To the port wine growers he protected, he was a patron; to the vineyard owners whose vines he ordered pulled up he was a scourge. And Pombal by and large achieved his objectives. His educational reforms made possible a flowering of Portuguese science and philosophy in the late eighteenth century. The merchants he favoured became the basis of a rich and opulent bourgeoisie. In both cases the state's role was crucial. As Dom Luis de Cunha had recommended, the liberty of the many was restricted to the benefit of the few.

Thus in Portugal, it was the state that created the bourgeoisie, not as in Anglo-America the bourgeoisie which restrained the state. This economic policy was logical in view of Portugal's position within the eighteenth-century international trading system. It relied on renewed state intervention within the entrepreneurial structure as well as in domestic and colonial markets, commerce and production. The policy protected mutually beneficial trade – the port wine trade, for instance – but it also sought to develop a powerful national class of businessmen with the capital resources and the commercial skills needed to compete with their foreign competitors. Far from being an imported policy it was a policy which grew out of the long tradition of Portuguese experimentation and debate dating back to the 1660s, and was based on a sophisticated assessment of the balance of social forces within Portuguese society. This nationalistic policy produced reactions inside Portugal, precisely because it intersected with other conflicts within Portuguese society – between old nobility and upstart businessmen, between modernisers of the educational system and defenders of tradition, between small and large entrepreneurs. Pombal dealt with the opposition ruthlessly. His reforms and his despotism are, therefore, inseparable. They were two sides of the same coin.

Hence, the Enlightenment, rationality and progress have a very different meaning in this context from that we have grown accustomed to. It is fundamentally the enhancement of state power we are speaking about, not the extension of individual freedoms. Pombal's actions were necessary, his apologists argued, to achieve progress. The problem with the idea of progress, especially for those deemed not to have progressed, was that it implied the stigma of backwardness, providing thereby a justification for actions which tradition, law or ethics had previously restrained. The interplay of these two notions – progress and backwardness – within the social, political and economic matrix of eighteenth-century Portugal is inextricably woven, consciously or not, into any interpretation of the age. Bolstered by rationalistic ideology which provided a convenient excuse for despotism, the contradictory images of eighteenth-century Portugal are in this perspective to a large degree dissolved. In a very real sense both images are accurate, because the other side of the coin of backwardness can be order, and the other side of progress can be tyranny. In neither was there much room for the rights of the individual. That space had been decisively pre-empted by the state.

4. Charles III of Spain

CHARLES C. NOEL

I

FROM the 1670s until its near-collapse in 1808, the Spanish old-regime monarchy, in an empire stretching from the western Mediterranean and North Africa across the Atlantic and most of the Americas to the Philippines, was subjected to a series of reforms. These aimed to recover Spain's military power and diplomatic position, to achieve economic growth, and to ensure domestic stability, and reached their peak during the reign of Charles III (1759–88). The late Habsburg and Bourbon reformers left almost no institutions or practices untouched, from industrial technology and the seigneurial system to book censorship and public health. Almost all were reformed, or at least subjected to critical scrutiny. Yet despite their energetic determination and talent, the reformers were rarely entirely successful. The age of what Spanish historians refer to as enlightened despotism – *despotismo ilustrado* – ended with many typical old regime abuses largely intact and with economic and political resurgence still elusive. Thus, not even Charles III, the most successful of the Bourbon kings, and his ministers were able to prevent the worst of all disasters, the eventual loss of almost all Spain's overseas empire after 1808.

The Monarchy – Spain and her overseas empire – was in many respects a typical old regime society. Men were divided into the privileged nobility and clergy, and the rest. But in this case the non-privileged rest consisted not only of peasants, workers of all kinds, and the commercial and professional middle classes, but also of millions of Indians, black slaves and freemen, and an almost endless variety of racial mixtures. Although in Spain the nobility dominated lucrative and influential church and government offices, they did not entirely monopolise them. In Spanish America, however, white, mainly non-noble creoles dominated commercial, cultural and political institutions, and were the rich landowners. Everywhere, leading urban economic institutions, like the *consulados* or merchant guilds, were controlled by the commercial, manufacturing and professional élite. But the nobility still controlled the municipal government of many Spanish cities and towns and dominated the seigneurial system. The latter, which affected about half the Spanish population, varied enormously from estate to estate, from one province to the next. By the mid-eighteenth century, seigneurial peasants usually had only nominal financial dues to pay their lords, but the latter still had jurisdictional rights to administer justice. These rights frequently annoyed peasants, but caused relatively few social confrontations.

The nobility and clergy thus dominated Spanish society through their control of national and local offices and the higher positions in the church, their manipulation of seigneurial justice, and their ownership of much of the landed and commercial wealth of Spain. Yet the authority of the nobles was not unlimited. They were divided amongst themselves into the rich grandees and aristocrats of the titled nobility, and the mass of untitled *hidalgos* – the famous dons who were usually country or city gentlemen of modest means. Often they were as poor as workmen or peasants. The two kinds of nobility confronted each other repeatedly during the century as rivals for influence and eventually some *hidalgo* intellectuals, like Gaspar Melchor de Jovellanos, led destructive attacks against the very nature of nobility as traditionally defined. Moreover, the nobility lacked significant representative bodies through which to express their will as an estate. There was nothing in the monarchy to compare with the British House of Lords. Only by holding government, church or judicial offices filled by the king could the nobility exercise influence over the state and protect their interests at the centre of political authority. By 1759, Spain had been united politically, though political uniformity in the peninsula and overseas was lacking. There was a single constitution for the Monarchy, with only Navarre in an anomalous position. Philip V (1700–46) had deprived Aragón, Catalonia and Valencia of their separate constitutions early in the eighteenth century. However, throughout the monarchy, officials – from the South American viceroys downward – often eluded full control from Madrid. For many reformers this lack of integration and control seemed particularly threatening.

The lack of integration was especially obvious in the relationship between Madrid and the overseas empire. For generations Spain had largely neglected colonial problems. In the first half of the eighteenth century Spanish trade with her American colonies was overwhelmingly in foreign hands, with its profits going above all to the French and British. Contraband trade was destroying legitimate commerce, corruption in the American administration was rife, and Indians were widely abused. Racism, slavery, intimidation and dishonesty had become the pillars of Spanish authority. Yet Spain profited little from this system.[1] By the 1740s some ministers and economic writers, like José de Carvajal, foreign minister of Ferdinand VI (1746–59) and José del Campillo had concluded that Spanish America needed drastic reform. The latter's *Nuevo sistema de gobierno económico para la America* (*New System of American Economy*) provided a key text for later reformers. Tighter administrative control, a stronger navy and a coherent trading system would benefit both Americans and Spaniards. By the 1760s it seemed clear to the reformers that America's resources and markets would provide the great opportunity for Spain's economic recovery from her seventeenth-century backwardness. 'America', as Carvajal wrote, 'is the soul of our greatness.'[2]

Despite her economic backwardness compared to Britain and France, the Spanish Monarchy was a vast, populous and potentially rich and powerful state. Its population stood at about 20–25 million in the 1760s – almost 10 million of which lived in Spain. Some American towns, like Mexico City, Lima and Havana, were beginning to rival major European cities in size and commercial vitality. American production of almost all significant agricultural and industrial goods was rising, often dramatically – from sugar and cacao to cotton cloth and Mexican silver, and Europeanised culture flourished. In Spain, too, the decay of the mid-seventeenth century was remote history by 1760. The peripheral regions of Catalonia, Valencia, Andalucia and Galicia were recovering especially well – faster than the central provinces of Old and New Castile and Aragón. Towns, particularly ports like Barcelona, Alicante and Málaga, were growing; the fishing industry was increasing; the technologically advanced Catalan cotton cloth industry was already flourishing. Even in Old and New Castile there was some progress with the establishment of textile, luxury wares and metal goods industries under Philip V and Ferdinand VI, usually owned or heavily subsidised by the crown. Agriculture was the crucial area of growth, providing support for the growing population. In Galicia, for example, sophisticated innovation by small and medium owners had, since the later seventeenth century, transformed the region into one of Europe's most agriculturally advanced regions. In Catalonia a similar process was underway, greatly enriching the wine growers and other owners, including Barcelona merchants who had invested in land for prestige and profit. Capital was being accumulated, some of which was ploughed into the cotton mills. In Valencia, Murcia and Andalucia new land was cultivated and new crops grown, including potatoes and maize. Not surprisingly, Roberto Fernández, introducing a collection of regional studies on the eighteenth century in Spain, concluded that the period from the late seventeenth century was one of marked demographic and economic expansion, ending only with the formidable crises of the 1790s.[3]

There were also signs of cultural and intellectual reinvigoration, inspired mainly by contact with recent French, Italian, English and German thought. As early as the 1680s in Valencia, reform-minded professors of medicine and anatomy had championed modern science, attacked scholastic monopolies of university teaching, and demanded modernisation of thought and attitude. By the first decades of the next century, they had been joined by thinkers like Manuel Martí, dean of Alicante, and Gregorio Mayáns y Siscar, an innovative and influential historian and philosopher who helped establish a vigorous and progressive Valencian provincial culture. Elsewhere, too, intellectuals had set out on the path of reform, and in Salamanca, Zaragoza, Seville and Madrid small groups, joined by scientists, moved in the same direction as those of Valencia. However, the outstanding propagandist of cultural renovation was the Benedictine monk, Benito Jerónimo Feijoo. In 1726 he began to publish his *Teatro crítico universal* (*Universal and Critical Essays*), a series which launched a massive and effective assault on Spain's intellectual and material backwardness. By 1760, he, Mayáns and a few other leading thinkers like Martín Sarmiento, had helped promote a limited but genuine intellectual reawakening among the educated, urban elite.[4]

By 1759, therefore, the monarchy possessed much of what could have supported an energetic reformist government – relatively strong institutions of central authority; a growing economy and population; and a degree of cultural liveliness and an awareness that modernisation was needed. What above all was lacking was determined reformist leadership in Madrid. Under Ferdinand VI, Charles III's dull-witted half-brother and predecessor, the monarchy had undergone a brief period of reform under the energetic guidance of the marquis of la Ensenada. With the co-operation of his fellow minister, Carvajal, Ensenada had undertaken a vast array of reforms – administrative, fiscal, educational, economic. He planned to transform the tax system of most of Spain by replacing many indirect taxes with his famous *única contribución*, ('single tax') to be levied on the wealth of all, including the nobility and church, thereby shifting the tax burden more effectively onto the backs of the wealthy. More significantly, the two ministers, determined to wrest domination of American commerce from British and French hands, had initiated colonial reforms and an effective shipbuilding and naval modernisation programme. However, in 1754 this push for recovery had been halted. Carvajal died and shortly after, alarmed by the naval build-up, the scheming British ambassador had engineered Ensenada's humiliating dismissal. Thenceforth, under Ferdinand's incompetent rule, the monarchy slipped into a kind of half-slumber. In 1758 a crisis arose when the King declined from mere incompetence into madness. Government came almost to a halt and Spanish America was left virtually unprotected from British interference. The widespread alarm caused by this collapse of authority ended only in August 1759, when Ferdinand died and Charles ascended the throne.[5]

II

By the 1750s the belief was widespread that drastic reform was needed. In 1745 Carvajal had written of the need for a Spanish 'resurrection', while at the time of Charles's accession, the Jesuit novelist and satirist, José Francisco de Isla, looked forward to the 'happy revolution' the new king would bring. To Manuel de Roda, one of Charles's outstanding ministers, the burden appeared enormous 'How much reform have we needed in Spain: So much that although we have tried, we have not known where to begin.'[6] There was general agreement among the reformers about many of the tasks awaiting them. Their primary aims were to make the Monarchy more powerful and to increase her wealth. These traditional aims were seen in a new light – the central role of her American colonies. Once protected from foreign interlopers,

Spanish America, through her commerce with the metropolis, would bolster Spain's economic recovery, providing raw materials like cotton, metals like silver, markets and tax receipts. Spanish merchants would have to learn how to supply these markets effectively; Spanish manufacturers to produce the goods the Americans wanted. The crown would legislate to secure and protect the exclusive Spanish monopoly of transatlantic commerce. Administrative and military reforms in America would enhance Madrid's authority there, safeguard coasts and ports, prevent or subdue Indian rebellions, send colonists into the borderlands of Texas and California to prevent Russian, French and British intrusions, and ensure a thorough integration of American policy with Spain's needs. Military – especially naval – strength was of paramount importance in protecting the empire from mainly British and French threats. Diplomatically, Spain would need always to negotiate from a strong position, to limit her commitments in Europe strictly in order to concentrate her efforts in America, and avoid war as a costly and uncertain gamble. No longer would Madrid pursue her old dreams of resurrecting her lost European empire.

This imperial policy, although strictly defensive, would demand vast efforts not only in America but in Spain. A powerful navy and effective army would be extremely expensive. Fiscal and administrative reforms in the peninsula would be needed to cut unnecessary expenses and increase tax revenues. More fundamentally, economic expansion of all kinds must be encouraged, partly as Charles himself and many other reformers saw it, to benefit his subjects, but principally to enhance the crown's resources. Spaniards' energies must be harnessed, their talents nourished and directed toward the monarchy's needs, their knowledge and skills developed. All, as Campomanes suggested, had to work together for Spain's benefit. Hence, clerics as well as laymen and the privileged elites, even women, would play their part. The body of the church and its hierarchy needed to be controlled, then mobilised to preach and support reform; the nobility and their capital brought in to provide leadership and material resources as well as educated talent in the effort; peasants' and artisans' habits needed reforming to make the land and industry more productive; and the middling groups of merchants, bankers, manufacturers and professional men had to be released from some of the restrictions which reformers believed had hindered their effectiveness. A culture of prosperity and modernity would be created, rejecting baroque obscurities and superstitions in favour of rationality, experimental science, political economy and critical history. Neo-classical 'good taste', as many reformers believed, would banish medieval barbarities in the theatre, bull-fighting and dress to make Spaniards more like other Europeans in thoughts and habits.

A defining characteristic of mid-eighteenth-century reformers was their conviction that compared to other European powers, especially her leading rivals, the British and French, the Spanish Monarchy was dangerously backward. Their task was to help Spain catch up, to modernise and to become more efficient. Being overwhelmingly pragmatic, they chose to act through the already available institutions of absolutist authority, convinced that only the latter could effect the kind of paternalistic, reformist government they valued. Strong royal authority would permit them, a self-conscious elite, to control the central political machinery in Madrid and to remould, where necessary, political institutions elsewhere. Ultimately, direction or at least inspiration would thus always emanate from the centre, integrating energies and institutions despite variations and local initiatives which reformers cautiously permitted. They assumed the need for social stability, the continuity of the existing social hierarchy, and the continued, even strengthened, command of social and political forces in the hands of clergy and nobility. Far from being revolutionaries, the reformers who dominated Charles's government were basically conservative renovators of traditional institutions.

In Spain, unlike in contemporary France, as Vicente Palacio Atard has suggested, 'reformers – the intelligentsia – and the crown – political power – identified with each other in a common enterprise'.[7] This was the case until the 1780s when some intellectuals began to question the need for absolutist government and the validity of noble and clerical leadership. Until then reformers, despite some quarrels about details or methods, and some conflicts of personality and rival ambitions, were largely united.

They were fortunate in the leadership given them by the king. Charles III was the most energetic and conscientious monarch to rule Spain from the seventeenth century until the mid-twentieth century. Though there have been some exceptions, most historians have credited Charles with the determination, experience, intelligence and humanity which helped make his reign as successful as it was. He impressed many contemporaries who knew him at first hand as honest, upright, good humoured, a man of 'good talents' who put much of his 'indefatigable' energies into personally transacting business regularly, despite his passion for hunting which took him into the countryside for hours every afternoon. With clear views and expectations, particularly in foreign policy, Charles directed his ministers, often taking a hand in the details, and reserving the final decision always to himself. Only as he aged, and relaxed into the skilled manipulation of his minister, the count of Floridablanca, after about 1777 or 1778, could he really be said to have shared final authority.[8] Charles's personal habits were admirable – he spoke, ate and dressed simply; he was relaxed and friendly with servants and workmen whom he often got to know well. Once his queen, Maria Amalia of Saxony, died in 1760, he was sexually abstemious. He was also extremely pious, though always suspicious of clerical power and above all of the Papacy and of his own prelates. On the other hand, he lacked genuine intellectual interests and was known seldom to have picked up a book once he had left the schoolroom. Nevertheless, he was an outstanding patron of painters and architects, whom he gathered about him in Naples and in Spain, earning his reputation as a discerning and generous supporter of refined baroque and neo-classical good taste.

When Charles ascended the Spanish throne in 1759 he was already an experienced ruler. The oldest son of Philip V, Spain's first Bourbon king, and his second wife, Elizabeth Farnese, Charles had left Spain for Italy in 1731 where at 15 he became duke of Parma. As the heir of the Farnese he inherited not only Parma, but an enthusiasm for the Farnese collections of sculptures and paintings. Then in 1734, with the help of Spanish forces, he conquered the kingdom of Naples from the Austrians and became the first Neapolitan and Sicilian monarch since the late middle ages to reside in his kingdoms. There, with the encouragement of some outstanding reformers, especially Bernardo Tanucci, perhaps Charles's greatest political mentor, Charles supervised a series of significant and often successful ecclesiastical, fiscal, administrative, economic and other reforms. Caroline Naples became a centre of innovative thought, above all in political economy, and the cultural capital of Mediterranean Europe. Charles therefore took with him to Spain in 1759 important lessons about the need for reform and methods of achieving this –

principally the need for vigorous direction from the throne, for control over the nobility and above all the church, for economic strength and for a pragmatic approach.

Among Charles' outstanding characteristics was his ability to attract talented and energetic advisers. His ministers were normally hard-working, honest and experienced. Occasionally, they were brilliant intellectuals, like Pedro Rodríguez Campomanes, a lawyer, economist and historian or, later in his reign, Gaspar Melchor de Jovellanos, one of the century's finest thinkers. Such men, holding a variety of ministerial, conciliar and other offices, formed the small core of activists who breathed the spirit of renewal into the machinery of Bourbon government. At the centre, besides Campomanes who made his office of *fiscal* of the Council of Castile into a focal point of reform, were the marquis of Esquilache, a humbly-born, tough and determined Italian whom Charles brought from Naples to reform royal finances: Manuel de Roda, minister of Grace and Justice from 1765, an outstanding regalist who helped control and reform ecclesiastical and educational policies; José Moñino, later count of Floridablanca, who started as Campomanes's colleague and co-*fiscal* in 1766, became a successful ambassador in Rome, and from 1777, as secretary of state, Charles's most powerful counsellor. Floridablanca, always a moderate reformer, came close to being a prime minister, dominating Charles's other advisers. The count of Aranda, one of the few grandees whom Charles trusted with power, was president of the Council of Castile (1766–73). Aranda was an Aragonese general, and although a cultivated, thoughtful individual acquired the reputation of being the strong man of the government. Others, like the radical reformer Pablo de Olavide, intendant of Seville; Francisco Pérez Bayer, an intellectual and university reformer; and José de Gálvez, the outstanding reformer of colonial administration, were primarily secondary figures who helped devise and implement reform, but seldom acquired power at the centre such as that held by Campomanes or Roda. Yet others encouraged Charles more informally – María Amalia until her early death was an enthusiast for some thoroughgoing changes, and helped influence the king; his Franciscan confessor, Fr. Eleta, frequently supported reforms, as did Charles's long-time companion, the duke of Losada.

The outer circles of reform were made up of ministers, councillors, secretaries, magistrates of the appeals courts who were also administrators, intendants, captains-general, and in

Spanish America, viceroys. New offices, too, were created to take reform into the corners of municipal administration – for example the *alcaldes de barrio* – or district aldermen who supervised small, neighbourhood jurisdictions in the larger towns and cities. While these officials were pressed to implement Madrid's policies, they also had the opportunity to inform and pressure the centre – frequently goading Charles's closest advisers to act. But the drivers of the machinery of power and thus of reform, the masters of the Monarchy, were almost exclusively from the paternalistic elite of middling and lower nobility. These *hidalgos*, often from peripheral provinces like Asturias or the Basque country, were also often from relatively poor families. Since they were making their careers as bureaucrats, professors or priests, they resented the power of rich grandees and the aristocratic domination of the complicated webs of patronage and influence which determined, from their point of view, too many fates. During the eighteenth century, in a tendency which reached its culmination under Charles III, the Bourbon kings had employed these ambitious *hidalgos* to buttress royal authority against the disruptive claims of titled aristocrats to share power at the centre. The domination of government by grandees and their allies, characteristic of later Habsburg Spain, was broken. Despite repeated, clumsy attempts to regain influence in the eighteenth century, the grandees remained excluded.[9] This was so even when ambitious magnates like the twelfth duke of Alba became protectors of rising *hidalgos* like Campomanes. Neither Charles nor the lower nobility were prepared to permit an aristocratic resurgence.

The frequently made and mistaken argument, on the other hand, that the reformers represented the thrust to power of the rising middle classes is belied by their social origins. The argument is further weakened, significantly, by the reformers' strong belief that the landed elite, reformed through education and propaganda, should continue to dominate Spain.[10] Even Campomanes, Olavide and Floridablanca saw the Monarchy's future in the hands of men with social origins like their own. As reformers they were the heirs of a long-established tradition of *hidalgo* service through public office, lay or ecclesiastical (or both). As magistrates or bureaucrats, *hidalgos* had dominated Spanish administration since the sixteenth century. The Caroline renovators continued this tradition and combined it with the equally traditional influence of *arbitrista* reform. The latter, promoted by sixteenth- and seventeenth-century writers like Sancho de Moncada, embodied both criticism of Habsburg policies and numerous recommendations for change. *Arbitrista* influence on the Caroline reformers was pronounced, particularly regarding economic and social policy. The result was that Charles's reforms had strong roots in Spain's past, in these twin traditions of *hidalgo* service and *arbitrista* reform.

There was a third source of inspiration amongst the reformers – the influence of the European Enlightenment. Enlightened thought had already affected some, like Feijoo and Campomanes, before the 1760s, and under Charles III the crown was to encourage some aspects of enlightened thinking still more. Eventually, a large number of intellectuals and social figures would be considered enlightened – writers like Jovellanos, Meléndez Valdés and Valentín de Foronda; economists like Francisco Cabarrús and León de Arroyal; journalists like Luis Cañuelo; and courtiers and aristocrats like the countess of Montijo and the duke of Villahermosa and others. Such men and women read Voltaire and Montesquieu and other French, British and Italian thinkers, frequently ignoring or bypassing Inquisitorial prohibitions on such works; they gathered in *tertulias* or salons hosted by either men or women; read and sometimes wrote for the slowly expanding and freer periodical press – a phenomenon encouraged by the crown as 'a useful instrument and a sign of modernity',[11] as in Charles's protection of the newspaper, *El Pensador (The Thinker)*. Eventually, the reformed universities, the new Economic Societies and some of the Academies were to become centres, located in numerous towns and cities even in America, of enlightened discussion and action.

At least until the final decade of Charles's reign, however, the Enlightenment in Spain accepted not only the political orthodoxy of absolutist government and social hierarchy, but the clerical orthodoxy of Catholicism as well. Until Charles's death there was no movement corresponding to deism in France, much less atheism. The men of the enlightenment – the *ilustrados* – although self-conscious of their role and their achievement, remained faithful Christians. Their reluctance to jettison these most fundamental beliefs allowed them also, too frequently, to accept many of the injustices of exploitation of Indians, black slavery and institutionalised racism against which they only occasionally spoke out. Their enlightenment, limited by their own hesitations as much as by the threats of the moribund, poverty stricken and intimidated Inquisition, has been called a Christian Enlightenment.[12] If such a hybrid could be said to exist anywhere, then it did so amongst these reformers whose demand for a secularised society founded on freedom was not voiced at all until the 1780s, and then only whispered. And almost never by the reformers who guided policy.[13] Instead of fighting for radical social and political change and against religiosity, reform in Spain saw the 'triumph of *proyectismo*' – the old *arbitrista* ideal of economic growth and limited social change through a search for rational solutions to the Monarchy's problems.[14]

The influence of enlightened thought in favour of critical rationality, against the grosser forms of religious superstition and secular ignorance, its support of the experimental and new social sciences, and its enthusiasm for material progress and humanitarian justice, were echoed, if often cautiously. The result was frequently opposition against many reforms, even sometimes the most cautious, against foreign ideas and against any serious interference with ecclesiastical authority and wealth. The power of the church was often perceived, by opponents of enlightenment and reform, as the great bulwark against atheism and social disaster. Traditionalists of many kinds, lay and clerical, looked to the church and some other institutions – the Inquisition, staffed partly by laymen, some of the university faculties, even occasionally to some of the councils in Madrid – to prevent assaults against their beliefs and interests. Some intellectuals, like Fernando de Zevallos in his *The False Philosophy* (1774–76) and Juan Pablo Forner in his *Philosophical discourse on man* (1787) became the paladins battling against the enlightenment. Others, like the popular preacher Diego José de Cádiz, took their campaign to the masses. But during Charles's reign books and sermons were less effective in preventing reform than manoeuvring within the government on many levels, or in local politics.

III

To the reformers, Charles's arrival in Spain was the homecoming of the hero who would inaugurate Isla's 'happy revolution'. Nor was his reputation in Spain, as a decisive reformer, belied by his first months on the throne. Having understood in Naples the perilous state of royal authority in Madrid, and of Spanish power on the Atlantic and in America, the King arrived determined to restore monarchical leadership and Spain's prestige and, above all, to protect Spanish shipping and her colonies from British 'extortions'. Initiating a six-year period of wide-ranging and intense reforms, Charles started by dismissing the finance minister he had inherited from Ferdinand VI, replacing him with Esquilache. With the latter's support, the encouragement of Queen María Amalia, and the co-operation of experienced ministers like Ricardo Wall, secretary of state since 1754, and Julián de Arriaga, the hard-working minister of the navy and Indies, the king demanded and secured more efficient administration, better co-operation from the bishops and other leading clerics, and the retreat into political quietude of the aristocrats, like the duke of Alba and the count of Oñate, who had taken advantage of the collapse of royal authority in 1758–59 to interfere in politics. As a symbol of his reformist intentions, Charles recalled the widely-respected Ensenada to court, and began gradually to fill important offices with the men who would help mastermind the expected renovation. In Madrid, he began a series of reforms of policing, administration and public amenities as well as beautification which transformed the capital from the mean and unbearably filthy city which greeted him in 1759.

On two fronts in particular, the reformers initiated policies which caused serious opposition. Esquilache and Campomanes undertook financial reforms which frightened and angered some clergy and other large landowners. The old proposal for the *única contribución* was revived and a committee set up to prepare for its imposition (1760). It was thereby planned to subject clerical and noble wealth to a proportional tax for the first time. Moreover, the old Council of Finance had almost all its authority removed and placed in the hands of the more energetic, reformist and more easily controlled ministry of finance, directly under Esquilache's supervision. In 1761, the crown took over the direct administration of the important ecclesiastical tax, the *excusado*, eliminating much of the systematic fraud the clergy had practiced to lighten its burden. By the end of the century, the yield of the *excusado* had increased six-fold. Finally, against the rich seigneurial nobility, Campomanes launched a programme of forced re-incorporation of their seigneurial rights and property into the crown, from which they had been alienated primarily in the sixteenth and seventeenth centuries. Though this policy was not greatly successful, some grandees, like the dukes of Alba and of

Arcos, lost both income and local political power.[15]

Campomanes, a brillant canon and civil lawyer, also helped strengthen and direct the crown's regalist policies. Regalism, with strong roots in sixteenth- and seventeenth-century Spain, was a complex of theology, and legal and historical theories and policies which aimed to enhance royal authority over the church and clergy in Spain; to reduce that of the Pope and the Roman curia; and by the mid-eighteenth century, to employ royal authority to force certain clerical reforms. Campomanes, formulator of the 'most radical and coherent regalism' of the later part of the century, had the support of a sizeable body of bishops and other clerics who shared regalist views and who were influenced by Jansenist reform ideas.[16] Jansenists, both laymen and clerics, wanted a more strictly spiritual Catholicism, purified of baroque excesses, worldliness and superstition. They valued a better educated clergy and laity and hoped, with episcopal and royal support, to mould a new enlightened Christianity founded on a rational piety. Since they and some of the reformers, like Campomanes and Roda, agreed on many aims, Jansenist-influenced clerics were favoured for leading church appointments, all of which were filled by the crown. Jansenists, including Felipe Bertrán and José Climent, were made bishops and employed their office to support Jansenist, regalist and other reforms. Jansenist fear and hatred of the Jesuits, again shared by most of Charles's reformers, as well as by the King himself, also endeared them to the crown. Helped by Wall and others, Campomanes pushed regalist reforms of the Inquisition and its relationship to the crown and to Rome. When the pious, well-intentioned but stubborn Inquisitor General resisted, he was exiled from Madrid. More significantly, Campomanes launched his plans to limit strictly the further acquisition of landed property by the church. His proposed law was introduced into the Council of Castile at the same time that his book, *Treatise on the Law of Amortisation* (1765), was circulated to justify a radical limitation of the church's freedom and wealth. The *fiscal's* law would have effectively 'amortised' all non-ecclesiastical lands, prohibiting their acquisition by the church except with specific royal permission.

There were other important reforms in these years. The grain trade was de-regulated, in a blow against the quasi-monopolistic grain merchants who had long victimised the consumer. Despite their frequent and successful evasions of the free grain trade policy, and the severe grain shortages of 1765 and 1766 – which naturally enhanced the merchants' profits as they increased their prices – the merchants were hostile to the reform. Closer to the throne, and as part of his own personal programme of reform, Charles was determined to break the influence that the six *colegios mayores* and their graduates, the *colegiales*, had over the bureaucratic, judicial and clerical machinery of the monarchy. The *colegios mayores*, established generations earlier to house poor university students, had been transformed into cliquish and exclusive clubs which offered easy comfort to their members for years, even decades, until fellow graduates holding posts of authority were able to infiltrate them into the plum posts of government and church. It was a corrupt, widely-resented system which brought many undeserving, ill-prepared men to office, protecting the narrow careerism of an elite within the elite. Charles, in order to destroy the stranglehold of these 'near masters of the monarchy' promoted their rivals, the *manteístas* among whom Campomanes was an outstanding spokesman. Though from much the same noble, landowning and office-holding background as the *colegiales*, the *manteístas*, excluded from the *colegios* and easy access to high office by their poor family or regional connections, helped the king in his attack. They were not only *revanchistes*, but convinced that vigorous and effective government would be impossible while the *colegiales* dominated. In 1764 Charles began to pry loose the latter's control over university professorships, and with the help of *manteístas* like Roda and Campomanes, systematically reduced their numbers in key bodies like the Council of Castile. Thereby, the crown began to free itself from this domination by clientage, patronage and *camarilla* groupings of the *hidalgo* elite.[17]

These early assaults against privilege and inefficiency were undertaken against the background of the Seven Years War and its aftermath. Already raging when Charles succeeded to the throne, the war, even before Spain reluctantly joined the belligerents in 1762, demanded that urgent attention be given the navy and army and to colonial policy. Esquilache's economic and fiscal reforms, like those of Campomanes and other ministers, were spurred on by the need to prepare for and effectively conduct a largely maritime and colonial war. Spain's poor performance during the war and above all the deep shock of seeing the British enemy capture and hold two important colonial capitals – Manila and Havana – in 1762, hardened the government's reformist resolve. More than ever, changes would be necessary to secure

Madrid's control over her empire and to block British intrusions into Spanish America.

Before further reforms could be undertaken, however, the Monarchy was confronted by the severest crisis of the reign – the uprising in Madrid (known as the uprising against Esquilache) and nearly 70 other Spanish cities and towns in the spring of 1766. The debate amongst historians about their causes and even their consequences continues, and no general consensus is available. However, most scholars accept that a number of factors encouraged the masses of Madrid, followed by cities like Zaragoza, Cuenca, Barcelona, Bilbao and Cádiz, to riot and demand a series of reversals of policy. High food and other prices; the newly-imposed dress regulations; xenophobic hostility to Esquilache and some of Charles's other Italian ministers – these had combined to cause widespread distrust and hatred of the Madrid reformers. Almost certainly, in Madrid if not elsewhere, some clerics and members of the nobility, some Jesuits and their allies, the clique commanded by Ensenada, took advantage of popular discontent. They may have manipulated the crowds against their target – Esquilache and his colleagues, the fomentors of feared and much-hated reforms.[18]

Whatever their aims and involvement, the opponents of reform gained little. Although Esquilache was dismissed as the crowds had demanded and Campomanes's law of amortisation was voted down in the Council of Castile following the Madrid rioting, most reforms proceeded. Indeed, the strong reaction against the disorders in Madrid and elsewhere strengthened the hand of Charles and his reforming ministers. Aranda was brought in as President of the Council of Castile, adding his support to Campomanes's endeavours. He also reinforced the military's presence in the capital, extending even further the Bourbon tendency to rely heavily on the captains-general of the provinces to maintain order and buttress royal authority. Moñino, appointed on Campomanes's recommendation as a fellow *fiscal*, bolstered the government's regalism while municipal administrative reforms in Madrid and elsewhere produced more carefully policed and rigidly supervised cities. The latter was achieved in the office of *alcalde de barrio*, first established in 1768. The *alcaldes* took royal authority into the corner of every home and shop with wide powers to intervene and arrest in cases ranging from family quarrels and drunkenness, to price regulation and marketing frauds. Though, surprisingly, elected by indirect universal male suffrage, they were the thinly-disguised intrusion of Bourbon absolutist demands for order into Madrid and other leading cities whose unruliness had so frightened the government in 1766.

From the reformers' point of view, their greatest triumph capped the crisis of 1766. This was the expulsion of the Jesuits from the Monarchy. To almost all the Caroline reformers the Jesuits appeared, along with the *colegiales* with whom they were closely united in a complex network of mutual support, the Monarchy's worst enemies – opponents of regalism, ultramontanists, authors and teachers of resistance to royal authority, subverters of reform, the manipulators who plotted the uprisings of 1766. The reformers found numerous allies among Jansenists, regalist bishops, the heads of religious orders like the Augustinians who hated the Jesuits for their own reasons, and the ambitious who saw the strength of the anti-Jesuit movement. Having filled important clerical posts since 1760 with anti-Jesuits, Campomanes, Roda and others found their plans went smoothly. With the support of the King, and in a masterstroke of Bourbon efficiency, the Jesuits were expelled from Spain beginning on 2 April 1767 and a few weeks later in Spanish America.

The departure of the Jesuits encouraged the reformers to turn to the institutions the Society had most thoroughly influenced – schools, the universities and the *colegios mayores*. Many reformers, like Jovellanos and Cabarrús, saw education as a national problem, the resolution of which would redeem the monarchy. University and other institutions of higher education needed to be integrated better into society to meet its economic and cultural requirements; they needed to reflect the best achievements of modern science and research; they needed to be taken out of the hands of the religious orders whose teaching was lax and backward. Thus the reformers in Madrid, led by Pérez Bayer and Roda, embarked on a wide ranging programme of administrative and curricular reform. But their real inspiration lay in the universities themselves. At Salamanca, Zaragoza and some other universities, a few reformist professors, like José Cevallos, professor of philosophy at Seville, were already demanding modernisation of courses and more conscientious teaching. From many sides, university renovators asked ministers to intervene to support them. They also widely agreed that elimination of *colegio mayor* influence in the universities was of fundamental importance. As Pérez Bayer informed the king, the *colegios mayores* were 'diametrically opposed to the public good of these kingdoms' because of their corruption of teaching and student life. In 1771 drastic reforms of the *colegios* began, and their influence over university teaching weakened. Finally, the *colegios* were entirely suppressed under Charles IV in 1798. To reform the universities themselves was an enormous task, considering the determined opposition from religious orders, some tenured professors who feared for their careers, and the lack of funds. Campomanes, Roda, Olavide and others expended much energy on the universities, but not always successfully. Decrees were issued repeatedly in the 1770s and 1780s requiring extensive administrative and curricular changes, particularly in natural sciences, medicine and law. But without the fullest co-operation of professors and the funds to build new scientific and other facilities, renewal was slow and piecemeal. Yet by the late 1780s much had been accomplished. Medical and science curricula had been brought largely up to date, and in the better universities, like Salamanca, Seville and Alcalá, a new spirit of curiosity and philosophical daring was widespread. If nothing else, the push for university reform had encouraged debate about new ideas, the nature of education and the role of the university. The movement had unleashed much radical talk which substantially affected intellectual and political life early in the next century.

Having only limited success with university reform, Charles's ministers turned elsewhere to find the means to modernise intellectual life and teach the scientific and technical skills they believed Spain needed. New, advanced secondary schools were opened in Madrid and elsewhere, directly by the crown or by other bodies with ministerial encouragement; surgical and pharmaceutical and other colleges and courses were established in Barcelona and other cities; the Madrid Seminary of the Nobility was organised to teach *hidalgos* scientific and technical subjects – to help them lead Spanish society more effectively; and some attempts were made to encourage cheap and even free primary education in Madrid and other cities. The great centrepiece of intellectual and technological modernisation, how-ever, was to be the Economic Societies of Friends of the Country. Their task, as their foremost promoter, Campomanes, defined it, was to bring together reform-minded gentlemen, clerics, ranking bureaucrats and other members of the propertied elite. In their societies they would inspire each other and outsiders to undertake the tasks of renovation. If the ministers in Madrid were the captains of reform, the members of the economic societies would provide its crew. The idea originated amongst a group of Basque dilettantes and reform-minded *hidalgos* in 1763, inspired by provincial academies and societies they had read about or knew from their travels abroad. The Basque model had little impact until Campomanes issued his call, through his essay *A Discourse on the Encouragement of Popular Industry* (1774), for patriotic men of the elite to establish such bodies throughout Spain. Campomanes expected them to be 'a public school of the theory and practice of political economy in all the provinces', to encourage the ideal of an educated, skilled elite serving Spain's economic needs. Their inspiration was to be not only foreign models, but the recommendations of the *arbitristas* and their own and the crown's perception of the Monarchy's needs. Members were expected to respond to local and provincial conditions and to act to safeguard and further the interests of agriculture, commerce and small-scale manufacturing. By combining their own and the state's economic interests they would further both, acquiring new, industrious habits and teaching them to their workers and peasants. The widespread and rapid response to Campomanes's invitation suggests that local elites may already have been thinking on much the same lines. By 1776, seven societies had been established, including one founded in Madrid by Campomanes and a few bureaucrat and merchant colleagues. By Charles's death over seventy had been established in Spain, and within a few years afterward, several founded in colonial cities like Manila, Lima and Havana. In some cities, like Barcelona, thriving older institutions – like the latter's Committee on Commerce – undertook the same tasks as the new societies. The societies, financed by membership dues and sometimes by royal or other grants, engaged in a wide range of educational and promotional programmes – establishing technical and primary schools, workshops and factories, museums and libraries; conducting scientific, agricultural and industrial surveys and experiments; reforming charitable institutions; and some, by accepting women members, integrating women into the provincial reform movement.

Their members were overwhelmingly from the same propertied elite who manned the government, church, army and municipal institutions of Spain and America. The inclusion of small numbers of merchants and manufacturers in this elite was no innovation – such individuals had infiltrated themselves since the late middle ages. The societies thus represented the traditionally powerful educating themselves and preparing to maintain their domination of a monarchy they themselves would help renovate. What was new was the release of provincial and local energies and talents, not only permitted but encouraged by the crown. As elsewhere in Europe, Spain's local elites were improving themselves, improving, but not transforming their world, cautiously sharing in the progressive culture of the Enlightenment.[19]

Many other important reforms were undertaken in Spain, especially in the 1760s and 1770s when the energies of both the King and his ministers were at their peak. Agrarian reforms were encouraged by forcing municipalities to sell lands they owned to peasants; by easing the process of enclosure; by establishing agricultural colonies in uninhabited areas – the famouse Sierra Morena colonies directed by Olavide are the best known example. Industry was encouraged with protective tariffs, subsidies and direct investment by the crown, and by encouraging greater respect for merchants and manufacturers by removing all legal stigma attached to traditionally 'ignoble' economic pursuits. Restrictive guild practices were limited. Charitable institutions were reformed, workhouses and municipal poor boards established, and the indiscriminate hand-outs of food and money by the clergy and laymen were discouraged. The movement of the clergy, their education, the relationship between the regular orders and Rome were all reformed and more carefully regulated. The Inquisition, despite the desire of some ministers drastically to reform or abolish it, was left in place. But, subjected to close ministerial scrutiny and the control of reformist inquisitor generals like Felipe Bertrán, it lost most of its threat for intellectuals before Charles III's death. Its revival under Charles IV was a reaction against revolutionary events abroad. Military and, more significantly, naval reforms were undertaken, revolutionising naval architecture, improving naval administration and modernising officer training, as well as building more and larger ships.

In America, and in America's relationship with Spain, significant reforms began to be undertaken within months of the conclusion of peace in 1763. With the sometimes hesitant support of Julián de Arriaga, administrative changes were initiated that would integrate more fully than ever Spanish and American institutions and policies and render them more effective in implementing further reforms. The crucial reform was the establishment, beginning in Cuba in 1764, of the American intendancies. With the encouragement of José de Gálvez, sent out as a trouble-shooting *visitador*, or Inspector, intendants had been sent to most parts of the colonies by 1788. They became key officials in achieving both integration and flexibility, with a wide competence which varied from province to province. Making themselves into the eyes and ears of the reformers in Madrid, they attacked many of the abuses and most of the corruption they found, sometimes successfully. Other *visitadores*, like José Antonio Areche in Peru, were despatched to investigate conditions and recommend reforms. Effective and conscientious viceroys like the general Alejandro O'Reilly were sent out, and new administrative and judicial districts established – all to tighten Spanish control. Colonial armies were also reformed and militias established. Not only to defend their land against British and other attacks, but to discourage or put down Indian rebellions which were a regular feature of the 1770s and 1780s. To help prevent further uprisings, the abusive *repartimiento* – forcing Indians to purchase goods at inflated prices – was prohibited, and levying of Indian labour reformed. So were taxes and fiscal administration, the slow postal service, and the all-important mining regulations and technology. Bureaucrats experienced in America were systematically given office in the Council of the Indies and in the Indies Ministry. In the latter, Gálvez, appointed to succeed Arriaga in 1776, brought greater determination, energy and knowledge. But perhaps the capstone of colonial reforms was the elimination of the unwieldy and counterproductive Cadiz monopoly of transatlantic trade. The monopoly had required all goods travelling between America and Europe, including those from other Spanish cities, to be funnelled through Cadiz. The system had discouraged commerce between Spaniards and the colonies and encouraged the contraband trade in foreign hands, the latter being the very fraud the monopoly was designed to prevent. To promote economic growth on both shores of the Atlantic, Gálvez and other reformers had the monopoly broken – against much resistance, in stages from 1765 to 1789. These and other reforms allowed most Spanish ports to trade directly and easily with their American counterparts. New mercantile connections grew uniting Spain and her empire as American trade through Spanish ports, and of Spanish manufactured and agricultural goods, expanded several-fold. At least one of the reformers' greatest aims was being achieved – Spanish producers were replacing their European rivals in the Spanish American economy. By Charles's death, the mark of success of American reform lay in rising prosperity, increasing population, growing silver production and soaring revenues going to Madrid.

IV

At the time of his death in December 1788, Charles and his ministers could have assessed his reign with some sense of achievement. Spain appeared stronger, especially with regard to the other maritime powers, than for several generations. The empire was larger, with the additions of Florida and Louisiana; the Monarchy more prosperous and better populated. New towns had been founded in Spain and in California, fresh lands brought into cultivation. The economic and administrative integration of the metropolis and her colonies had been increased, and throughout the empire governing political institutions were better articulated, and somewhat better adapted to implementing reform. Moreover, much talent had been assembled in both central and provincial government, in Spain and America. In other institutions, like the universities and the economic societies, numbers of reform-minded subjects (as many as 10,000 by 1821, according to one estmate of economic society members)[20] were being assembled to be offered assistance by and to give their support to the Madrid reformers. The crown had strengthened its control over the clergy and the Inquisition, and reformist prelates effectively promoted. The grandees had been excluded from real political power and the seigneurial regime, at least in Andalucia, was slowly being undermined. The army and navy helped make imperial control more secure and acquitted themselves honourably in the war of 1779–83. And a few hesitant steps had been taken towards the emancipation of women, under the aegis of Campomanes, Floridablanca and the King himself, in the economic societies.

But in 1788, the flourishing condition in which the Monarchy seemed to be was partly illusory. In any case growth was largely the result of economic developments rooted in earlier times and over which Caroline reformers had little direct influence. Despite the American reforms, contraband trade continued to exist, protected partly by corruption and abuses. Frenchmen and other foreigners were still prominent in the legitimate transatlantic commerce via Spanish ports. The borderlands of California, the south-west and Louisiana were too vast, their Indian inhabitants much too scattered and still too badly organised for the Monarchy to profit from her tenuous authority there. Worse, no fundamental accommodation of creole needs and interests had been achieved overseas, and indeed, the tightened control from Madrid increasingly exasperated the creole elite. The latter's pessimistic bitterness and their growing economic maturity fed their rebellions of the early nineteenth century. Thus, the logic of the imperial dilemma was fulfilled: reform supported economic growth, which in turn promoted political maturity, and maturity demanded – and secured independence.

There were other failures in Spain itself. Her economy, despite its expansion, failed to leap ahead as had those of Britain or France by 1790. Campomanes's vision of an entirely regalist church was never attained and in the 1790s clerical arrogance and obscurantism were to revive with disastrous consequences in the nineteenth century. Road and canal transport, the publishing industry, guild regulations, technical education, though they had all been touched by the reformers, remained backward. Even pet projects, like the economic societies, were widely considered near failures in the later 1780s.

Why had the reformers not achieved more solid results? To begin with, their task was enormous and the material resources of the crown, stretched over the vast empire, insufficient. Thus, the hopes which some ministers had for a vigorous primary educational reform were frustrated by lack of funds. So were many of the favoured projects of the economic societies, as their members acknowledged in a survey conducted toward the end of the reign. Spain, with its underdeveloped economy was not rich enough to sustain the needed imperial reforms. Moreover, the Spanish and American middle classes seemed reluctant to support reform actively. This was perhaps because, except for a few commercial policies, Caroline reform, determined by the traditional elite, had comparatively little to offer. Instead, the middle classes of Cadiz, Mexico City, and other towns tended to take refuge in their *consulados*, often resisting reform. Even the flourishing Barcelona bourgeoisie remained surprisingly passive, while for the poor and uneducated the reforms brought few if any advantages and aroused much hostility. Thus, the latter absented themselves from the neighbourhood elections of their *alcaldes de barrio* and from the technical schools set up by the economic societies. They hated the work houses established to replace free charity, opposed the free grain trade, and disliked having to pay to have their streets lit at night and their garbage collected weekly. The reformers were met with widespread indifference and opposition. They were keenly aware of their opponents' strength and it often aroused in them a marked pessimism. Roda once wrote to his old friend, José Nicolás Azara, a Spanish diplomat in Rome: 'At every step we face a thousand obstacles because superstition and ignorance reign in all the States of the Republic. . . . I will not tell you of the projects which are being frustrated nor the reasons, not wishing to bore you nor to take your time, there are so many.'[21] Opposition from the clergy and the great landowners helped scuttle Campomanes's law of amortisation and the *única contribución*, which though decreed in effect in 1770, was in practice dropped a few years later. Moreover, Olavide was destroyed by his clerical opponents, while in Mexico the viceroy, Bucareli, resisted the reforms of Gálvez for years with the support of well-entrenched local interests.

But perhaps the greatest causes of the reformers' disappointments lay in their own limitations and failures. The reformers frequently lacked diplomacy and made tactical errors, like Gálvez when confronted by Bucareli; or as when Charles permitted some of the richest grandees of the monarchy to sit on the committee preparing the *única contribución*. More significantly, the reformers lacked a clear programme. As Roda said, they did not know where to begin. Lacking any firm ideological base – they espoused mercantilist as well as liberal and physiocratic ideas in a pragmatic jumble – they never set clear priorities that could be widely agreed on and which could provide true insight into the needs of the Monarchy. Therefore their efforts were too diffuse, their energies too easily dissipated. This made them frequently passive – reacting to demands for university reform and to the ingenious idea of the economic societies, both proposed by citizens not ministers. Neither were they particularly well informed, nor expert enough in many of their endeavours to provide the skilled leadership necessary.

Ultimately, however, their greatest weakness was their refusal to accept the need for more than mere reform and renovation. The Monarchy could only have survived if it had been transformed, politically and socially. By the 1780s, Valentín de Foronda, Cabarrús, and a few other radical observers saw just that. But the Caroline reformers who determined policy in those years refused to jettison absolutism and, above all, the continued

leadership of the landowning and office-holding *hidalgo* elite.
They trusted only these men like themselves, of property and
substance, and perhaps like-minded bourgeois where appropriate
and necessary. They feared the masses – more after 1766 than
before, undoubtedly – and excluded artisans and peasants from
even the economic societies. Their reforms were fundamentally
self-interested, designed to perpetuate elite control over a more
vigorous Monarchy and a more prosperous, but submissive
artisanry and peasantry. Hence the political isolation which so
distressed Roda.

5. Reform in the Habsburg Monarchy, 1740–90

H. M. SCOTT

I

HISTORIANS in search of textbook specimens of enlightened absolutism have been accustomed to dwell on Joseph II's reforming achievements during the 1780s. The Emperor has often appeared the very exemplar of a monarch inspired by the Enlightenment. His religious, educational and, above all, agrarian reforms were certainly noteworthy, and he even attempted to reform the institution of serfdom. Yet the opposition and, ultimately, rebellion which his policies aroused, and the withdrawal of many of the more controversial innovations in the final months of his life or immediately after his death in February 1790, have sometimes made Joseph II appear a paradigm for the failure and even irrelevance of enlightened absolutism.

The Emperor's reputation has experienced similar vicissitudes within the narrower arena of historical writing about the eighteenth-century Habsburg Monarchy. It has proved difficult to maintain any balance between the sharply contrasting figures of Maria Theresa, ruler from 1740 until 1780, and Joseph II, co-regent after the death of his father, the Emperor Francis Stephen, in 1765 and sole ruler after 1780. Admiration for the heroic and attractive Empress has often been accompanied by distaste for her impulsive and, at times, obnoxious son, and the Emperor's historical reputation has suffered accordingly. To be 'for' Maria Theresa has usually meant to be 'against' Joseph II. The massive life-and-times published during the 1860s and 1870s by the Empress's first important historian, the imperial archivist Alfred Ritter von Arneth, has had an enduring impact.[1] Arneth's preferences were clear: he was an unqualified admirer of Maria Theresa, who saved the Monarchy during the 1740s and ruled it wisely thereafter, and a stern critic of her wayward and

145

headstrong son. This preference extended to the selective utilisation – and occasionally the suppression – of documents to support his over-arching view of the period. For a generation, Joseph II was without a historical champion to dispute the field with Arneth, until the appearance early in the twentieth century of a remarkable monograph by the Russian Paul von Mitrofanov.[2] This was the first scholarly study of the Emperor and it did something to correct the balance. Yet Mitrofanov's book was essentially a study of Joseph II's reforms during the 1780s and this, together with the continuing influence of Arneth's massive volumes, has ensured that the Empress has usually received more than her scholarly due and her son rather less.

Until the 1940s and 1950s, historical writing remained overwhelmingly monarchical in focus. But during the past generation a more sophisticated historiography has developed and continues to flourish. A bitter feud over the remarkable religious reforms encouraged detailed investigation of many other areas of government policy, and there is now a substantial literature on the wide-ranging innovations introduced between the 1740s and the 1780s. This has made clear that these decades saw the most radical programme of reform from above in later-eighteenth-century Europe. More was attempted and achieved in the Habsburg Monarchy than in any other state, as Vienna came to intervene, sometimes decisively, in areas where central government had not previously sought to exert influence.

The extent of continuity between the two reigns and the importance of Maria Theresa's rule for the reform programme have also been demonstrated. The discussions and initiatives during the Empress's lifetime did much to prepare the ground, but the pace of reform quickened appreciably after her death in 1780. Joseph II's personal rule was the high-water mark of Habsburg reformism and the more radical measures were mainly delayed until he secured full authority. There was also a sharp contrast in the approach and mental outlook of the two rulers. Maria Theresa's actions were rooted in her traditional sense of Christian responsibility for all her subjects. She was always politically conservative and reluctant to undertake reform, though once convinced it was essential, she was capable of surprisingly radical measures. Innovation was for her always a matter of pragmatism and necessity. Joseph II's pursuit of change was more principled, even doctrinaire, and revealed his impatient personality and reluctance to acknowledge any obstacles to his policies. The Emperor's sense of responsibility for his subjects depended less on Christianity than on natural law theories. It found expression in his conviction that it was his duty to promote the 'general good' or 'general best'.

This utilitarian approach, together with a desire to strengthen the state and to extend his authority, provided the central strands in his reforms. Joseph II's actions were often deeply unpopular and could sometimes appear ludicrous, as in the case of his unsuccessful attempt to introduce wooden coffins with hinged floors, which could be re-used once the corpse, trussed up in a sack, had been dropped into the ground. Contemporaries were aware of the new spirit which the Emperor brought to government after his mother's death. The very next year, censorship was substantially relaxed (the first tentative steps had been taken before 1780) and this opened the way for a flood of publications, many critical of the Emperor's own actions. The 1780s saw remarkably free political debate, principally in Vienna, though some restrictions were re-imposed during the Emperor's difficult final years. The new tone which Joseph II brought to government was also apparent in his immediate escape from the imperial palace at Schönbrunn and his adoption of a relatively austere life-style.

Recent scholarship has also established the impossibility of explaining enlightened absolutism in the Habsburg Monarchy in terms of the French Enlightenment. This had some impact among a very small number of nobles and state officials, but it often co-existed beside more traditional deductive philosophy. The State Chancellor, Kaunitz, was a particularly good example of this blend of the contrasting teachings of Christian Wolff and of the *philosophes*. In any case, Italian reformist ideas and cameralism were always far more important sources of government policy.

II

The lands of the Monarchy were uniquely far-flung: no other eighteenth-century state was involved at so many points on the map of Europe. These possessions lay principally in Central Europe, but the Habsburgs also governed the distant Austrian Netherlands (most of present-day Belgium and Luxembourg) and territories in northern Italy, where Lombardy was administered directly from Vienna and Tuscany was ruled personally by Francis Stephen until his death in 1765 and thereafter by Joseph II's brother, Leopold. The Monarchy's Central European heartlands consisted of three contiguous blocks of territory: the various Austrian provinces ruled from Vienna which were the family lands of the Habsburgs and the traditional source of their power, principally Upper and Lower Austria, Styria, Carinthia, Carniola, the Tyrol and Vorarlberg; secondly, the so-called 'Lands of the Bohemian Crown' (Bohemia, Moravia and Silesia), which had been acquired in 1526; and, finally, the Kingdom of Hungary, most of which had been under Turkish control for much of the sixteenth and seventeenth centuries and had been reconquered in the decades around 1700, along with the neighbouring and small (but separate) principality of Transylvania, which was ruled from Vienna. Hungary clearly constitutes a special case: it is discussed separately by Dr Evans (see Chapter 6). The principal focus of this chapter is the 'Hereditary Lands', that is to say the Bohemian and Austrian provinces. During the half-century after 1740, several significant territorial changes took place. Above all, most of Silesia was lost to Prussia during the first decade of Maria Theresa's reign; while Galicia (strictly speaking only East Galicia) was acquired in the first partition of Poland (1772), the Bukovina was seized from the Ottoman Empire (1775) and the Innviertel from Bavaria in 1779.

The Monarchy's geographical extent was also reflected in the diversity of its subjects and territories. The bewildering mixture of social, economic and geographical conditions, religious and racial groups and languages and dialects contained within its borders was also quite unique, and this enormously complicated the task of government and the implementation of reforms. The territories that made up Maria Theresa's inheritance had been acquired at separate times and in different ways. There was no uniform system of government; on the contrary, even after the administrative reforms of 1749–61 (discussed below, pp. 152–9), the Monarchy's subjects continued to be ruled through a variety of institutions and Vienna's authority varied significantly from one province to another. The freedom of action enjoyed, for example, in Lombardy was in sharp contrast to the very real limitations acknowledged by the Habsburgs as rulers of Hungary. In most provinces, the traditional local authorities remained involved in government and were a further restriction on Vienna. This, together with the often bewildering variety of local circumstances, ensured that innovations were usually introduced on a provincial basis. Joseph II, whose extensive travels gave him a remarkable knowledge of his territories, generally insisted that reform should proceed province by province, with due allowance usually made for particular local circumstances. Yet his ultimate aim, particularly apparent after 1780, was uniformity throughout the Monarchy in administrative, military, legal and financial matters.

The varied and complex ways in which the Habsburg Monarchy was governed had one obvious benefit for the historian. Much of this administration was conducted on paper, through innumerable reports and *vota*, as departments struggled for ascendancy. Though there are some important gaps in the sources (above all, the proceedings of the Council of State have been destroyed), the surviving archival material is enormously rich – it may even be richer than that for any other later-eighteenth-century state – and this enables the genesis and implementation of particular measures to be studied with some precision. In particular, the role of individual ministers and officials in the reforms is often clearer than it would be in other countries.

The scattered nature of the Monarchy's possessions was also crucial for the varied intellectual environment which produced the reforms. The Habsburg Lands were unusually and perhaps uniquely open to the intellectual currents of the day, though their full circulation was hindered by the strict censorship controlled by the Jesuits until the 1750s. Thereafter, the new state Censorship Commission was more willing to allow advanced ideas to circulate, at least in educated circles. The Habsburg possessions in northern Italy were particularly important for the religious reforms, providing a conduit for Italian reformist ideas and for examples of successful innovations to pass into the Monarchy's Central European heartlands. Another important source of the ecclesiastical legislation was the Jansenism which came to Vienna through the distant Austrian Netherlands and also from Italy. The established links with Germany, where the Habsburgs' position as Emperors was a source of prestige and even power, enabled natural law and cameralist theories to circulate within the Monarchy. Austria herself had her own cameralists during the final decades of the seventeenth century, above all men such as Hörnigk and Becher, and their influence remained strong. But it was principally in the Protestant universities of north Germany that the subject of cameralism had become established and had developed so spectacularly during the first half of the eighteenth century.

One point which is frequently made about enlightened absolutism in the major states is the inhibiting effect of international rivalry and of the large armies this demanded. It is also true, though less remarked on, that such rivalry could provoke reforms. In the Habsburg Monarchy, the principal stimulus to innovation was the serious defeats inflicted by the Turks during the 1730s and by Prussia during the 1740s, and the subsequent failure to recover Silesia during the Seven Years War (1756–63). International failure and Austria's sense of relative decline made her rulers and ministers accept the absolute priority of radical measures. The sight of the upstart Prussian state catching and then overtaking the Monarchy was particularly painful, and made it willing to embrace root-and-branch reforms which might not otherwise have been attempted. In a general sense, it made the modernisation of Habsburg government and the economic advancement of its subjects and therefore of the state the main aim from the later 1740s and particularly after 1763. The hope that a strengthened and prosperous Monarchy would, one day, be a match for Prussia lay behind much of the reforming activity. Fiscal and military considerations directly influenced and even dictated certain initiatives. Yet these innovations were not simply the product of military and political failure. Prussia's emergence and her victories convinced Maria Theresa and her advisers that reform was essential, but the actual policies adopted were the product of varied intellectual forces and the responses they dictated to the Monarchy's problems. Enlightened absolutism here, as elsewhere, resulted from an attempt to apply the best recent theories to a particular set of circumstances. At least until 1780 it was not a programme but a reaction, shaped by many of the intellectual currents of the day, to events and to the problems facing Vienna. Only during Joseph II's personal rule did government policy become more rigid and even doctrinaire.

Many reforms were foreign in inspiration. This was not simply a matter of the influence of ideas from outside the Monarchy, important as this undoubtedly was. Actual institutions and policies were also studied, often by sending high-ranking officials abroad to examine them *in situ*, and then applied where appropriate. The extent of Prussian influence on the Habsburg reformers was considerable and surprising in view of the enduring hostility between the two states. The Prussian military system naturally inspired emulation: the modernisation of the Austrian army after 1748 owed much to the Hohenzollern model, while the introduction of conscription and of recruiting districts in the Hereditary Lands in the early 1770s was copied from Prussia's celebrated cantonal system. Haugwitz's administrative reforms of 1749 were rooted in his admiration of Prussian administration and even extended to a proposal to style the new agency of central government the [Austrian] General Directory. Even more remarkably, the efforts to improve primary education were inspired by Prussia's celebrated Pietist schools. Frederick the Great's state was the principal foreign model, but important parts of the religious legislation looked towards earlier reforms in the smaller German and Italian states.

The link between defeat abroad and reform at home was particularly clear in economic and commercial policy. The loss of Silesia deprived the Habsburgs of a major source of revenue and of its most important manufacturing and exporting sector, the Silesian linen industry. This, together with the enormous cost of the Seven Years War and the Monarchy's financial plight thereafter, caused a determined and sustained attempt to encourage prosperity and to stimulate economic development along the lines indicated by cameralist thinking. In 1745–46, the General Commercial Directorate (*Universalcommerzdirektion*) was established in Vienna to co-ordinate these initiatives, and a start was made before the Seven Years War by raising tariffs against foreign goods and lowering duties at the Hungarian frontier, in order to encourage internal trade within the Monarchy. These initiatives were the origin of the attempt after 1763 to promote economic unity among the various Habsburg provinces, with the goal of creating a huge free trade area within which Hungary was to be the agricultural centre and a market for the manufactured goods produced in the western territories. In 1775, the Hereditary Lands were turned into a common market, with the abolition of internal tariffs (only the Tyrol and Vorarlberg were excluded).

Better communications were sought in order to encourage trade and to advance economic integration. During the 1770s, attempts were made to speed up river transportation in Hungary, while there was an expansion of road building, particularly in the Bohemian Lands and, during the 1780s, the great Arlberg highway was built. These structural improvements were accompanied by measures to improve agriculture (mainly by the foundation of Agrarian Societies to act as a channel for new farming ideas), to stimulate trade and manufacturing and to remedy Habsburg economic backwardness. The immigration of skilled foreign artisans and the circulation of the latest technology were encouraged, while the monopoly position of guilds (seen as a barrier to economic progress) was slowly undermined. State subsidies were given for new manufacturing enterprises and – particularly after 1780 – domestic production was protected by high import tariffs against foreign goods.

Many of these initiatives had been anticipated before 1740 and were part of classical mercantilism and cameralism. While much remains obscure about the reign of Charles VI (1711–40), it seems clear that continuity was particularly strong in this area of state activity. Charles VI's efforts to undermine the position of the guilds, to promote commerce especially through Trieste and Fiume, to remove restrictions on the grain trade and to encourage manufacturing were built on during the reigns of Maria Theresa and Joseph II. The short-term impact of this government intervention was limited, but it was important in preparing the way for the Monarchy's nineteenth-century economic advance.

III

The first series of administrative reforms, carried out immediately after the Peace of Aix-la-Chapelle, were the work of Count Friedrich Wilhelm Haugwitz, supported at a crucial moment by the Empress and by Francis Stephen. A member of the provincial administration of Silesia since 1725, Haugwitz was a disciple of the Austrian cameralist Wilhelm von Schröder and shared his hostility towards the territorial Estates. These assemblies were to be found in most of the Habsburg provinces, were dominated by the nobility and retained significant powers, above all over taxation. Haugwitz had been impressed by the new administration in Prussian Silesia and especially its immediate and striking fiscal success. Within a few years, Frederick the Great's government had vastly increased the amount of taxation collected, almost doubling the total revenue from 3.9 million florins to slightly over 7 million florins.[3] This had been brought about by the more efficient administration provided by Prussian officials and by abolishing both the Estates' right to approve taxation and their involvement in its collection. Prussia's success in Silesia highlighted the relative inefficiency of Austrian government, above all where taxes were concerned.

The main flaw was central authority's evident weakness throughout the Monarchy. There was not one system of administration but two, as the provincial Estates maintained their own parallel institutions of government. In the various Habsburg territories, royal officials were less powerful and less numerous than their local counterparts. Taxation could not be imposed but had to be bargained for, and once the provincial Estates had agreed to the level of the Contribution (the main military tax), their officials – and not Vienna's – collected it. This was both time-consuming and, from the viewpoint of central government, inefficient and frustrating. Its financial consequences had become glaringly apparent during the wars of 1733–48 which had made clear that this ramshackle system of government was incapable of supporting Austria's great power position. In the other major European states, and principally France and Prussia, during the seventeenth and eighteenth centuries, there had been an expansion of central authority and particularly fiscal power, but no such evolution had taken place in the Habsburg Lands between 1600 and 1740.

The Estates' enduring influence was a consequence of the Monarchy's distinctive political evolution during the early modern period and the piecemeal nature of territorial consolidation. The rise of the Austrian Habsburgs since the sixteenth century had been based not on increasing central intervention at the local level, but on an alliance between the dynasty and the twin forces of the nobility and the Catholic Church.[4] In the Austrian duchies and the Bohemian Lands, as in the Kingdom of Hungary, the secular and ecclesiastical magnates ruled in the name of the Habsburgs and implemented – or circumvented – the orders received from far-away Vienna. The weakness of central authority was reinforced by the social and political dominance of the territorial aristocracies in the various provinces, by the powers of the individual Estates and by the separate legal systems and distinctive laws which each maintained and defended. Such unity as existed had been provided by the dynasty itself, by the Court in Vienna, by the army, by the triumphant Counter Reformation and by a distinctive culture, that of the Catholic baroque. This particularism and provincialism had been strengthened by the development of separate systems of government for each group of territories, headed by the Austrian and Bohemian Chancelleries which, by the eighteenth century, were located in Vienna. Dominated by the territorial magnates, these chancelleries often defended their own sectional interests, rather than advancing those of Vienna.

This was the system, exposed as deficient particularly during the 1740s, which Haugwitz set out to reorganise. He had begun to introduce piecemeal reforms even while the fighting was going on, first in Austrian Silesia (1744) and then as governor of Inner Austria after 1746, but the full-scale reconstruction was delayed until peace was signed. The over-riding aim was a peacetime standing army of some 108,000 men, paid for by a substantial increase (negotiated in 1748) in the Contribution from the Hereditary Lands. This was to be facilitated by the centralisation of political and fiscal power, and by its separation from judicial matters, now handled by a new High Court (*Oberste Justizstelle*) for the Hereditary Lands.

The proposed changes were vociferously opposed by the magnates, who dominated both the Estates and Habsburg provincial government and who feared a reduction of their own power. This led Maria Theresa to implement them 'almost as a *coup d'état*'.[5] A series of royal proclamations on 2 May 1749 simply dissolved the separate Austrian and Bohemian chancelleries. A measure of unitary control over the Hereditary Lands was provided by the creation of the *Conferenz in Internis*, which replaced the *Hofdeputation*. This was supported by the establishment of the *Directorium in Publicis et Cameralibus*, the centrepiece of the new arrangements. The *Directorium*, as it quickly became known, was designed to execute internal policy and was directly modelled on Prussia's famous General Directory. With Haugwitz as its influential first president, it sought to give unity and direction in domestic affairs and to govern the Hereditary Lands as a single unit. During the following decade, it both expanded and extended its authority, and also gave birth to a series of subordinate departments handling particular types of business.

These changes were accompanied by measures to extend Vienna's control and especially its fiscal power at the local level. A series of bodies was established in the various provinces, known first as the Deputations (*Deputationen*) and then, after 1749, as the *Representationen und Cammern*. Staffed increasingly by royal nominees, and paid by and reporting to Vienna, these were the local arms of the *Directorium* and sought in particular to control taxation. They were part of Haugwitz's over-riding strategy of pushing back the authority of the Estates and advancing that of the central government. Yet the reform of local administration was only a partial success. Control of taxation was not wrested from the Estates, which still consented to the Contribution and continued to collect it in some provinces.

A well-placed contemporary observer, the court-chamberlain and noted diarist Prince Joseph Khevenhüller-Metsch, described the 1749 reforms as a 'revolution in government'. Though this is certainly true of the way they were discussed in secret and implemented without warning, his verdict can only be applied with difficulty to the actual innovations. Haugwitz's measures were essentially pragmatic and were shaped by fiscal and military imperatives. The ending of the separate systems of government for the Austrian duchies and the Bohemian lands was crucial, but the *Directorium's* new president was unable to achieve the centralisation and unitary control he hoped. The reforms' scope was also consciously limited. Hungary and Transylvania, along with the less important Austrian Netherlands and Lombardy, were explicitly excluded. The Catholic Church's enormous power remained intact, while the nobility's influence and authority proved to be less seriously affected than the magnates feared.

Haugwitz's reforms proved to be both flawed and imper-manent: they were undermined by the Seven Years War. He himself always argued, particularly when his system came under attack during the conflict, that the changes had been intended only for peacetime. This seems disingenuous. These were clearly the product of humiliation and desire for revenge on Prussia and the recovery of Silesia. The administrative changes were accompanied by a series of military reforms, intended to modern-ise the Habsburg army and make it a match for the formidable Prussian forces, and by a prolonged debate on future foreign policy. Indeed, the strengthening of the core Habsburg territories at which the 1749 reforms had aimed was linked to the new diplomatic orientation urged by Wenzel Anton von Kaunitz, who controlled Habsburg foreign policy as State Chancellor after 1753. Kaunitz believed that Prussia, not France, was now the Habsburgs' greatest enemy, and he proposed a *rapprochement* with Versailles, which would have the incidental benefit of depriving Frederick the Great of his established ally. The consequent signature of an Austro-French alliance in May 1756 (the centre-piece of the celebrated Diplomatic Revolution of that year) was the prelude to the war to recover Silesia. The Chancellor's skilful diplomacy had assembled an imposing alliance of Austria, Russia and France, together with the faded power of Sweden and contingents of soldiers from the Holy Roman Empire, and Prussia's rapid defeat was anticipated. That it did not materialise was due to several factors. Foremost among these were Prussia's own resilience and France's diminishing enthusiasm and reduced commitment to the continental struggle, as she concentrated on her own separate war with Britain overseas. But the collapse of the Habsburg home-front and thus of the 1749 reforms contributed to Austria's failure, and this was quickly seized on by Haugwitz's arch-rival, Kaunitz.

A serious defeat at Torgau in November 1760, following hard on an earlier reverse at Liegnitz that summer, made clear to the Empress and her advisers that Habsburg armies were incapable of defeating Prussia and recovering Silesia. Austrian government had all but disintegrated and the finances were in chaos. The annual cost of the Seven Years War had risen to a staggering 40 million florins and Haugwitz's much-vaunted changes could not support expenditure of this order. This was exploited by the ambitious Chancellor, who perhaps saw a way of explaining away the failure of his own foreign policy by the fiscal and administrative shortcomings of his rival's system. Kaunitz had earlier carried out a noted re-organisation along modern lines of the State Chancellery (which controlled foreign policy) in 1753. He believed that Haugwitz's aim of unity and centralisation in internal administration had not, in practice, been secured. His remedy was a new Council of State (*Staatsrat*) advising Maria Theresa about all aspects of domestic policy and providing coherence and direction from the centre. Kaunitz had already proposed this in August 1758, and he again urged its creation in December 1760 (the month after Torgau), when it was accepted by the Empress.

The Council of State was formally established in January 1761, with a membership of six and a wide remit. Though Hungarian, Italian and Belgian affairs were all formally excluded from its competence, from the outset it discussed Hungary as well as other aspects of internal policy. Formally it had no executive authority, but it quickly exploited its potential influence and came to play a crucial role in the reforms after the Seven Years War. Its first task was to conduct a wide-ranging investigation into Habsburg government. This lasted throughout 1761 and was accompanied by a struggle for power between Haugwitz and Kaunitz. By December, a decision had been reached and central government was again reconstructed. The underlying principle of the 1749 reforms – the union between administration and fiscal power – was now reversed and a more traditional separation imposed between taxation and political authority. This was a death-sentence on Haugwitz's *Directorium*, which lost most of its military and financial functions (with the significant exception of the Contribution). It was renamed the 'Bohemian and Austrian Court Chancellery' and now discharged a limited range of business. Power was instead concentrated in the new Council of State, which the rulers sometimes attended. In the event, the Council was unable to cope with the mountain of paper directed to it, but it proved to be far more enduring than the earlier reorganisation, partly because of Kaunitz's political and personal longevity.

The overhaul of central government was accompanied by a restructuring of local administration, carried out after peace had been concluded in 1763. The absence of uniformity in the way the Hereditary Lands were governed was apparent in the local arrangements now established. The *Representationen* established a dozen years before were re-named the *Gubernien* (literally: 'governments'), though in Moravia this body was known as the *Landes-Gouverno*. These became the principal agencies of local government, inheriting the duties and responsibilities earlier discharged by the *Representationen*.

The success of the administrative reforms was incomplete. The participation of the local Estates in government was diminished, but it was certainly not ended. There was no such intention on the part of Haugwitz or Kaunitz, and indeed the enormous financial burdens of the Seven Years War may have increased their importance, since they continued to approve taxation. In any case, a crudely adversarial view of the relationship between crown and Estates must be avoided. The fiscal appetite of central government was certainly far greater than the willingness of the territorial assemblies to grant taxation, or the capacity of the individual provinces to pay it. But such disputes were over the precise level of taxation, rather than the principle itself. In many – perhaps most – areas of internal government, crown and Estates were partners rather than opponents. Many of the later reforms were discussed and implemented jointly between royal officials and the influential permanent committees of the Estates.[6]

Government policy until 1790 and beyond was partly implemented at the local level by agents of the provincial assemblies and
of individual lordships and towns. However, though the officials
of central government were and remained less numerous than
their local counterparts, they increased substantially in number
as a result of the reforms of 1749–61. At the accession of Maria
Theresa, there were some 4,000 royal officials (excluding those
in the Court itself) in central and local government in the
Hereditary Lands; by 1763, the comparable figure had risen to
around 6,000.[7]

This was the broader significance of the administrative changes. Though far from totally successful, they increased centralisation, broke the Estates' dominance over local government and
created an extra tier of administration more responsive to
Vienna's direction. Between 1763 and 1790, the *Gubernien* were
employed by Vienna to impose its social, economic, educational
and agrarian reforms over the heads of narrow sectional interests.
Their authority – like that of all local administration in later-
eighteenth-century Europe – was incomplete and could be very
limited indeed. At times, the new bodies obstructed Vienna's
aims. However, the rapid expansion of government made Maria
Theresa and her advisers more aware of the situation in the
various provinces and of the real problems which existed. This
was to be particularly important for agrarian reform. The new
state officials also provided a *cadre* of administrators willing
and able to implement the measures decided on by central
government. The early administrative changes made possible the
later reforms and, in some measure, shaped them as well. In
a similar way, the re-modelled central government provided
direction and sometimes initiative. The new Council of State was
to play a crucial role in drawing up reforms and in pressing for
their implementation. The reigns of Maria Theresa and Joseph II
saw a dramatic increase in the pace of all government activity,
as Vienna's aims expanded rapidly and its policies became
increasingly interventionist. This is apparent in the average
annual number of published royal edicts, which grew from 36 in
the 1740s to 96 by the 1770s and to an astonishing 690 in the
1780s, the decade of Joseph II's hectic personal rule.[8]

Much remains elusive about these new officials, though their
role was crucial. Their often obscure origins and incomplete
personal biographies make any prosopographical analysis impossible. It is clear, however, that the old idea that the reforms of

1749–61 opened up public service in the Habsburg Monarchy to an aspiring middle class which provided professionalism in government must be abandoned. In the short term, the changes broke the nobility's dominant position in Vienna and in the provinces. At the local level, Maria Theresa's later years witnessed an aristocratic recovery, but no parallel process can be detected in central government. The new element in administration during the second half of the eighteenth century was less a matter of social origins than of education and receptiveness to new ideas. Many – and perhaps an increasing number – of the enlarged corps of royal officials had either studied the new science of cameralism at universities or colleges or had a second-hand acquaintance with such ideas.

These efforts to advance administrative unification and centralisation were accompanied by initiatives to create a single, state-imposed system of public law. The existing situation was confusing in the extreme, with numerous jurisdictions competing for supremacy. In the aftermath of the 1749 reforms, a start was made on the codification of civil and criminal law. This made slow progress, and encountered considerable opposition from vested interests anxious to defend their own legal privileges. After lengthy deliberations, the *Codex Theresianus* was produced in 1766. This resembled the administrative changes of 1749–61 in that it was a compromise between the established corporate interests (nobility, Estates, Church and towns) and the new rationalism and natural law theories. Widely unpopular, it was nevertheless the first hesitant step towards the imposition of a modern system of civil law throughout the Monarchy. Three years later, the Criminal Code (the so-called *Nemesis Theresiana*) was issued. This was primarily a codification, rather than a reform, and it represented a similar compromise between old interests and new theories. Though it retained numerous barbaric penalties and even contained a series of pictures showing how they were applied, it brought about some rationalisation in the administration of justice. This excited lively debate, and seven years later in 1776 torture was abolished, thanks to the effective pressure of a group headed by Joseph von Sonnenfels, with the significant support of the Emperor.

Joseph II pressed on during the 1780s with the task of imposing uniformity in legal administration and making justice swifter and less expensive. He acted first to curtail the jurisdiction of the manorial courts, which was drastically limited in 1781. These efforts enjoyed some success. This was more than can be said for the codification of the civil law begun under Maria Theresa, which made no real progress during the Emperor's personal rule. Joseph II was sympathetic to enlightened legal opinion, as represented by Beccaria, Sonnenfels and Martini, and his policies during the 1780s partly reflected their influence. by 1784 he had decided that the criminal law needed to be modernised and a commission was established. The Emperor's close personal interest ensured that it reported within three years. The outcome was the Criminal Code (*Allgemeines Gesetzbuch*) of 1787. This was a general reform of the criminal law. It abolished the death penalty (save in a few instances where military discipline and state security were involved) but worsened the conditions under which hard labour was performed. Witchcraft, black magic and heresy ceased to be crimes, and the principle of equality before the law was upheld. This modernisation of the criminal law, together with the tentative improvements in civil law, were an important step forward and not the least important legacy of these decades.

IV

The remarkable religious reforms have given rise to a fierce debate over what has come to be known as 'Josephinism' or 'Josephism'. By this is meant the series of measures, climaxing during the 1780s and therefore permanently associated with Joseph II, which transformed the position and the practices of the Roman Catholic church and also influenced policies towards education, censorship and poor relief. State control in the ecclesiastical sphere was increased, ceremony and doctrine were reformed, the parish system was thoroughly overhauled and – most remarkably of all – a degree of toleration was introduced for Protestants, Greek Orthodox and even Jews. The fierce historical controversy around these measures which has raged since the 1940s is due to two principal circumstances. The surviving material bearing on the religious changes is very large, and the prejudices which many historians have brought to the subject of Josephinism are even more apparent. More important, however, has been the truly remarkable nature of the reforms themselves.

The Habsburg Monarchy was firmly, indeed ostentatiously,

Catholic. Its Catholicism, moreover, was firmly rooted in the Counter Reformation. The support of the Roman Catholic church had been one of the twin pillars on which the Austrian Habsburgs had risen to prominence during the seventeenth and early eighteenth centuries. This was epitomised by the power wielded by the Jesuits, who effectively controlled education, particularly at the university level, and also monopolised the key posts of confessors to the ruling dynasty. The imperial family's own identification with the Church Militant was complete, and symbolised by its active participation in numerous religious processions in Vienna such as that of Corpus Christi, which ended with the celebration of the Eucharist. The Habsburgs themselves participated enthusiastically in many of the ceremonies of that baroque piety (*Pietas Austriaca*) particularly characteristic of the period from the later-seventeenth century onwards: religious processions, pilgrimages, crib-blessings, and so on. Yet during the generation after *c.* 1760, *Pietas Austriaca* was dismantled, the Jesuits were suppressed and the government introduced the most remarkable series of ecclesiastical reforms in later eighteenth-century Europe. The sense of incongruity, even paradox, is increased by the fact that many of these measures were initiated – and the decisive breach made – before 1780, during the reign of the pious, conservative Maria Theresa. Though the Empress was not unaffected by the more progressive religious currents of her day, she was always a reluctant reformer, particularly where the church was concerned. Though much remains controversial about 'Josephinism', it is clear that the religious changes are largely to be explained in terms of two developments: the international movement of reform within the eighteenth-century Catholic church and its decisive influence in the Habsburg Monarchy; and, secondly, the momentum provided, from the 1760s, by progressive officials, headed by the Chancellor, Kaunitz, and with the full support, particularly and decisively after 1780, of the Emperor himself.

The suppression of the Society of Jesus in 1773 was crucial, but Austria – alone among the major Catholic states – played no direct part in bringing it about. Maria Theresa did – and could do – nothing to support the Society, whose demise resulted from concerted diplomatic pressure on the Papacy by France, Spain, Portugal and Naples. Yet the Empress's territories were remarkably open to the ideas now described as 'Reform Catholicism'. Jansenism entered primarily through the Austrian Nether-

lands, Febronianism came from the *Reich*, while the teachings of
Muratori, which were to prove crucial, circulated from the
family's possessions in the Italian peninsula. Josephinism was
an amalgam of all these currents of thought, and of rationalism,
too.

The ideas of Ludovico Antonio Muratori (1672–1750), the
leading figure in the early Italian Enlightenment, were particularly influential and were perhaps the principal source of reform.
A polymath, Muratori's most enduring scholarly achievement
was as a historian. He assembled and published many of the
sources for Italian history and was an important figure in the
process by which historical scholarship came to be based primarily on manuscripts and archival research. His contemporary
importance came principally from a book published towards the
end of his life, *Della regolata divozione dei Cristiani* (*On a well-ordered
devotion*) (1747), which became a handbook for Catholic reformers.
In it, Muratori brought together his belief in the crucial importance of pastoral work, his own brief experiences as a parish
priest, the distaste this had instilled in him for the superstitious
practices which now passed for Catholicism at the popular level
and in particular the excessive veneration of saints and their
miracles, and his own search for a truer, more inward-looking
faith. For the baroque piety of the Counter Reformation, Muratori
wanted to substitute a simpler, more internalised Christianity,
shorn of many of its characteristic practices (the excessive number
of pilgrimages and feast days was a particular target) and far
fewer monks and nuns, whose numbers were now grossly swollen
in most Catholic countries. Instead, he advocated renewed
emphasis on pastoral work, and a sustained effort to bring the
common people to a truer faith through the use of the vernacular
in services and by closer contact with the Scriptures. Faith
rather than works should once again be the path to salvation.

Muratori's plea for a simpler Christianity was widely influential
in its own day. Though he himself was clear that he was no
Jansenist, his ideas were taken up by this parallel movement for
Catholic reform. Eighteenth-century Jansenism is an elusive
topic: it was always a movement rather than a party. It retained
loose and rather vague links with its origins in seventeenth-
century France and with the ideas of Cornelius Jansen. The
characteristic spiritual concerns of the Jansenism of Port-Royal –
grace, justification, predestination, a reduction in the frequency
of communion services – were now less important than its

opposition to many aspects of Catholicism. Its targets were very largely those of Muratori: the excessive monastic population, the poor state of pastoral care, baroque piety, the power of the Jesuits and of the papacy. Jansenism also aimed to substitute a simpler liturgy and service, and to improve pastoral care, and it was, in general, less inward looking and much more assertive than Muratori.

By the 1730s and 1740s, Jansenism was acquiring shape and momentum as a movement for reform. Influential in the universities of the southern Netherlands and northern Italy, it passed easily from these Habsburg territories into the Central European lands of the Monarchy. So, too, did the teachings of Muratori, which came to be widely influential within the Habsburg episcopate. Though Jansenism and the ideas of Muratori had quite separate origins, the overlap between them, in their targets and their objectives, soon ensured that this distinction was blurred and came to be merged into the broader movement of 'Reform Catholicism'.

The third precise source of reforming ideas was a timely pamphlet which appeared in 1763, as the movement for religious reform was gaining ground. This was *De Statu Ecclesiae* (*The State of the Church*), published under the pseudonym 'Febronius' by the suffragan Bishop of Trier, J. C. N. von Hontheim. Though this was eirenical in intention and looked to a re-unification of the Protestants with the Catholic Church, its contemporary importance was due to its revival of the old controversy about the extent of papal power. Rejecting ultramontanism (the doctrine now particularly associated with the Jesuits, that the Catholic church was everywhere subject to control from Rome), Febronius instead asserted that individual bishops should possess these powers and that secular rulers should assert their own authority and use it to carry through reforms. His ideas were immediately and widely influential within Germany and inspired efforts to extend state control over the church, though their impact in the Habsburg Monarchy was delayed.

The final and most elusive influence on the religious reforms was the French Enlightenment. The rationalism of the *philosophes* was affronted by many of the practices of mid-eighteenth-century Catholicism. Very few of the reformers were outright atheists, but many were critical of church wealth and the numbers of monks and nuns, as well as of baroque piety. There was also widespread hostility among educated opinion throughout the

Monarchy towards the Jesuits and their influence in education and at Court. Against this must be set the undoubted religious conservatism of the vast majority of the Catholic population, devoted to the traditional practices, and sceptical and often simply hostile towards the ecclesiastical changes.

Recent scholarship has located the origins of Josephinism as early as the 1750s, and that decade certainly saw some important developments. The appointment of the reform-minded Trautson as Archbishop of Vienna in 1750 brought Muratori's ideas to the centre of the ecclesiastical stage. Trautson's successor after 1757, Migazzi, at this point shared these views and wrote an introduction to the German translation of *On a Well-Ordered Devotion* which appeared in the Austrian capital in 1762. Though he later became an influential religious conservative, during the 1750s and 1760s Migazzi was a supporter of moderate reforms. In the period immediately before the outbreak of the Seven Years War, a remarkable plan was drawn up which foreshadowed the later attack on the monasteries. In order to halt the advance of Protestantism in Hungary, an improved parish system was to be set up, funded by a levy on the monasteries and the church and with the resulting income administered by a central religious fund. Crucially, however, unilateral action was not attempted. Maria Theresa instead sought papal approval, which was not forthcoming, and the whole scheme was overtaken by the outbreak of the Seven Years War. Too much tends to be made of these plans, though the hostility they reveal towards monasteries and their wealth was certainly significant for the future. The influence of Josephinism has been pre-dated as a result of the search for its origins. The 1750s, and the 1760s, too, were important more because of developments which were essential preliminaries to the later reforms. In particular, the power of the Jesuits waned and that of the Jansenists and other reformers waxed, while Lombardy was used as a laboratory for a series of religious changes that were subsequently introduced into the Central European territories.

The influence of the Jesuits in the Monarchy declined rapidly after the early 1750s, as their entrenched position was dismantled. Their dominance in higher education was broken (see below, p. 173), while the establishment of a state-controlled system of censorship deprived them of a crucial source of their power. This control was destroyed in 1759, and the last Jesuit left the new Censorship Commission in 1764. The moving spirit behind the new body's more liberal attitude towards the circulation of books was Maria Theresa's family physician, the Dutchman Gerard van Swieten, who was also the Court Librarian. Van Swieten was a convinced Jansenist and his patronage enabled a series of like-minded reformers to move into positions of influence – men such as Martini, the noted legal reformer and fellow-Jansenist. The final act was the appointment of the Jansenist Ignaz Müller as extraordinary confessor to the Empress in 1767, at Joseph II's instigation. Hitherto, this Jesuit monopoly had been almost unbroken and Müller's appointment symbolised the way the tide was running against the Society, in the Monarchy as in Catholic Europe generally. Maria Theresa was herself sympathetic to many of the ideas of Jansenism, and this was strengthened both by the influence of her new confessor and by her own heightened religiosity after the death of her beloved husband, Francis Stephen, in 1765.

Simultaneously Kaunitz, who was the principal advocate of religious reform at least until 1780, was advancing his *protégés* and supporters into positions of influence within the state administration. The Council of State was under his control and could usually be relied upon to support progressive measures, while men such as the influential Heinke, Greiner, Gebler, Brukenthal and Karl Zinzendorf provided a team of reform-minded ministers in key positions. Simultaneously, the near-bankruptcy of the Habsburg state after the costly and destructive Seven Years War provided a more pragmatic motive to extend state control, for the church's apparent wealth was in sharp contrast to the government's poverty. Traditionally, the church had been more heavily taxed in the Habsburg Monarchy than in any other Catholic country. Taxation had been imposed after 1716 to pay for the Turkish Wars and for the repair of key Hungarian fortresses. The church in the Hereditary Lands (but not in Hungary) paid the Contribution after 1748 on its property. After 1763, at Maria Theresa's instigation, a valuation of all religious property was carried out. When the papacy resisted suggestions that it should allow an increase in taxation, the government unilaterally imposed a levy in December 1768.

This highlighted the extent to which the papacy's traditional claims were being questioned in Vienna. It was even more apparent where developments in Lombardy were concerned. Austria's Italian province was, by the 1760s, effectively ruled by Kaunitz, who enjoyed near-dictatorial powers and was assisted

by a significant number of reform-minded officials. This enabled
Lombardy to be used as a laboratory for experiments which, if
successful, could then be applied to the Monarchy's heartland. It
was the case with fiscal policy, and even more where ecclesiastical
legislation was concerned. Lombardy's large number of monasteries made a particularly inviting target and, at the end of the
1760s, some were dissolved. More important for the development
of Josephinism, Kaunitz and his supporters in the administration
developed arguments to justify state supremacy over the church
in all but questions of doctrine and religious practice, and these
theories were subsequently employed in the Hereditary Lands.

Kaunitz and, to a lesser extent, Joseph II provided much of
the momentum in ecclesiastical policy during the Co-Regency.
Maria Theresa, though she endorsed many of the stirring
declarations of principle emanating from the Chancellor, always
hesitated when actual changes were proposed. During the 1770s,
however, she was brought to agree to a series of measures which,
though individually not particularly significant, cumulatively
represented a real attack on the traditional position of the
Catholic church. This legislation followed the lines sketched out
during earlier debates within the government and at court, and
was underpinned by an assertion of state control over the church
and of its right to tax the clergy.

Many of these measures appear insignificant to a modern eye.
The earlier attempt (1754) to reduce the number of public
holidays on church festivals (often saints' days) was extended in
1772 and efforts made to limit the extent of such celebrations.
Religious festivals were an obvious target both for Catholic
reformers and for cameralists, who emphasised the consequent
lost production. In the same year, restrictions were placed on
pilgrimages. A series of measures sought to limit any further
expansion of religious, and especially monastic, wealth by prohibiting the taking of vows before the age of twenty-five, limiting the
'dowry' a novice could bring, and restricting gifts of land to the
church. Ecclesiastical jurisdictions were attacked by the abolition
of separate church and monastic prisons and by limitations on
the extent of asylum conferred on criminals on church property.
The reform of universities (see p. 173) included a thorough
overhaul of the teaching given to prospective priests, who were
required to spend some time studying in a faculty of theology,
where the government's views now dominated the curriculum.
These measures, and the assertion of state control over large

areas of the church's activities, were in themselves dramatic and significant for the future. Yet surprisingly little was actually done before 1780 to curb monasteries (only in Lombardy and Galicia were any dissolved), to improve the parish system or, above all, to grant a measure of toleration to the Monarchy's religious minorities. In each of these areas, reform was to be delayed until Joseph II assumed sole power, in itself a revealing comment on the very real limits which Maria Theresa placed on any ecclesiastical changes.

This was especially true where religious toleration was concerned. The Monarchy's heterogeneity was evident in the substantial numbers of non-Catholics to be found within its borders. This was particularly so in Hungary and Transylvania, with their large populations of Protestants, Greek Orthodox and Jews. Bohemia contained a substantial Jewish minority, and another, larger, minority came with the acquisition of Galicia, while Protestantism also flourished in Bohemia, Moravia and especially the remnant of Silesia, and even in some pockets in the Austrian duchies. The numbers of non-Catholics in practice obliged Vienna to allow a measure of toleration, or at least to set limits to the extent of discrimination, usually in return for special taxes. Yet official policy upheld Catholic orthodoxy, and all these groups were barred from public offices and experienced some form of discrimination, in the form of restrictions on their religious services, on the building of churches and synagogues or, in the case of the Jews, an obligation to wear distinctive clothing. Toleration made next-to-no progress as long as Maria Theresa was alive. Her notorious and virulent anti-Semitism – shared by the vast majority of her subjects – made Jewish emancipation unthinkable, while her strict Catholic orthodoxy led her to sponsor initiatives against recalcitrant pockets of Protestantism within Austria in particular. The Empress was fully prepared to deport such heretics forcibly to the furthest corners of her realms, as when some 2,700 Protestants were expelled from Upper Austria, Styria and Carinthia and sent to Transylvania in the period before the Seven Years War. Towards the very end of her reign, Maria Theresa was prepared to concede some limited extension to the toleration in practice extended to the Moravian Protestants, but even this was denied the Jews.

Real reform only came after 1780. The new tone apparent during the personal rule was evident in Joseph II's cancellation of the Corpus Christi procession and in his choice of a simple parish priest as confessor. The Emperor's belief in the supremacy of the state over the church quickly became apparent in a series of restrictions on Rome's authority within the Habsburg Monarchy. Within two years, he had created what was in effect a national church throughout his territories which owed minimal obedience to the Pope. This new power was the basis of the ecclesiastical changes on which Joseph II now embarked. He had frequently complained at the slow and even non-existent progress of reform during the 1770s, and the momentum for change was considerable after the beginning of his personal rule. The agenda was far wider, and it was headed by toleration. The Emperor had himself always favoured a measure of religious freedom. This was not because he was indifferent to religion; he was, on the contrary, a sincere and devout Catholic. But he recognised the impracticality and the inadvisibility of discriminating against the large non-Catholic minorities, particularly when such discrimination would weaken the state economically. His characteristic attitude was apparent in his remark in 1777 that 'toleration means only that in purely temporal matters, I would, without taking account of religion, employ and allow to own lands, enter trades and become citizens those who are competent and who would bring advantage and industry to the [Monarchy]'.[9]

Within a year of the Empress's death, the Toleration Patent of 1781 conferred the right to worship in private on Calvinists, Lutherans and adherents of the Greek Orthodox faith. Subject to certain restrictions, they could build their own churches and open their own schools. A measure of civil equality was granted: Protestants and Orthodox could now go to university, enter the civil service and buy and sell land. This toleration, however, was subject to real limitations. It did not extend to members of radical religious sects. Joseph was perfectly prepared to deport a group of deists to Transylvania and, when this failed, to impose physical punishment on them. Yet it was a decisive step forward and brought about a real improvement in the position of the Monarchy's Protestant and Orthodox subjects. As in the later grant of toleration to Jews, Joseph II's motives were a mixture of humanitarianism and economic self-interest. The Enlightenment doctrine of natural rights involved the idea that men had an inalienable right to worship according to their own beliefs, however mistaken these might seem. This did not involve an acceptance of different faiths, merely a recognition that these existed and that little could be done to prevent it. This was reinforced by the practical consideration, central to cameralist teaching, that a state should not weaken itself by excluding groups of subjects from economic activity.

The initial intention had been to include Jews in the Toleration Patent of 1781, but the resulting controversy prevented this being done. The position and even more the numbers of Jews varied enormously from province to province. They were numerically least significant but economically most influential in Vienna itself, where their role as government financiers and army contractors made them particularly prominent. The importance of Jewish finance to the Monarchy declined after 1740, though some revival was apparent during the 1760s and 1770s. Maria Theresa's anti-Semitism apparently extended to being prepared to borrow from Protestant financiers rather than from the Jewish houses that had been so prominent under Charles VI. Indeed, the Empress had even attempted to expel the Jews from the Bohemian Lands in 1744–45, in retaliation for their financial support to the Bavarian Emperor Charles VII after his invasion and occupation of Prague. She desisted only when the consequent damage to Bohemia's economic life was made clear, and settled instead for an increased toleration tax. Most of the Monarchy's Jews lived in Hungary, the Bohemian Lands and, above all, the new province of Galicia. Everywhere they were subject to heavy toleration taxes – an important source of revenue for the hard-pressed Habsburg treasury – and to restrictions on their worship, dress, household and commercial activities. This discrimination had its origins in medieval sumptuary legislation intended to highlight their separateness from the rest of the community. The repressive legislation had been periodically renewed, most recently by the Jewish Ordinance (*Judenordnung*) of 1764.

Jewish toleration was embodied in a series of patents for individual provinces in 1782–83. These removed the more obnoxious social humiliations, such as the obligation on men to wear yellow armbands and women and children to have yellow strips in their hair. Married and widowed Jewish men no longer had to wear beards, while they could now go out before midday on Sundays. The *Leibmaut*, a tax usually on livestock but which was also imposed on Jews travelling through the Monarchy, was abolished. The same freedom to worship was conferred as on Protestants and Greek Orthodox and with it the right to enter all educational establishments and attend public entertainments.

This legislation had twin goals: emancipation and assimilation. To break down the barriers between Christian and Jewish communities, Joseph II decreed that after a period of two years, all Jews must employ the German language in business transactions and that Hebrew and Yiddish could only be used in religious services or in family and social gatherings. He also opened up all trades and professions to Jews, with the clear intention that they might make a greater economic contribution. Yet these advances, remarkable as they were, did not amount to complete emancipation or civil and religious equality. This was never the Emperor's intention: he aimed only to enable Jews to become good and useful subjects. Joseph was, in any case, forced to retreat from some of his more ambitious plans by opposition within the Council of State and by fears of popular anti-Semitic violence. Even such established progressives as Kaunitz hesitated in the fact of the undoubed unpopularity of the measures.

The limitations were clearly apparent in the Toleration Patent for Lower Austria (January 1782). Although this abolished the traditional restrictions and gave Jews the right to live anywhere in Vienna and not merely within their own ghetto, they were not allowed to build a synagogue and no increase was to be permitted in the number of families allowed to reside in the capital. The restrictions on the numbers of Jews living in the Bohemian Lands were maintained, and fiscal exploitation continued. Nevertheless, Joseph II's Jewish reforms, carried through in the face of vociferous official and popular opposition and against the hostility of Jewish leaders fearful that the traditional community would be destroyed, brought about real if limited improvements. In the context of near-universal anti-Semitism – the Emperor himself was not entirely free of such prejudices – they were remarkably far-sighted and humanitarian measures, and they earned more applause from his enlightened contemporaries than any other reform.

Joseph himself wished to further the process of assimilation. His radical programme is apparent from his measures in Galicia during the 1780s. The Jewish community there was large (around 225,000), relatively poor and internally divided, and was in all respects a suitable target for social engineering designed to break down the barriers between the communities and to lead to eventual civil equality for the Jews. Throughout the Emperor's personal rule, a determined attempt was made to settle poor, rural Jews on Crown and vacant lands, with financial assistance and fiscal incentives from the state. Though the scale of the problem was far larger than the resources available, these agricultural colonies had some success. The growing radicalism of his policy reached its climax in the Galician Jewish Ordinance, issued in 1789. This sought to further assimilation by the dramatic step of allowing only those Jews who became farmers or artisans to remain resident in Galicia, and also attempted unsuccessfully to transfer the costs of the colonisation scheme to the Jewish community itself. This measure was even less successful than some of the Emperor's other reforms, but it does reveal his radical approach. To his detractors, Joseph II was the 'Emperor of the Jews', a sobriquet which is also a testimony to his real contribution to the removal of discrimination and to eventual emancipation and assimilation.

Joseph II's established hostility towards the contemplative orders and his concern to improve the quality and quantity of priests, and to make pastoral work more effective, dictated the other major ecclesiastical reforms of the 1780s. In Upper and Lower Austria and Styria, the wealth of the religious houses was particularly evident, with around 40 per cent of taxable revenue in ecclesiastical hands; in the Bohemian Lands, the figure was perhaps half this. Such wealth was an obvious target for a reforming monarch in search of money to pay for his planned religious changes and, in 1781, the dissolution of some monasteries began. At the beginning of Joseph's personal rule, there were slightly over 2,000 monasteries in all the Habsburg territories, with some 40,000 monks and nuns. By 1790, these totals had fallen to around 1,250 and over 27,000 respectively. Slightly over a third of all monastic foundations were dissolved and rather less than a third of their inhabitants expelled. The majority of these suppressions took place within the Monarchy's Central European territories, with 55 per cent of all religious houses suppressed in the Hereditary Lands and 75 per cent in Hungary.[10]

The scale of the dissolution was considerable and greater than that attempted in any other Catholic country, though many of the larger monasteries survived and it was far short of a total suppression. The revenue, along with that from some ex-Jesuit properties, was applied partly to paying pensions to some former monks and nuns, and rather more to funding Joseph II's education reforms (below, pp. 176–7) and his ecclesiastical measures. Foremost among these was the radical overhaul of the diocesan system carried out in 1782–83. Demographic changes meant that, by the second half of the eighteenth century, parish boundaries no longer corresponded with the distribution of population. New dioceses and parishes (each with its own school) were set up unilaterally, the state took responsibility for paying priests, and a parochial system was established which was to be an important source of the strength and vitality of the Catholic church in the nineteenth century. Simultaneously, the Emperor sought to improve the training given to parish priests by setting up six 'general seminaries', which replaced the diocesan seminaries. These were unpopular with the bishops and arch-bishops, who resented their own loss of influence, and the experiment failed, being abandoned shortly after Joseph II's death. The Emperor's determined assault on popular superstitions was also only partially successful. He sought in particular to reduce the number of pilgrimages and religious processions, but these initiatives were deeply unpopular and were stubbornly resisted by the faithful, who remained firmly attached to the old ways. Overall, however, Joseph II's religious reforms, and in particular the measure of toleration introduced and the parochial reorganisation carried out, were surprisingly successful. Humanitarianism and Reform Catholicism together did much to improve the position and the practices of Catholicism and non-Catholic religions within the Monarchy.

V

The educational reforms were the most successful and enduring, particularly the initiatives in primary and, to a lesser extent, secondary schooling during the 1770s and 1780s. These were intimately connected with the religious changes. This was because the Jesuits had dominated education for a century before Maria Theresa's accession, controlling most *Gymnasien* (Latin secondary schools) and in particular enjoying a stranglehold over the universities, where all professors in the dominant faculties of theology and philosophy were members of the Society. Their pre-eminence ensured that the Monarchy's universities, and particularly the leading one in Vienna, continued to provide a very traditional, scholastic and even arcane education, and they were incapable of providing the training needed by administrators in the expanding state apparatus. The lower nobility and the sons of the aristocracy had, particularly after the 1730s, shunned these institutions and instead sought the more practical training essential for government service either in the noble academies which sprang up in the various Habsburg provinces or abroad. They went to the Benedictine University in neighbouring Salzburg, to Leiden in the Dutch Republic, and increasingly to the Protestant universities of northen Germany, above all Halle, Jena, Leipzig and Göttingen. There they secured a more useful education in public law (especially that of the Holy Roman Empire), natural law, history and, above all, the new science of cameralism. A generation later, three out of six members of the influential Council of State had been students of a northern German Protestant university, most notably Kaunitz, who had studied at Leipzig.

The shortage of trained recruits for government service produced a series of initiatives during the 1740s and 1750s. The *Theresianum* was established in 1746 to train future civil servants and a chair in cameral science was founded, though its usefulness was limited by its subsequent transfer to Jesuit control; a military academy was set up at Wiener Neustadt in 1751–52 to prepare young noblemen for the officer *corps*; and in 1754, the Oriental Academy was established to educate future diplomats. The frontal assault on the position of the Jesuits in the universities came during the 1750s. The Dutch Jansenist, Gerard van Swieten, carried through a reform of Vienna's Faculty of Medicine and, in 1759, this was extended to the all-important theology and philosophy faculties. The Jesuits lost their previous monopoly, being compensated with control of the *Theresianum*, and a series of reform-minded individuals was appointed to chairs, notably the celebrated jurist, Martini and the canon lawyer, Riegger. Curricula were brought up to date and chairs of cameralism were founded. The first such professor at Vienna, appointed in 1763, was Joseph von Sonnenfels.

The impact of these initiatives was less than had been hoped. The new administrative elite which it had been intended would be trained in the Monarchy's own universities did not at first materialise and prospective bureaucrats continued for some time to look to north Germany for practical training. Higher education was resistant to reform and progress was slow and incomplete. It was further hampered after 1780 by Joseph II's insistence that only severely practical subjects should flourish, above all theology, law, medicine, languages and cameral science. The Emperor's scepticism about the value of intellectual activity for its own sake distorted development, though he did ensure that all teaching should now be in German. Less practical subjects, above all philosophy, were neglected in favour of a severely functionalist approach to higher education. Nevertheless, the first steps had been taken towards the creation of a more up-to-date university system. These initiatives were, in any case, far less important than the rapid expansion of primary education.

Until the reign of Maria Theresa, schooling was very largely in the hands of the Catholic church and, to a lesser extent, of town authorities and individual noblemen in the countryside.[11] The reigns of the Empress and her son, and particularly the years after 1765, saw increasing state involvement in education and an assertion of the government's overall responsibility for schooling. The motives for this intervention were mixed. A few of the more advanced figures in government, men such as Kaunitz and Gebler, believed that education was a good in itself and a means by which individuals might realise their potential. Such enlightened ideas were certainly one element in the formulation of policy, but were less influential than economic and religious considerations. Cameralist teaching embodied the idea that universal elementary education could encourage obedience and diligence in the agrarian and manufacturing work force, though further study was to be discouraged. The capacity to read and write could also, in a distinctly utilitarian way, make the population better and more useful subjects of the state. Such practical considerations were important, but they were less influential than purely religious considerations.

The educational reforms were to a large extent an extension of the Josephinist religious measures. Reform Catholicism came to embody the idea that basic literacy, and the study of the Bible this would make possible, would assist the development of a simpler, purer faith. The first half of Maria Theresa's reign had seen growing official interest – an Educational Commission (*Studienhofkommission*) had been set up in 1760 – and sporadic initiatives, but it was only in 1769 that school reform was placed firmly on the agenda in Vienna. A report from Firmian, prince-bishop of Passau, spoke of the march of 'heresy and unbelief' in Lower and Upper Austria. This focused the Empress's established concern with the decline of Christian morality and Catholic orthodoxy. The maintenance of these was the principal task of the established network of parish schools, which they were clearly failing to fulfill. Maria Theresa now ordered the provincial governments of Lower and Upper Austria to prepare schemes for the improvement of primary schooling, and the whole matter was also taken up by the central government.

The debate which followed was dominated by the ideas of the pre-eminent Catholic educational reformer, Johann Ignaz Felbiger. He was the Augustinian abbot of Sagan in Prussian Silesia and had remained in the province after its conquest by Frederick the Great. Felbiger's responsibility for the neighbouring Catholic schools had been the source of his enduring interest in education. His own reforms at Sagan were closely modelled on the schools and shaped by the ideas of the German Pietists, who were the leaders in educational theory and practice. Pietism was the reform wing of German Protestantism and its advocates were particularly influential within Prussia. Their educational reforms and especially the famous Normal Schools (*Normalschulen*) were the source of Felbiger's innovations, though he also adapted some of the more impractical Pietist ideas to fit in with the realities of rural education. These were to provide a model for schools established in the surrounding area and were also to train primary teachers for these institutions.

The advocates of educational reform at Vienna were initially blocked by the Court Chancellery, headed by the influential conservative Chotek. But the cause was then taken up by the Council of State and some progress was made. In January 1771, a Normal School was created out of the St Stephen's town school in Vienna, with the aid of a government subsidy. An even more radical scheme to secularise education entirely throughout the Monarchy was put forward in 1770, but it was rejected because of its immense cost and because it was feared that the destruction of the infrastructure provided by the church would make it impracticable: the fact that it would place Protestant educational reformers in control also made it anathema to Maria Theresa. The vetoing of this scheme early in 1772 brought plans for reform shuddering to a halt while, simultaneously, the Vienna Normal School was in confusion and apparent decline.

The cause of school reform was saved by the suppression of the Society of Jesus in the following year, which created an educational vacuum. More important, it provided the wealth essential for any wide-ranging reform. The Empress quickly secured control over all former Jesuit property, and part of the income from this was employed to establish the new network of primary schools. Two months after the suppression of the Society, an Education Commission was established in Vienna. It was dominated by Reform Catholics and *Aufklärer* (men such as Müller, Greiner and Martini) – the same partnership that had contributed to the reform of popular religious practices, censorship and the universities. By the end of 1773, Martini had drawn up a report which advocated that universal compulsory schooling on Felbiger's model should be introduced, financed by former Jesuit wealth.

Felbiger himself was 'borrowed' from Frederick the Great and now brought to Vienna to supervise this reform in person. After his arrival in May 1774, he was given near dictatorial authority, and he dominated educational policy until Maria Theresa's death. His first task was to draft the General School Ordinance (*Allgemeine Schulordnung*) published in the following December. This established compulsory schooling between the ages of six and twelve for five days a week throughout part of the year at least. It embodied the principle of compulsory education and was to be enforced by fines on parents whose children failed to attend. The state provided free textbooks, and often buildings as well for the new schools that were established, but expected even the poorest parents to pay fees and made individual parishes responsible for the salaries of teachers. This system was quickly extended to the other provinces of the Monarchy. Hungary, for example, was the subject of a separate ordinance (the famous *Ratio Educationis*) in 1777, and here again central government enjoyed a considerable measure of success.

Joseph II's personal rule saw a massive increase in state expenditure on education, directed by the influential Gottfried van Swieten as head of the Education Commission and made possible by the confiscated monastic wealth. This was partly used to finance a rapid increase in the number of primary schools, to some extent at the expense of the secondary sector. The success of these reforms was striking and they provided the basis for the nineteenth-century advance in public education. By 1790, as many as two-thirds of children in the Bohemian Lands were attending primary schools. These provinces had traditionally enjoyed a considerable measure of educational provision, and the church and the aristocracy both actively supported the measures of the 1770s and 1780s. This was not everywhere the case. Conservative, clerical resistance was widespread and did something to hinder the implementation of these reforms. Nevertheless, by 1790, the extent of elementary education provision was greater than in any other major continental state, and it was the most enduring legacy of the reform era. Even more remarkably, educational opportunities were in theory to be as open to girls as to boys. Though at first male children formed the bulk of the new school population, there is some evidence that by the end of the eighteenth century, girls were beginning to catch up.

VI

The final area of reforming activity, agrarian life, proved the most explosive. This was because, in a society as overwhelmingly rural as the Monarchy, any interference in the relationship between nobleman and peasant would be deeply unpopular and widely resented. Nevertheless, the agrarian reforms of Maria Theresa and Joseph II were among their more remarkable achievements, and only in Denmark (see pp. 245–63) was more actually done to improve the lot of the peasant. Neither the Empress nor her son was to be wholly consistent in their approach to this problem and, at least until 1780, the mother was often more radical, willing to undertake a root-and-branch reform which, at this stage, alarmed the Emperor. The agrarian problem was the central issue in domestic policy throughout the 1770s and 1780s and aroused more serious opposition than any other measure.

The rural world of the Habsburg Monarchy in the later-eighteenth century was an extremely diverse society, with wide variations not merely between different provinces but on different and sometimes neighbouring estates. Most land was in the possession of the nobility and the Church, from whom the peasantry held the fields and strips they cultivated. The terms on which they held this land and the extent and nature of the obligations they owed their lord became the main focus of Vienna's concern. At the beginning of the early modern period, the Central European peasantry had largely been free, but thereafter its status had declined sharply. The sixteenth-century Price Revolution and the opportunities it created for profitable export, above all of grain and grain-based products, had been the source of this deterioration. This had led the nobility to expand the demesne (the lands which they themselves cultivated) and, in order to diminish labour costs and therefore to protect their own profits, to require compulsory and usually unpaid work from their peasants. Their efforts to impose labour services had been assisted by the political upheavals of the seventeenth century and finally consolidated by the mounting fiscal demands of central government during Austria's wars against Louis XIV and the Turks.

The enserfment of the peasantry is an established theme in the history of early modern Eastern Europe. Its fate in the Habsburg territories was in some respects less severe than in other countries and, above all, in Poland and Russia. There was also considerable variation in the rural picture. In the Alpine provinces of Tyrol and Vorarlberg, the peasants resisted efforts to impose labour services and remained free in the later-eighteenth century. By contrast, the Bohemian Lands and the Kingdom of Hungary had become serf economies, where individual noblemen on what were sometimes very large estates farmed their own demesnes with the aid of substantial labour services from a depressed and impoverished peasantry. These labour services were usually known as *robot* (after the Czech *robota*, work), though sometimes in the Austrian territories as *Frondienst*. They were everywhere accompanied by a series of further restrictions on the serf's life and person. He was not allowed to move from the estate, or marry, or take up any trade without his lord's permission. He could be obliged to sell his own produce at artificially deflated prices to the seigneur, and sometimes to buy the lord's produce and use his mill as well. He could also be forced to perform additional labour services, often without any pay, while his children could be made to work for the seigneur between the ages of 14 and 18.

The operation of this system and the actual burdens on the serf varied considerably. In the former Polish province of Galicia, a peculiarly oppressive regime operated where in summer a peasant could be forced to perform as much as six days of labour services a week. In the Kingdom of Bohemia, the figure was usually three days, and in some of the Austrian duchies it was even lower. In some respects, the Austrian peasant was worse off than the Bohemian or Hungarian serf, though he was likely to own his own land. This was because the extent of his *robot* was undefined, and he also had to pay an annual rent. There was considerable divergence between the various Austrian territories. Labour services were usually lighter in Upper Austria (and often commuted for payments in cash or kind) but relatively heavy in

Lower Austria and Styria. Their extent came to concern the government, and efforts were made to regulate them, but it was always the plight of the Bohemian and Hungarian serf which received most attention. In 1767, an attempt was made in the *Urbarium* to limit labour services to one day per week in Hungary, but it was widely ignored. Instead, from *c.* 1770 onwards, Vienna's attention came to be focused on the plight of the Bohemian serf.

Serfdom was not simply an economic system or even a legal status, but was also a form of social organisation. The labour services which the serf performed and the restrictions on his mobility were part of a wider subordination. The nobleman also exercised patrimonial jurisdiction in civil and sometimes criminal cases and, even after the administrative reforms of 1749–61, might well be involved in collecting the Contribution. The lord's estate and his authority in all secular matters established the parameters of a serf's life, which was little affected by the central government. The nobleman and his bailiff were far more immediate and potent influences than the tax collector or recruiting officer. This meant that the condition of the enserfed peasantry would ultimately depend on his lord.

The concern of central government at the serf's plight did not begin under Maria Theresa but extended back into the seventeenth century. The peasant bore the principal burden of direct taxation and it was therefore in Vienna's interest that his obligations to his lord should not become oppressive, since this would obviously hinder his ability to pay his taxes. The government also wanted to ensure a regular supply of fit conscripts for the army, which might be imperilled by a too-oppressive seigneurial regime. There was finally fear of insurrection, should the serf *régime* become too severe, and a traditional Christian sense of responsibility for all the Monarchy's subjects. Efforts had been made, by *Robotpatenten* in 1680, 1717 and 1738, to limit labour services to three days a week, where agreements between lords and their tenants did not already exist, but it seems clear that these edicts had been widely ignored. One by-product of the expansion of government during the first half of the Empress's reign was an increasing awareness of the serf's plight. This was being exacerbated by the rapid rise in population, which increased the number of landless peasants and put pressure on the existing arable land. The particular demographic causes were not apparent to contemporaries, but its consequences certainly were.

In 1770 and again in 1771, torrential rains destroyed the harvest throughout Central Europe and particularly in the Kingdom of Bohemia. The famine which ensued, along with the collapse of Bohemia's manorial economy which deprived the peasants of wages to buy food, made the situation critical, and as many as a quarter of a million people may have died in these two years alone. The situation was exacerbated by the actions of many Bohemian noblemen who, instead of distributing the grain in their stores to their own starving peasants, sold it at considerable profit in neighbouring Saxony and Prussia, equally devastated by famine. The government in Vienna was already aware of the delicate situation which existed, particularly after the celebrated Mansfeld case when a leading Bohemian nobleman was fined and had his estates confiscated for two years for particularly blatant mistreatment of his serfs. In 1769, Maria Theresa had sought to limit the scope of noble jurisdiction over the peasants on their estates, though this edict's impact was limited. The next year, an equally unsuccessful effort had been made to limit the worst excesses of serfdom such as child labour and the forced sale of produce. Her government was driven to intervene by the Bohemian famine of 1771–72 and by the effective pressure exerted by her son.

Joseph II had travelled through the Kingdom of Bohemia late in 1771 and had been appalled by the wretchedness, disease and starvation everywhere. The conventional measures adopted in times of famine, and particularly government price-fixing and encouragement of internal trade in grain, had failed to cope with the crisis, while the local authorities in Bohemia were all-but-paralysed. The Emperor's impassioned arguments, and the special authority which his travels and first-hand knowledge of the situation gave him, forced Vienna to consider the position of the Bohemian serfs, but his influence was not yet sufficient to bring about effective action. That was accomplished only by a particularly serious Bohemian peasant revolt.

Since 1771, a commission had been arguing inconclusively over the terms of a new Bohemian *Urbarium*. In 1774, in an attempt to cut through all the debate, Maria Theresa had proposed nothing less than the abolition of serfdom. Her son demurred, fearful of the social and fiscal consequences, and instead argued for local solutions to particular problems. His proposal for individual *Urbaria* to be negotiated between lord and serfs on every estate was accepted and work commenced on these. When it became apparent that the Bohemian nobility were preventing progress, a serf rebellion began. During the first few months of 1775, it swept across Bohemia, with hints of class war in the attacks on manor houses, religious foundations and individual landlords. The Protestantism which had existed underground since the defeat of the Bohemian rebellion of 1618–20 and which had revived notably in the eighteenth century, also came into the open, and the serfs even established their own 'government'. The rising contained many echoes of the great Cossack-peasant rebellion in Russia led by Pugachev (1773–75) and this may have increased Vienna's anxieties. Eventually, the Bohemian serfs were defeated, by an army of 40,000 infantry and four regiments of cavalry. The government adopted a very lenient policy towards the rebels and now introduced the long-delayed reform.

Though Vienna was finally forced into action by the Bohemian rebellion, the measures introduced arose from more motives than fear. Some of these aims were traditional, above all the established concern that landlord oppression should not damage the state's fiscal income. This was particularly important in the context of the struggle with Prussia, which had driven up government borrowing and the Contribution to new levels. Such thinking was reinforced by the cameralist doctrines which were now widely influential in government circles. The notion that a state's prosperity ultimately depended on that of its subjects led to serfdom being condemned as economically inefficient. The absence of peasant security of tenure was highlighted as a disincentive to agricultural improvements and thus to economic progress. The cameralist ideal was a free peasantry, owning their own land or with security of tenure, whose prosperity and increasing numbers would be the basis of the state's well-being. Maria Theresa had spoken in 1769 of the need to 'sustain the peasantry, as the most numerous class of subjects and the foundation and greatest strength of the state'.[12] This was an ambitious programme and one whose implementation would be difficult and might be expected to arouse the formidable opposition of the nobility.

Joseph II's approach to agrarian questions was certainly influenced by cameralist ideas and by concern for the state's fiscal income. It was also affected by more purely military considerations. In 1771, after a long struggle, the Emperor and his principal military adviser, Marshal Lacy, had introduced recruiting districts, modelled on the famous Prussian cantonal system, into the Hereditary Lands. Their intention was to solve the army's manpower problems permanently by establishing territorial levies. This required a regular supply of healthy peasants, and this was threatened by the misery of the serf population and by the oppressive nature of labour services in many provinces. The state and the seigneur were here competitors. Joseph II had an established interest in promoting agricultural improvement, epitomised by the celebrated episode in August 1769 when he drove a peasant's plough while travelling in Moravia. He was also influenced by more enlightened, humanitarian motives. The doctrine of natural rights was particularly important. This established that a serf was a human being with certain inalienable rights and not simply a source of labour and taxation. The view was pressed in particular by the influential official, Franz Anton von Blanc, who was the real source of Joseph's proposal for a new Bohemian *Urbarium* in 1775. Maria Theresa's undoubted humanitarianism was more grounded in traditional Christianity than the Emperor's, but it was deeply rooted and sincere, and at this point her approach was considerably more radical.

In August and September 1775, a new *Robotpatent* was promulgated for Bohemia and Moravia. This fixed an upper limit of three days per week for labour services, where other agreements were not already in operation, with proportionately less for serfs with less than a full holding and for landless peasants. It was a determined attempt to define and to limit serf obligations and to protect them against the worst excesses of seigneurial oppression. The 1775 decree seems to have brought about a limited but real improvement in the position of the Bohemian serfs. There were problems of enforcement, and the bulk of the nobility opposed the edict vigorously and with some success. But the *Robotpatent* did bring about a reduction in the burdens on the peasantry and some improvement in the serfs' condition. The same can be said about the other principal agrarian legislation of Maria Theresa's final decade: the reform of labour services in the Austrian territories. These were limited to two days a week in Lower Austria (1772) and three in Styria (1778).

The 1770s had seen the state intervene – albeit partially and hesitantly – in landlord–peasant relations. Maria Theresa would certainly have gone further, arguing that the situation was so tense that only a full abolition could resolve it. Her natural conservatism led her to hesitate for many years before intervening, but once persuaded of the need to act, her radicalism was real. Joseph II's position at this time was more cautious and broadly similar to that adopted by Frederick the Great in Prussia (see pp. 280–1) and Catherine the Great in Russia (see pp. 307–10). Though recognising the drawbacks of serfdom and the strong humanitarian arguments against it, the Emperor also believed that immediate, outright abolition involved an unacceptable risk of social and economic disruption and might weaken the state, which depended on taxation and on noble support. He was particularly afraid of the potential damage to the military system and to fiscal income. This gradualist approach was also apparent in his support for the *Raab* reforms.

The power of the Bohemian nobility made it difficult and perhaps impossible to turn their serfs into small peasant proprietors. In the mid-1770s, however, this was attempted on the cameral (Crown) lands and on the ex-Jesuit estates in Bohemia. This scheme was put forward by Franz Anton von Raab and, with the influential support of Maria Theresa and Joseph II, he began to implement his ideas on a few selected estates at the beginning of 1776. The cameralist axiom that a prosperous, landowning peasantry would be the best foundation for a strong state was the source of this initiative, which divided up manors and sought to substitute a land-holding yeomanry for serfdom. To achieve this end, agreements were drawn up for individual estates by which labour services were abolished and peasants were turned into hereditary lease-holders, paying moderate cash rents; the fact that the state and not a nobleman was the landlord enabled this to be done without compensation. A secondary aim was to increase population and, to further this end, peasants were brought in from outside Bohemia and settled on the partitioned estates. The success of Raab's work was considerable, but it was necessarily small-scale. The principal landowners in Bohemia remained the nobility and the church, and they were resistant to any suggestion that they might imitate the *Raab* system and divide their own estates. During the 1780s, Joseph II sought to widen the scope of this initiative to include town, church and even noble lands, but this encountered severe opposition, not least from within the state administration, and these efforts were far less successful than those of the 1770s. Behind the *Raab* system were the twin goals of agricultural progress and a free peasantry enjoying secure possession of their land. These remained the central objectives of Joseph II's agrarian reforms after 1780.

A series of measures during the Emperor's early years as sole ruler sought to improve the position and legal status of the serf, but left labour services intact. In November 1781, the Bohemian peasants were 'emancipated' and this was extended to Galicia in the following April. This meant that the serfs were declared to have certain basic civil rights which they had not hitherto possessed. Theoretically they could now marry, move or enter a trade provided they secured their lord's formal permission, which was not to be unreasonably withheld. An elementary system of legal aid was instituted for peasants seeking redress in the courts against noble oppression, while child labour was formally abolished. A second Patent in November 1781 enabled serfs in Bohemia to secure at modest price greater security of tenure and the right to dispose of their land as they wished. These were followed by a further series of measures extending the reforms of the 1770s to other provinces. In 1782 labour services were limited to three days a week in Carniola and, in 1784, the same figure was laid down for Galicia, while Joseph II's legal reforms and especially the new Penal Law of 1781 were additional blows against the lords' patrimonial jurisdiction, imposing the obligation to employ professional, university-trained judges. Most remarkably of all, Joseph II attempted to extend agrarian reform to Hungary in 1785. The standard limit of three days a week was decreed for labour services and the Hungarian serf was declared to have the same civil rights given to his Bohemian counterpart four years earlier. The real limitations on Habsburg authority in the Kingdom ensured that these measures were widely and flagrantly disregarded and usually remained a dead letter, though they excited noble resentment and opposition. Elsewhere, Joseph II's early agrarian reforms were resisted and, at times, ignored, but they did bring about some amelioration in the peasants' condition and furthered the improvements begun in the 1770s.

These reforms had deliberately ignored the central problem, that of labour services. Joseph II was determined to attack these, but wished to avoid damaging destruction to the fabric of society and state. The central conundrum was how to protect tax revenue and noble income. On cameral and ex-Jesuit estates in Bohemia, no compensation had been necessary. This was impossible on noble and church land if and when the *Raab* system was extended to private estates throughout the Monarchy, which was the Emperor's eventual aim. The existing landholders would have to be compensated on a massive scale, and neither the over-stretched state budget nor the newly-free peasantry could afford such substantial sums. By about 1783, Joseph II had devised his solution. This linked the curtailment of labour services to a drastic revision of the tax system, which envisaged one single tax on land reminiscent of the physiocratic *impôt unique*. The Emperor's scheme for root-and-branch reform encountered fierce and determined opposition in the Council of State and in the administration, and this slowed its implementation. A more important source of delay was the need to compile a new *cadastre* listing all land liable to be taxed. The shortage of qualified surveyors, bureaucratic inertia and even resistance, and noble opposition together delayed its compilation, but by the final year of the Emperor's life, the *cadastre* was ready.

In February 1789, the radical Tax and Agrarian Regulation was promulgated, though its implementation was postponed until the following November. This decreed that a peasant, freed of all labour services, would henceforth retain 70 per cent of his annual income to support his family, to improve his own small-holding, and to pay his dues to the church and to his village. The remaining 30 per cent was to be divided between the state as taxation and the nobleman as compensation for the loss of labour services. With characteristic precision, Joseph laid down that $12\frac{2}{9}$ per cent would be paid as taxation and the remaining $17\frac{7}{9}$ per cent as a permanent commutation payment to the nobleman. This change was to apply only to rustical serfs, that is to say peasants holding lands which were not part of the lord's demesne (dominical peasants would continue to rent their lands with the new security in theory provided by the earlier legislation) and in Bohemia at least only a minority would have benefited immediately. But the wide implications which the measure possessed, and the direct threat to the nobility's social position and economic well-being aroused immediate and vocal protests, and in Hungary contributed to the rebellion which threatened by the end of 1789. Joseph himelf was forced to postpone implementation until the autumn of 1790, and after his own death in February of that year, his brother and successor Leopold II withdrew it altogether.

The most dramatic and remarkable of the Josephine reforms, and the culmination of Habsburg agrarian legislation, thus

proved to be stillborn: the serfs in the Monarchy would everywhere wait until 1848 to be emancipated. The entrenched
opposition of the seigneurs proved too strong and labour services
remained in place for another two generations. Yet the agrarian
reforms were far from a complete failure. In the context of the
later-eighteenth century, it is the extent of what was achieved,
rather than its obvious limitations, that is striking. The Monarchy's peasants enjoyed greater security of tenure, a measure of
equality before the law and improved material conditions as a
result of the reforms of the 1770s and 1780s. Only in Hungary
was resistance complete and successful, and the condition of the
serf as wretched as before.

Joseph II's agrarian reforms were one reason for the growing
unpopularity of his government and its planned changes. Administrative centralisation was particularly resented in the Austrian
Netherlands, where his reign ended amidst a rebellion against
Habsburg rule, and in the Kingdom of Hungary, where open
resistance was very close by 1790. The Monarchy's involvement
in a disastrous Turkish war after 1787 was a further source of
unrest and criticism, and in the final years of his life, the Emperor
was driven to adopt severely repressive measures. Joseph II's
policies, and especially the emphasis on German as the official
language throughout the Central European lands, could also
have unexpected and even paradoxical results, as R. J. W. Evans
makes clear (see Chapter 7). Yet the undoubted problems by the
later 1780s qualify rather than overturn Joseph II's reputation
as a radical reformer. Many of his measures – such as the
introduction of the idea of marriage as a civil contract, rather
than a sacrament, and the removal of legal disabilities from
illegitimacy – were very radical indeed and anticipated nineteenth-century developments.

A strong and wealthy state, with a formidable army, was
always a central aim of Habsburg policies during the half
century after 1740. Yet, particularly under Joseph II, there was
also real humanitarianism and concern with the fate of the poor
and the destitute. The Emperor not only claimed to rule in the
interests of all the people; he actually tried to do so. A series of
measures during the 1780s extended the provision of poor relief
and established primitive welfare services, particularly in the
towns. Medical education and provisions were improved,
hospitals – notably the famous Vienna General Hospital – were
established, giving free treatment to paupers, and homes for

orphans and unmarried mothers were set up. Such initiatives were always a minor part of Joseph II's wide-ranging programme, but they are clear evidence of the radical nature of his policies and confirm his claim to be considered one of the most remarkable reformers of the later eighteenth century.

6. Maria Theresa and Hungary

R. J. W. EVANS

MODERN Habsburg history begins in 1740, with the accession of Maria Theresa as ruler over a group of realms which, lacking allies themselves, were immediately beleaguered by a hostile alliance of five European states. Over the next decades, especially after the temporary peace of 1748, and then between the final peace of 1763 and up to her death in 1780, her measures were piecemeal, but remarkably concrete and far-reaching, modernising and consolidating those realms for the future. The modern Austro-Hungarian relationship was unveiled in those same decades. We find one aspect of it in the drama of September 1741, as the young queen, Maria Theresa, arrived to seek succour from a typically tempestuous and obstructive coronation diet at Pressburg. Attired in Hungarian style (white, with gold braid and blue floral decoration), she made a direct and emotional appeal to the assembled nobles, and evoked their famous offer of physical sacrifice, amid the general cry of '*vitam nostram et sanguinem consecramus*'.[1] At the other extreme lies the hardly less dramatic diet of 1764–5, which concluded in bitterness and deadlock, with Maria Theresa pointedly absent, and was followed by fifteen years of direct rule from Vienna. Thus the two lines of policy alternately pursued in Habsburg governance of Hungary right down to 1918 – by turns essentially respectful or disrespectful of constitutional tradition – were now first clearly elaborated in the context of the Empress's transformation of the whole Monarchy.

Yet the subject is little known to historians; perhaps because on the surface Hungary was the one area under Habsburg rule which did not change much. Its feudal economy and society and its corporate privileges, especially that of noble tax-exemption, survived; a *modus vivendi* appeared to have been reached between the energetic but careful ruler and the loyal but unbending constitution. Indeed, what happened in 'Austria' (I shall use that term as shorthand for the rest of the Habsburg lands) was crucially important for Hungary too, precisely because it happened there alone: the development of an effective absolutism outside Hungary, and its associated realms of Croatia and Transylvania, increased the gap between the two halves of the Monarchy, and foreshadowed the 'Dualism' of a later age. Thus much of the reform programme had only an indirect impact on Hungary. For example, the new economic management, which impinged mainly through its effect on tariff policy and through the efforts at 'impopulation', or the settlement of foreign colonists upon lands recently recovered from the Turks; or measures in the fields of justice, policing and local administration.[2] Nevertheless Maria Theresa's government – and this is the most significant point of all – did gradually modify the two great pillars of earlier Habsburg authority, as well as introducing a new one, throughout her realms; and in these last respects Hungary was hardly less affected than the rest.

In briefest summary, the structural changes initiated from the 1740s onward contained a political-social, a cultural-spiritual and a military element. In the first place, aristocratic and provincial oligarchy was supplemented by, and in good measure reorientated as, a centrally-directed and standardised state service, with written forms and bureaucratic norms. Secondly, the Catholic church and religious precept were supplemented by, and in good measure reorientated as, a deliberate patriotism and state identity, based on notions of the common good – that terminology of 'general good' ('*allgemeines Wohlsein*') or 'the general best' ('*das allgemeine Beste*') or similar, which constantly recurs in the enactments of Habsburg absolutism. Finally, the army was harnessed by the mid-eighteenth century to fulfil a new role as unifying and integrative factor in the defence of the state. Let us consider these changes in turn.

Even in Austria the reform movement needed to work through traditional notables, particularly, of course, reasonably enlightened and progressive ones. That was the more true in Hungary, where office remained restricted to the noble nation and many of the gentry were unwilling or ill-equipped to serve anyway. Resort could be had, in the first instance, to the established aristocratic families who at the start of the reign maintained the three chief organs of executive authority under the Crown: the Chancellery, the Lieutenant Council, and the

Chamber. In each of them an expansion of business under Maria Theresa went with more professional management.

The dutiful and diligent Hungarian Chancellor for over twenty years from 1762 was Count Ferenc Eszterházy, who belonged to the senior branch of the richest and most powerful clan in the country (he was second cousin to Prince Nicholas, the celebrated *maecenas* and patron of Joseph Haydn). Having been debated at the Chancellery in Vienna, whose senior personnel rose from 24 to 75 in this period, the sovereign's decisions then passed for their implementation in Hungary to the Lieutenancy Council, under the Palatine (royal viceroy) or his deputy. There too the workload increased markedly: about 2,500 letters were received annually during the 1740s, over 10,000 by the 1770s; and the number of officials swelled from under 50 to 122 to handle it. There too magnates held sway, like Count Lajos Batthyány, Palatine from 1751 to 1765 (and a previous Chancellor), who headed the second most prominent Hungarian family.[3]

Such men, besides being aristocrats, were also favourites and trusted agents of the Empress: Eszterházy seems to have featured in court entertainments under the sobriquet of 'Quinquin', as well as being a gaming partner of her husband Francis; Batthyány's younger brother Károly acted as tutor to the young Joseph II. Alternatively the ruler could reverse the strategy, by infiltrating preferred individuals into the aristocracy. One such was Ferenc Balassa, who became a key figure in the Lieutenancy Council for thirty years from 1756. More often similar elevations took place through the Hungarian Chamber, or treasury (*Ungarische Hofkammer*), which experienced still greater expansion under Maria Theresa: thanks to extra responsibilities, its receipts grew tenfold and its higher-grade employees trebled in number. Antal Grassalkovich, who had risen from obscurity to a dominant position on the Court of Appeal through his legal prowess, took over the presidency of the Chamber from 1748 until his death in 1771, and built up the Crown's and his personal revenue in dramatic fashion. In the next generation Pál Festetics, son of another acquisitive judge, made himself as indispensable for the management of the 1764 diet as had Grassalkovich in 1741, and became senior adviser simultaneously to both the Viennese and the Hungarian Chambers.[4]

Balassa, Grassalkovich, and Festetics may have bettered themselves substantially in office, but all entered it as native and noble-born Hungarians, in accordance with the provisions of the constitution. The same was true of the Chief Justice, György Fekete, and his brother-in-law, the formidable chancellery council-lor Kristóf Niczky, of Antal Brunswick, Ferenc Xaver Koller, and others who acquired the title of count for their services to Maria Theresa and whose biographies, if ascertainable (in Koller's case even the elementary information is confused) would tell us much about her style of government in Hungary. At the top there appears to be only one exception: the Saxon prince, Albrecht, son-in-law to the Empress, and again a favourite, if of rather a different sort, who took over the duties of Palatine after Batthyány's death, but only as royal Lieutenant and without the statutory election at a diet. Lower down, especially in the Chamber, we can certainly find a few Germans or Austrians, such as Anton Cothmann, a shadowy but manifestly influential immigrant from Westphalia; but again these are exceptional. More typical in his background, albeit remarkable in his gifts, was the cameral official Wolfgang (or Farkas) Kempelen, born at Pressburg in 1734. Basically a reformist administrator with special responsibility for the fabric of Habsburg palaces, for impopulation, and for the vital royal salt monopoly, Kempelen's memory lives on, when his seniors are long forgotten, as the inventor of water projectiles, notably the giant fountains at Schönbrunn, of a chess automaton which fascinated monarchs from Maria Theresa via Frederick the Great to Napoleon, and of an artificial head able to reproduce the sounds of the human voice.[5]

The acid test of these intensified Hungarian administrative arrangements came with the first of only two major enactments by Maria Theresa for the country: the *Urbarium*, or peasant regulation of 1767. Whereas political change there generally limped behind the pace set in the rest of the Monarchy, the agrarian nettle was grasped earlier in Hungary than elsewhere. In fact the two issues are closely connected, for it was precisely Hungarian recalcitrance in other matters which precipitated confrontation on the peasant question. Hard bar-gaining at and between diets – because of war only two could be summoned during the first half of the reign – left the still untaxed noble nation looking out of line. The new Council of State (*Staatsrat*) established in Vienna in 1761 as a supreme advisory body immediately turned its attention to obtaining relief for Her Majesty's Hungarian taxpayers, who also provided the rank and file in her Hungarian regiments; its more forthright members, like the State Chancellor, Kaunitz, and the Swabian-born Borié, said much in private about the iniquity of their burdens.[6]

In 1764, faced with near-bankruptcy at the end of the Seven Years War, Maria Theresa made one more attempt to win the diet's approval for a higher level of voluntary grant. But when the estates, thoroughly mistrustful by now of 'Austrian' methods of government, insisted on the redress of their own grievances – 228 separate items – in return, she ordered the material condition of the peasantry to be placed on the active agenda. Now came a catalyst, in the form of rural disturbance in western Hungary (perhaps encouraged from Vienna). This allowed the government to dictate solutions, first piecemeal, then in the form of a national standard for dues and obligations adjusted to local conditions, and to impose them against much resistance in the counties. On crown lands and estates which escheated to the ruler under the ancient Hungarian inheritance law of *aviticitas*, Maria Theresa tried to go further still in the direction of tenants' guarantees. Moreover, the impopulation activity which likewise reached its climax in the 1760s and 1770s shared some of the same ulterior motive.[7]

This severest crisis in the Queen's Hungarian policies, which caused her to suspend the office of Palatine, as we have already seen, and to prorogue the diet indefinitely, also proved a severe embarrassment for her leading servants there. The unease of their collective response was exacerbated by the fact that many of their own peasants stood in the forefront of the rebellion. Were they actually the most exploited by their masters? Certainly Maria Theresa was ready to believe so: 'It is only the cruelty of their lords which has driven these wretched people to such extremes. Would you believe that the late palatine [Lajos Batthyány], prince [Károly] batyani, the esterhazi are the most hardened oppressors: that makes one shudder.'[8] At all events they *felt* most exploited, because of the development over recent decades of demesne farming and long-distance trading to supply neighbouring Austrian markets, which demanded extra inputs of their labour; and they were better able, by some contact with a more cultured West, to articulate their discontents.

Yet on the whole the administration succeeded in accommodating or modifying these policies. An elaborate memorandum by Eszterházy persuaded the Queen to respect constitutional channels as far as possible. The most important practical advice was tendered by two of the new men: by Festetics, who looked to set up contractual relations on the land, enforced by ordinance as necessary, and whose intimate links with Maria Theresa and her cabinet secretary, Neny, in those years are documented in his correspondence; and by Balassa, who appears to have been the brains behind the actual mechanism adopted, of a sliding scale based on the size of an optimum peasant holding in regions of different fertility. On balance, in other words, the new administrative cadre mediated between Vienna (especially the Council of State) and the nation, sat on the mixed Austro-Hungarian urbarial commission which settled the details, then acted as royal commissioners to implement them; just as it was men like Grassalkovich, Cothmann and Kempelen who supervised the process of impopulation on both public and private estates.[9]

By the 1770s, as the storm abated, a plethora of other measures could be co-ordinated by the Lieutenancy Council and issued to counties and other localities as so-called *normalia* (we should not, strictly speaking, describe them as legislation, since they lacked the sanction of the diet). The *homo regius*, Archduke Albrecht, the director of whose chancellery was Kempelen, proved reasonably sympathetic to domestic interests; while aristocrats old and new, as royal-appointed High Sheriffs (*Supremi Comites*) in the counties, could do much to ensure that *normalia* on commerce, justice, finance, health, communications and the rest would be obeyed. Especially significant were the many enactments in the field of religion and education, to which we shall now turn as evidence of the cultural and spiritual preoccupations of reforming absolutism.

For some 150 years before 1740 the intellectual underpinning to the authority of the Austrian Habsburgs had been Roman Catholic orthodoxy as defined by the Council of Trent. The Prussian challenge – the challenge of a more enlightened Protestantism as well as of a more effective governmental machine – soon made it clear to Maria Theresa that this orthodoxy, like its long-standing administrative forms, could no longer sustain the Monarchy. How then should it be adapted or appropriated to the renewal of the state?

Not, it must be stressed at the outset, by any abdication from the absolute spiritual claims of the Roman Catholic church *vis-à-vis* other Christians or unbelievers. Maria Theresa firmly believed in her rule as a monarch by divine right, and in her duties as guardian of Catholic values. That applied most particularly in her Hungarian domains, where the whole Counter-

Reformation only generated belated fervour during the eighteenth century, and which still harboured the Monarchy's only large populations of non-Catholics. 'Religion yet has much good to do in Hungary', as she commented in the early 1750s.[10] Thus the assault on other churches continued unabated throughout the reign, or at least until its very last years. Evictions, especially of Calvinist congregations, occurred alongside the forced resettlement of Austrian Lutherans in Hungary; continuous pressure was exerted on towns – and again Calvinists in royal boroughs like Debrecen fared worst – to elect Catholic councillors or observe Catholic feast-days. The ideological struggle against the Heidelberg Catechism and against Protestant editions of the Bible went with active proselytising, often Jesuit-led, on a broad front.[11] Campaigns against the Eastern Orthodox to found and extend Uniate churches among Rumanians, Ruthenes and Serbs tell their own story: the regular frustration and miscalculation of the authorities only highlight their wildly doctrinaire presuppositions in imagining that they could impose such solutions.[12]

Intolerant outreach was matched by official exclusiveness. Protestant élites, the most important aspect of Hungarian cultural distinctiveness, had no chance to participate in government or administration, however loyal they might be, apart from the special case of Transylvania (and even there the small minority of Catholics tended to be favoured). The Roman Catholic hierarchy, on the other hand, remained formally involved both in senior royal counsels and in the local executive, most significantly through the person of József Batthyány, second son of Lajos, a long-serving member of the Lieutenancy Council during the years when he rose to become archbishop, cardinal and primate of the country. For that reason, among others, the Hungarian authorities proved less responsive over confessional grievance in this period than the Council of State, to which Protestants could appeal through an agent stationed in Vienna.[13]

Yet this ecclesiastical establishment in fact changed markedly during the reign of Maria Theresa. In the first place, the balance of forces between church and state tilted in the direction of the latter. While Austria experienced major Erastian developments from the 1750s, as the state took a tighter hold on church revenues, restricted the freedoms of monasteries, curbed the powers of the papacy, etc. (those measures which first gave rise to the term 'Josephinism', though they are now seen as having been initiated by Kaunitz and the Empress), parallel developments took place in Hungary. In 1758 Maria Theresa – 'as an adornment for the kingdom of Hungary' – revived the title, originally granted to St Stephen, of 'apostolic' ruler, which carried an implication of the monarch's enhanced authority in spiritual matters.[14] During the early 1760s she went a step further in refurbishing royal claims by encouraging her librarian, Adam Franz Kollár, himself a Hungarian, to publish two books whose titles sufficiently indicate their line of argument: *Historia diplomatica iuris patronatus Apostolicorum Hungariae Regum* ('Documentary History of the Rights of [Ecclesiastical] Patronage of the Apostolic Kings of Hungary') and *De Originibus et usu perpetuo potestatis legislatoriae circa sacra Apostolicorum Regum Ungariae* ('On the Origins of Legislative Power in the Ecclesiastical Sphere and its Perpetual Exercise by the Apostolic Kings of Hungary').

Kollár's work created a furore at the diet of 1764, despite the inoffensive manner of its presentation (with the sting in his long Latin footnotes!): the Hungarian opposition could see that the court library lay not far from the cabinet office. Not only were clerical privileges more intact in Hungary than in Austria; an attack on them could also more clearly be associated with an attack on noble privilege – we may discern, *mutatis mutandis*, some similarities with the clash a few years earlier between Machault d'Arnouville and the French episcopate. The battle proved fierce but inconclusive; it held up the diet's proceedings for a whole month, and led to the withdrawal, but not the recantation, of Kollár's theses. The wildest diatribe against the Crown, a pamphlet entitled *Vexatio dat intellectum* by one of the archbishop's chaplains, was burned on the orders of Maria Theresa.[15]

The Kollár affair, with its mildly anti-establishment overtones, coincided with, and formed part of, the struggle over the *Urbarium*; yet – more than that – it contained a large element of shadow-boxing, since the cultural attitudes of the Hungarian Catholic hierarchy were also shifting, *pari passu* with those of the reformist regime. Thus other changes, like a substantial diocesan reorganisation, could proceed with active local support – even that of the author of the *Vexatio* – although they had not been approved by Rome.[16] Most important of these changes was the second fundamental piece of Maria-Theresan 'legislation' for Hungary: the *Ratio Educationis* of 1777. Like the *Urbarium* ten years earlier, this major pedagogical initiative originated fortuitously in part (it followed the Pope's decision to dissolve the Jesuit order), but then bore the fruit of intense administrative planning: by

Brunswick and his younger colleague, Ürményi; by Festetics, who had studied at Leipzig with Gottsched, and Niczky, an alumnus of the Theresianum in Vienna; by Kollár and others.[17]

The result was a long and detailed programme for Catholic schooling at all levels, in a framework of vigorous state control, with a stress on training for citizenship (Protestant educational institutions, while not directly addressed, were evidently expected to follow suit). And the *Ratio* represented merely the capstone on a series of more piecemeal measures which culminated in the 1770s, ranging from the foundation of a mining academy with European catchment at Schemnitz (Selmecbánya) in the north, to an intensive effort to inculcate the rudiments of knowledge in backward Orthodox peoples of the Banat in the south.[18] Educational reform was thus made possible by co-operation between Maria Theresa's government – launching similar initiatives elsewhere in the Monarchy, though without quite such a clear blueprint – and her Hungarian administrators. It articulated much of their own cultural orientation: a moderate receptivity to enlightened influences from abroad, but anchored in local Latinate traditions, and with an explicit but subordinate place for domestic vernaculars, especially German. It enjoined more teaching of Hungarian subjects, particularly law, so long as they were suitably glossed in a royalist sense – since the mid-1760s that had been the task, at Maria Theresa's personal request, of Festetics, whose portrait in Budapest's National Gallery depicts him with the *Corpus Juris Hungarici* at his elbow.[19]

Many of these people, clergy hardly less than laity, were cultivated, even learned, in a highly cosmopolitan way. We can see that in their libraries and collections: Eszterházy's at Cseklész, or Festetics's concentrated upon coins (which are likewise shown on his portrait); József Batthyány *primus inter pares* among the prelates, expending vast sums to acquire source materials for Hungarian history. We can see it too in their linguistic practice. Official usage still firmly insisted on Latin, which appears dominant in all political and administrative contexts, except in the business of the Chamber and in dealings with the army. There and elsewhere it was being chipped at by German, and much less by Magyar, though the latter had a vital role in the management of landed estates (as a Slav vernacular might also have, especially in Croatia or in the north). Magyar formed the regular language of conversation and correspondence within the élite; but German was needed, not just for Austrian contacts, but as the leading medium in many Hungarian towns; while Latin remained in frequent private use too, a living tongue as well as a formalised *lingua franca*. French was increasingly preferred for many holograph letters, as an index of culture and status: the Batthyány family correspond in it among themselves by the 1740s; Festetics corresponds with his father in Hungarian, but with his own son in French; and it is surprising how many clerics use it too, among them the conservative but worldly-wise Migazzi, Archbishop of Vienna and a diocesan in Hungary as well.[20]

This is the bewildering world of a ruling group often quadrilingual in Magyar, Latin, German and French, whose members might also learn some Italian and remember the rudiments of a local Slavonic tongue acquired in youth. (But we should note that many Hungarian lesser nobles, parish priests or merchants were themselves at least trilingual.) What contexts or preferences determined the choice of one tongue rather than another? That question would provide scope for a separate investigation. Here we should rather seek to place such linguistic cosmopolitanism in the framework of a wider problem of identity posed by contemporary attempts from Vienna to build not only a new Austrian state, but with it a new loyalty to Austrian statehood. The latter has strangely been much less discussed than the former, yet it is a purpose arguably implicit in the governmental ideal of bureaucratic service: the wash of official self-justification in terms of the 'common good' and similar phrases seems to demand as its corollary a concept of state patriotism, which would likewise warrant a separate investigation to do it justice. Essentially the endeavour was a practical rather than a theoretical one, its most important intellectual formulation appearing with Joseph von Sonnenfels, whose book (to which we shall return) about the love of the fatherland (*Vaterland*) – another vogue word of the period – looks to a 'nation of patriots' who cultivate citizenship by such means as economic development and the codification of law.[21]

But which territory was the '*Vaterland*'? Did it include or exclude Hungary; and what implications did the notion carry for the government's backers there? Sonnenfels remained vague, but he seemed to be looking to the *Gesamtmonarchie*, the Habsburg lands as a whole. So was the Council of State, which explicitly condemned the Hungarian constitution as a front for '*Nationalismus*', in an interesting early use of that term. The *Staatsrat*, in fact, particularly Kaunitz and Borié, embodied an anti-Hungarian ethos which, still feeding on the experience of civil wars between the 1680s and 1711, stressed the Crown's rights of conquest, stamped local discontents as treachery, condemned 'barbarian' practices, and looked to a civilising mission. It was committed to subsuming regionalisms in some 'Austrian' totality.[22]

So, a fortiori, was Maria Theresa, whom the Staatsrat served in a still dynastic enterprise. Yet the matter is not so clean-edged. She was certainly 'Austrian' in one respect, as her predecessors had not been: though the wife and mother of emperors, she found herself divorced from the traditions of the German Empire, both before and after the death of her husband, in different ways. The identification with Austria could only be enhanced by the very defence of it against a Prussian- and Bavarian-led 'Germany', which formed a novelty in Habsburg history. By the same token, however, this identification did not necessarily work against the status of Hungary. Indeed, it might enhance that too: the Hungarian crown was the only part of Maria Theresa's territories in 1740 which had always stood outside the Holy Roman Empire, and it provided her chief royal title, even after Francis's imperial election in 1745. It also provided the Habsburgs with their ostensible claims to Galicia and Dalmatia – Kollár (again!) and the Croat canon, Krčelić, were pressed into service to justify those claims in print.[23] In the event only thirteen towns of the Zips (Szepesség), pledged to Poland since the early fifteenth century, returned to Hungary at the time of the First Partition (and Venice still clung to Dalmatia); but Maria Theresa did take important steps by the end of her reign towards re-creating the integral kingdom of St Stephen demanded by patriots. On the one hand she incorporated parts of the Military Frontier and then (1778) the whole of the Banat of Temesvár, which had been administered directly from Vienna since the expulsion of the Turks. That initiative had been urged upon her by Hungarians of all shadings, even quite forcibly by Archduke Albrecht; and her trusted bureaucrats, especially Niczky, took the lead in implementing it. On the other hand Maria Theresa granted Hungary a seaport at Fiume on the Adriatic (1776), then abolished the separate Lieutenancy Council for Croatia in 1779.[24] At the same time she allowed more domestic control over the fiscal implications of these changes through the expansion of the Hungarian Chamber which we have already encountered.

These policies are not easy to interpret. They were not merely concessions: the timing does not fit, since opposition had reached its peak in the mid-1760s; but nor were they consistently followed. Over the largest territorial grievance, Transylvania, Habsburg government pursued a fairly deliberate strategy of divide and rule. That is evident from the discussions surrounding its decision to underline Transylvania's two and more centuries of separation with a decree, in the significant year 1765, elevating it to the rank of Grand Principality. There the military frontier was actually extended and a new taxation system introduced, while the authorities courted the Orthodox population (at least in its Greek-Catholic guise) and favoured the local Saxons, notably Samuel Brukenthal, as a counterweight to the Magyars.[25] Moreover, in the case of Hungary proper, the acknowledgement of autonomous status could serve as a ruse: a revealing debate took place at the end of the 1760s on Joseph II's proposition that Festetics should join the all-Austrian *Staatsrat*, but the conclusion was reached that it would be better not to advertise thereby that body's *de facto* handling of Hungarian business.[26]

No doubt even the proven loyalist Festetics aroused misgivings among some in high places in Vienna. Yet the decision was also designed to spare Hungarian sensibilities: Eszterházy had argued strongly *against* Festetics's appointment. It seems a fair assumption that enough pluralism of identity existed on both sides to admit of real ambiguity. A proper analysis of the issue would demand close attention to such things as linguistic usage. For example did German-speakers refer to Hungary (when they did not call her a kingdom) as a *Reich*, with that word's strongly autonomous implications, or as a *Land*, which carried overtones of the merely provincial? Did they describe her diet as a *Reichstag* or as a *Landtag*? Maria Theresa, Kaunitz and others do not seem to have been able to make up their minds. In one letter of 1766 the Empress writes to Festetics that his revision of the *Corpus Juris Hungarici* should serve 'the general good of the *Reich*' – which the context reveals to be the whole Habsburg Monarchy – provided it eliminates aspects 'which are detrimental not only to the King [herself], but to the *Reich* [of Hungary] and its estates themselves'.[27]

What of Hungarian perceptions of Austria? It seems clear that those who rallied to the cause of modernisation and rose in the royal service were also often those with some first-hand impressions of life and culture across the western frontier.

Conventional wisdom recognises that many more Hungarians lived in Austria, mainly in Vienna, during the eighteenth century. But no concerted study has been made of them, and it appears to be assumed that they stayed either as visitors, in which case the experience formed merely an episode (albeit perhaps an extended one, like that of the 117 Hungarians educated at the newly-founded Theresianum between 1749 and 1774),[28] or as long-term residents who were simply 'denationalised' and assimilated. In fact we find very few Hungarians in purely 'Austrian' jobs – that was the *quid pro quo* for the fairly strict protection of Hungarian posts (formally through the requirement of *indigenatus*, or naturalisation; informally through the practical need for local legal knowledge). Hardly any served in the civil administration, except for some dual membership of Austrian and Hungarian Chambers; and whereas Koller took over one important central office from 1767 as head of the Illyrian Deputation, his work in the Banat manifestly had a Hungarian flavour too (the more so in retrospect). A handful became ambassadors – Miklós Eszterházy, Ferenc's brother, in Spain and Russia, then the great Orientalist and bibliophile, Károly Reviczky in Poland and England – or attained prominent positions at court. Károly Batthyány gained the coveted position of preceptor to Archduke Joseph; Kollár succeeded Gerard van Swieten as court librarian.

Kollár came from simple circumstances in one of the Slovak-settled counties of northern Hungary. To describe him for that reason, as his recent biographer has done, as 'the most successful Slovak in the history of the Habsburg Monarchy' is patently anachronistic and smacks of misplaced national zeal. Yet the assertion contains a grain of truth: Kollár does betray – in certain other of his scholarly passages – a pride in his Slav ancestry.[29] More generally and significantly, he continued to display a loyalty to the Hungarian realm, for all that he owed his preferment squarely to imperial patronage. So much the more did the rest of his fellow-countrymen, even the permanent officials of the Chancellery, who, despite their acceptance of the Austrian environment, seem to have remained on the fringes of real power in Vienna. Those who operated from within the country, concentrated in the little city of Pressburg just across the border, had still weaker links with the court, though their correspondence with Viennese patrons suggests how much they had come to look to it for favours.[30] Among the greatest favours it could bestow was the new Order of St Stephen in its various gradations, established by Maria Theresa in 1764 to reward especially the activity of Hungarians in the public service both of their native land and of the Monarchy as a whole.

One further key institution demands investigation in this connection (although as yet we know little enough about it): the army. The Austrian army furnished the third pillar of Habsburg power in the eighteenth century, and the only novel one. Now it became for the first time a permanent, large and coherent force, closely associated with the state apparatus and drawn from all parts of the Monarchy, including Hungary.[31] A law of 1715 authorised the recruitment of Hungarian contingents for the standing army; the wars of the 1740s first witnessed a great body of Hungarians fighting outside their own theatre of battle. They displayed considerable verve and enjoyed marked success: from that time dates the European reputation of the hussar cavalry and of pandours, heyducks and other irregular foot-soldiers.

Hungary provided a dozen or so generals in the 1740s, more later: from the veteran János Pálffy, already too corpulent to mount his horse, through Károly Batthyány and Ferenc Nádasdy, who distinguished themselves in the defence of the Austrian Netherlands and the conquest of Bavaria, to András Hadik, commander of the troops which sacked Berlin in 1757. Their exploits of course go far to explain Maria Theresa's gratitude towards their country, which naturally included personal favours too: Pálffy was raised to the dignity of Palatine at her instance in 1741; Batthyány secured his prized but ticklish assignment as tutor; Nádasdy fully restored his family to the status forfeited when his grandfather of the same name had been executed for treason in 1671. Thus the army also began to function as the only normal channel through which Hungarians could rise outside their own hierarchy: the self-made Hadik, for example, became successively President of the Imperial War Council, governor of Transylvania, and then of Galicia.

Army discipline and long postings in other parts of the Monarchy, even if dictated by military preparedness rather than by any grand political plan, evidently represented a good school for Austrian loyalty, and scope for advancement depended on that. Yet there was much more to eighteenth-century military life, not least the considerable opportunity for wider experience and freer thought that it offered (at least to higher ranks). Enlightenment in the Habsburg lands was often promoted by soldiers or ex-soldiers: Petrasch, Kinsky, even Sonnenfels are outstanding examples, and plenty of Hungarian cases could be adduced too.[32] Equally the army did not necessarily extinguish Hungarian sentiment in its officers, and might have quite the reverse effect. The new Hungarian Royal Bodyguard, founded in 1760 as an élite corps open to provincial nobles, including Protestants (but with the opulent Prince Miklós Eszterházy as its Captain) soon demonstrated that, even as it bound socially, it could loose culturally: several of its members swiftly turned to the emergent cause of vernacular Magyar literature.[33]

Pride in military achievements could therefore be a force which strengthened both Austrian consciousness in a wider sense and Hungarian consciousness within it. There is an interesting vignette to illustrate this in a curious little piece of doggerel entitled *Lettre à Voltaire ou plainte d'un hongrois*, written in 1764. The 'letter' consists of verses by Baron Lőrinc Orczy, a general, and by his younger fellow-officer, János Fekete, son of the Chief Justice, György Fekete, who was a leading bureaucrat and neo-aristocrat. A *jeu d'esprit* by two enlightened Francophiles, the text upbraids Voltaire for ignoring Hungary in his work, and points significantly to the army and military virtues as the chief factors which not only sustain Maria Theresa's rule, but help to spread civilisation there and elsewhere.[34]

Of course the army was also a liability. It could not bear the full weight of the centralising programme. Neither did it remain entirely immune from conflicting interpretations of patriotism. In 1790 several Hungarian officers would present a petition calling for national regiments to be stationed on national soil and employ the national, Magyar language: they were led by the son of Festetics, and their sentiments echoed the demands of the political opposition for which János Fekete had now become a chief spokesman.[35] Moreover, it was mainly the rising cost of the army which lay at the root of the unresolved constitutional impasse over state taxation and the related issues of tariff policy and the peasant question, and which thus led to regular deadlock at the diet in 1741, 1751 and 1764–65. This represented a crucial problem in the long run, as any student of the nineteenth-century Habsburg Empire will be aware; yet it was one which could remain latent for extended periods. The fragility of material and cultural solutions reached on other fronts, with the Hungarian nation underpaying and underrepresented at the same time, could, in other words, always be dramatised as a dispute over the army; but that was not because military aims as such formed the highest priority of absolutist government, or because those aims – as yet, at least – stood at variance with the perceived political priorities which the Hungarian nation set itself.

I hope I have said enough to imply the essential patriotism of Hungary's royal servants under Maria Theresa, however much it might be combined with personal ambition or chicanery. With very few exceptions they counselled respect for national traditions and liberties, albeit in different ways: the defensive Eszterházy, the empire-building Grassalkovich, the clandestine Festetics, the suave and adroit Cardinal Batthyány, the shrewd but ostentatious Brukenthal in Transylvania. (Balassa's often provocatively unconstitutional opinions perhaps prove the rule, but he appears exceptional in other respects too, as a cross between the extremely unscrupulous careerist and the indefatigable do-gooder.) In their private lives they were supporters of national causes, such as schools, colleges, or collections, sometimes with public access, although the full development of that came in the next generation (Brukenthal bequeathed his uniquely valuable museum for the purpose). This was not, of course, the xenophobic patriotism of the county gentry, the greatest obstacle to fuller Habsburg control over the country. Yet in some ways foreign contacts might heighten elements of national awareness. The active pursuit of change, through the only channels actually capable of promoting it, required both Hungarian (or in Brukenthal's case Transylvanian) patriotism and solid Austrian loyalty as overlapping, even complementary, sentiments.

The belated reform movement engineered by members of the domestic political nation in Hungary under the auspices of Maria Theresa reached its climax in the last years of her reign. They were encouraged by the reintegration of the rich plainlands of the Banat; by the acquisition of Fiume, a window on the West and training-ground for a rising generation of entrepreneurs, or at least of enterprising administrators; by the transfer of the country's one university from an isolated provincial town to the historic metropolis of Buda.[36] Such fruitful co-operation expunged for many contemporaries the memory of earlier squabbles. Maria Theresa herself declared, in connection with her decision on the Banat: 'I am a good Hungarian. My heart is full of gratefulness (*Erkenntlichkeit*) towards this nation.' There was correspondingly widespread and genuine regret at her death, and the afterglow of that mood illumines the pages of nineteenth-century historians like Mailáth, Horváth and Marczali. The first of these, Mailáth, who could still draw on personal reminiscence (his father was the first Governor of Fiume), concluded that 'in Hungary too the constitution would have slumbered and the principle of absolute monarchy triumphed, if only the Empress had lived longer'.[37] The judgment may be exaggerated; but the endemic oppositional spirit within the body politic had rarely appeared feebler.

Loyalists looked to the 1780s to continue the momentum. Signs of admiration for the new ruler Joseph II are plentiful. Young Festetics writes to his father in late 1781 about 'the wise Ordinances of our Philosophical and Enlightened Monarch'. Count Sámuel Teleki, a Calvinist admittedly, but before the Patent of Toleration, writes to his wife that he 'cannot praise enough the qualities of His Highness: we are proud to have him as our ruler'; by 1783, after the Patent, 'just to look at him gives me consolation'.[38] Joseph's early moves could foster illusions further: he actually cut the Gordian knot over Transylvania, uniting its administration to that of Hungary proper; he centralised the officials of the Lieutenancy Council and the Chamber in Buda. His first anticlerical measures could count on considerable support; so could the Toleration Patent, and not merely among Protestants, long the most discontented party in the kingdom, for whom it opened up the prospect of office and a part in the national mission, as well as pecuniary and social advantage. One of its first beneficiaries was Teleki who, back in 1770, though a well-endowed magnate in Transylvania, had sought the ear of a court protector to release him from 'the sad necessity of languishing in the shameful inactivity of an obscure and miserable life'.[39] His words, however inflated by rhetoric, reveal much about the motivation of the kind of enlightened administrators we have been examining.

From 1784–5 disenchantment was all the greater; alienation from Joseph's system, and its consequent collapse, were all the more rapid. The Emperor's tearaway assault on the whole Hungarian constitution and nation seems to have been compounded of three factors. The first was his intemperate personality, which inherited Maria Theresa's abruptness and insistency without her charm, or caution, or favours. Secondly, Joseph embodied the 'Austrian' prejudice against Hungary in its most drastic form. Finally, and most importantly in the present context, he appears to have worked off a frustration at the country's protected status under his mother: his role there had raised political problems and – even though he was proclaimed co-regent in Hungary as elsewhere (with a formula drafted by Festetics) – left him largely impotent right up until 1780 in the territory where Maria Theresa least trusted his judgment. Thus he could nurse ever fiercer resentment at the myriad trammels on royal authority within the crown of St Stephen.[40]

However we weigh those factors, the chaos by Joseph's death in 1790 was correspondingly most complete in Hungary. And the failure of 'Josephinism' there manifests a deeper truth. Historians' terminology has obscured the fact that the bureaucratic reform movement conventionally described nowadays as 'Josephinist', in Hungary actually almost always stood closer to a 'Maria-Theresanism'. The mentality of the leading Hungarian *Aufklärer*, like that of their Queen, was French in fashion but not in philosophy, and little affected by advanced criticism either of the Church (the country bred hardly any Jansenists) or of society (hardly any non-nobles were involved at this stage). It was highly resistant to the claims of German national culture, which Joseph actively fostered, however involuntarily, not least by his theatrical enterprises; and it remained highly ambivalent about the unnecessary use of an army whose Hungarian component, while increasingly important, was still comparatively modest. Above all, it looked to progress by means of the creative moulding of tradition, especially via a careful extension of the royal prerogative. If Maria Theresa was 'an active feudal monarch', to cite Dr P. G. M. Dickson's recent assessment,[41] Batthyánys, Eszterházys, and Festeticses represented the vanguard of a kind of patriotic feudal *ralliement*.

In the end Hungary and Austria did not grow together, as they and Maria Theresa had hoped. For that reason historians have largely failed, besides their vague plaudits, to investigate the nature of the symbiosis in her reign, especially in its prominent personalities, before Joseph II almost laid it waste again, and exposed it to a backlash from both sides during the years after 1789. The Queen's enactments were saved, with the return under Leopold II to the *status quo ante*, and some of the earlier impetus could be recovered, frequently through later generations of families which had risen to official and social eminence under Maria Theresa.[42] But the amphibious Austro-Hungarian patriotism of her day could not be saved. Most fatefully for the future, the linguistic nationalism which progressively succeeded it, while opening up membership of the nation to the population as a whole, at least to the Magyar part of the population, would also cut Hungarians off from any positive identification with the rest of the Monarchy. That outcome, as the next chapter will indicate, forms part of a larger evolution among the peoples of the Habsburg realms during the age of absolutism.

7. Joseph II and Nationality in the Habsburg Lands

R. J. W. EVANS

To examine the relation between the mighty programme of eighteenth-century Austrian state-building and reform which culminated in the rule of Joseph II during the 1780s, and the origins of that dominant nineteenth-century issue which ultimately destroyed the Habsburg Monarchy, might seem an obvious and by now otiose duty. There is a huge literature, much of it recent and perceptive, on both Josephinism and nationality. Yet with the squarely metropolitan focus of the former, and the disparate, centrifugal treatments of the latter, the two sets of problems have remained largely unrelated to each other. I cannot evidently begin to do justice to their juxtaposition now: instead I offer the barest sketch of the terrain, a kind of preliminary land-survey, whose fate (like that of Joseph's) may be instant combustion.

Of course, for Josephinism's chief protagonist my subject would have been a non-subject. Nationality was one thing the Emperor did not try to embrace in his statistical investigations. He recognised only the demands of creating a modern, centralised society, towards which he expected all right-thinking citizens to display an equal commitment, but which raised opposition from (as he saw it) traditional sectional habits and loyalties, among them those of province or region.

Historians have tended to turn these assumptions on their head. The outcome of Joseph's policies, they have been inclined to argue, was on the one hand broad socio-economic changes, better education, more mobile and advanced populations ripe for coherence and integration, but on a national rather than a dynastic base; while on the other hand the very pressure for uniformity generated, in the realm of ideas, precisely those resentments which would be articulated in the guise of nationalism. There is much evidence to support a story of action and reaction. The general decision to impose German as language of

209

administration and schooling was notoriously ill-fated; while particular ordinances, like the insistence on propriety at Serbian funerals (no kissing the body of the departed, etc.) certainly raised a storm of protest which could be channelled by leaders of the ethnic community.[1]

Yet that is not the whole story. Josephinism and national consciousness had a common term in the Austrian *Aufklärung*, whose origins were older than either of theirs. The movement of enlightened concern and critique in Habsburg Central Europe (dating back to the mid-century and earlier) encountered doctrinaire state-building and rival national claims rather as if they formed two prismatic eye-pieces in a field-glass, ready to deflect it, but only gradually, in two opposite directions. Josephinism captured the ideas somewhat earlier, nationality somewhat later. And the consequent parting of the ways can be located fairly precisely in the years around 1785. It is highlighted by a series of individual rifts between the Emperor and leading officials and writers (Brukenthal, Pelzl, Rautenstrauch, Skerlecz, Széchenyi, Zinzendorf . . .) who often turned away from the reality of Josephinism in proportion as they came into direct contact with the Emperor's wishes, his brusque manner, and his lack of 'philosophical' commitment. But then, as scholars ever since Mitrofanov have pointed out from time to time, Joseph was not necessarily the model 'Joesphinist'.[2]

Until that point was reached (even a little afterwards) the opportunity existed for a genuine interplay. The *Aufklärung*, so the poet Blumauer quaintly put it in 1782, at the height of its influence in Austria, 'proceeds by rubbing *Geist* against *Geist*, as American Indians rub stick against stick to make fire'.[3] Its overwhelming practical concern for welfare and improvement created two particular areas where statecraft and national interest might thus burn in the same flame.

The first is an attention to vernacular language. Evidently the government favoured the German tongue throughout the Empire, and there were good neutral reasons for that. The spread of German to accompany the changing tastes and horizons of higher society from the 1750s into the 1780s evoked hardly any objection. Hungarian officials in the 1770s and even later positively welcomed its advance at the expense of Latin. But other vernaculars were readily drawn in *pari passu* at a lower level, especially for popular education and instruction. It became possible to study Czech, albeit (logically enough) more in Austria than in Bohemia (since enough people knew it there anyway). Slovaks successfully lobbied for their own school at Pressburg. The Hungarian *Ratio Educationis* even allowed for academic dissertations to be submitted in any of the country's languages. Thus lesser vernaculars were actually felt necessary for state purposes, and the new *Normalschulen* encouraged learning of them out of hours.[4]

Alongside this went a great increase in their use for the printed word. New journals were launched in the 1780s with government approval in Magyar, Czech and Slovak. Nearly 400 official documents appeared in Rumanian during the same decade. The authorities promoted not just textbooks, but a flood of primers on non-curricular activities: from military matters to the prevention of fires, and above all questions of peasant tenure and husbandry.[5] Let us take just one instance, since I wish to refer back to it at the end of this chapter. In the early 1770s Maria Theresa summoned an illiterate Slovene bee-keeper called Janša to teach his craft in Vienna. Janša's book on apiculture not only appeared in German, but it passed through five editions in Czech by the end of the century, while prominent national awakeners like the Slovak Fándly and the Rumanian Piuariu-Molnár made the same subject available to their own people.[6]

Concern for vernaculars answered a deepening profession of faith among *Aufklärer*. We can already find it with the most significant East-Central European precursors of their popular mission, the Transylvanian Puritans. 'A people which derives everything from foreign languages is most unfortunate and most worthy of sympathy among all nations. ... I have firmly resolved that, God willing, I may before I die, communicate all knowledge to Hungarians in the Hungarian tongue.' Thus Apáczai in the preface to his *Encyclopaedia* of 1655. His appeal was at last widely echoed by 1781, when Bessenyei (founder figure of several genres in modern Hungarian literature) could claim that 'the basis and instrument of a country's welfare is culture. ... The key to culture is a national language ... and the cultivation of that language is the first duty of the nation.'[7]

Such a strident call necessarily implies tensions to come. The very strength of German linguistic and literary development between the 1760s and 1780s carried other vernaculars along with it, and of itself upset the neutral programme of Joseph II, which suddenly came to appear Germanising, and to deserve resistance.[8] The Emperor, for all his dealings with the Viennese Burgtheater, seems to have remained blissfully ignorant of that possibility. Besides, how far should popular education go? The restrictive response of many government officials was not shared by national *Aufklärer*.[9] Yet for the time being a good deal of symbiosis was achieved in the propagation of living languages. The whole process – and its implications – might perhaps be compared to the replacement of Persian in India from the 1830s, under Bentinck and Macaulay, by English at the upper level of state and judiciary, and by local tongues in lower courts.

The second area of common ground is more imponderable: the whole concept of citizenship, or *Staatsbürgertum*, and its various levels in the eighteenth-century Austrian Monarchy. The matter deserves proper semantic study. Enlightened language about *allgemeine Glückseligkeit* and *Wohlfahrt*, or *das allgemeine Beste*, allowed ambiguity, not about the general ends of public policy (since evidently the citizenry at large is meant), but about the extension of any particular community within the 'state'. As elements of feudal privilege survived, so might elements of ethnic and cultural identity be accommodated: witness the case of Maria-Theresan Hungary which we have just examined. Josephinism assimilated the nation to the state; but it allowed notions of patriotism, even encouraged them. The best evidence for that (as for so much else in the dynastic reform movement) is provided by Joseph von Sonnenfels.

In 1771 Sonnenfels issued a brief treatise *Ueber die Liebe des Vaterlandes*, which begins with a lament for the demise of the virtue announced in his title. 'To our ears the word Fatherland is a meaningless noise; but for the Romans and Greeks it rang out like the name of a loved one.' Nowadays, he tells us, a tract on taste will be reprinted six times in a year, then again with plates and vignettes; but any discourse on patriotism will surely be shredded. *Patria* is the country where one lives, its laws, its form of government, its citizens. Love of it is positive attachment and pride.[10] And so on. But where, *concretely*, was the *patria* of the Habsburgs' own subjects? Sonnenfels fills his text with classical example and allusion; he mentions several other European states, China and India, Persians and Mongols, even Samojeds and Hottentots, but Austria never! An expanded version of the work as late as 1785 remained essentially the same. Even when citing – right out of the blue – the explosive slogan 'Unity of language is unity of nationhood', Sonnenfels could only give a facetious and uncomprehending gloss, deriving from a passage in F. M. Pelzl's freshly-published history of Bohemia.[11]

Pelzl was in fact one of those enlightened intellectuals, with a cast of mind not very different from Sonnenfels's, to whom more restrictive, though overlapping interpretations of *patria* had begun to occur by the mid-1780s. That is evident especially in *Aufklärung* historiography, a field which merits far more attention in this context than historians have so far accorded it. We need more work like the recent account of the Bohemian school, with its Czech overtones, by A. S. Myl'nikov.[12] I have space here merely to indicate three aspects which illustrate the incorporation of national heritages within a wider cosmopolitan stance. One is the recovery of the Central European Renaissance, as critics of the Baroque establishment invoked older traditions to serve their purpose, among them the humanist version of patriotic values.[13] Another is the vogue discipline of literary history. While De Luca's *Das gelehrte Oesterreich* of 1776–78, 'dem Vaterlande gewid-met', represents a lone attempt at a bio-bibliography of Austria, others simultaneously established the contours of a national pantheon for Bohemia (Voigt, Pelzl, Ungar . . .), Hungary (Bod, Weszprémi, Wallaszky, Horányi . . .) and elsewhere.[14] The third is the consciously modern discipline of statistics (in the broad contemporary sense of the term), with the semi-official local loyalties engendered in author and reader of such work as de Luca's (again) for Austria, Korabinsky's for Hungary, or the younger Riegger's for Bohemia.[15]

Even the Austrian government acknowledged 'nations', ex-plicitly enough, in the secondary sense of peoples. Maria Theresa recognised that usage; so sometimes did Joseph. They did so especially with reference to the Orthodox populations of southern Hungary, where the Banat actually had its *Nationalisten*, and the subject was immediately related, through the issue of Serbian Church versus Hungarian counties, to constitutional questions about the nature of citizenship.[16] Thus alternative patriotic loyalties became increasingly available, and loosely conjoined with the new prominence of the vernacular issue, though a larger Austrian *Landespatriotismus* could subsist with them, at least until the 1790s, when the regime's mistrust of the latter gave a clear advantage to the former. In fact the terms of the interrelationship, if not the balance within it, remained recognisably similar: from the precocious Bél in the earlier eighteenth century, who described himself as 'by language a Slav, by nation a Hungarian, by erudition a German'; to the Croatian late-Latinist Derkos in his thoughtful little treatise of

1832 entitled *Genius patriae super dormientibus suis filiis*.[17]

Manifestly this kind of fruitful impact of the Josephinist state upon ethnic nations through the vernacular and the idea of a community of *Staatsbürger* did not operate in a vacuum. It depended on a sliding scale of responses within the given territorial, religious and social framework, as also (perhaps) on the actual degree of contact with an impersonal and arbitrary state machine.

Things worked least well with the Magyars, who afford the most obvious evidence of enlightened patriots joining conservative ones in bitter opposition to Joseph, and thereby redirecting the traditional constitution towards modern national goals. Even that was no foregone conclusion: rising forms of national sentiment did not in themselves destroy the Emperor's position, at least until the very end (that was accomplished by other factors). He retained the support of some patriots throughout, and the language ordinances of 1784 enjoyed some success (even if their deadlines could not be kept), while the decayed state of the Magyar language was admitted by progressives to be the nation's fault, not Joseph's.[18] Meanwhile the whole citizenship issue broadened and modified Hungarian claims, as demands for membership of the feudal nation began to give way to demands for membership of a cultural and territorial nation gaining self-consciousness precisely through awareness of an overlapping Austrian identity.

Still, by the mid-1780s the fragile harmony between Josephinism and *Aufklärung* among Magyars, even loyalist ones, had been largely vitiated. It is most instructive to observe the Protestant dimension of this.[19] Joseph suffered the special historical misfortune that those most accessible to Enlightenment in Hungary, the Protestants, were already deeply alienated from any Habsburg solutions (or at least needed far more careful handling than he could muster). They exhibited some gratitude of course for the Patent of Toleration, which in 1781 largely removed the legal disabilities on them; they also showed a sincere commitment to the progress of popular instruction, which would lead Kazinczy, for example, to play a prominent role in the new educational system throughout the 1780s, Gergely Berzeviczy to join the judiciary in 1788, Sámuel Teleki (the bibliophile) to function as district commissioner during the same years, and his nephew József to become a district director of schools. A Calvinist physician, Sámuel Benkő devoted a whole book in 1787 to general welfare, and entitled it – with an obvious debt to Sonnenfels – *Tentamen Philopatriae*.[20]

But resentment won the day: about Vienna's economic policies; about wilful controls on freemasonic lodges; especially about interference in the ecclesiastical sphere. Consider József Podmaniczky, one of the country's most powerful intellects, Governor of Fiume, member of Joseph's Lieutenancy Council, posting off critical reports on government policy to his mentor Schlözer at Göttingen, for the edification of European opinion; or the Calvinist and Lutheran leaders of Upper Hungary, Gábor Prónay and István Vay; or the councillors of the city of Debrecen, digging in their collective heels.[21] The thoughtful diarist Keresztesi, no reactionary, who actually resurrected Protestant worship in Nagyvárad and had to appeal for Joseph to stop the local bishop from serving cheap wine in the vicinity of the large tent where he held services, still turned against the Emperor, and describes how his flock actually tore up the Toleration Patent, amid Catholic shouts of '*vivat*'.[22]

Among other Hungarian Protestants we encounter a similar evolution, more restrained, but in the same general direction. Germans were readiest to acknowledge a significant measure of Austrian loyalty, notably Johann Genersich, who in 1793 compiled a young person's guide to Sonnenfelsian patriotism (and would, significantly, end his life as one of the first professors at the new Lutheran theological faculty in Vienna).[23] But the Saxons of Transylvania felt mortally offended by Joseph's highhandedness, and their leader Brukenthal, that man of high enlightened culture and Western taste, earned the revealing rebuff from his ungrateful Emperor that he was 'no Transylvanian, but a Nationalist'.[24] For their part the Slovak Lutherans clearly owed much in an indirect way to official encouragement. Pedagogues like Institoris-Mossóczy at Pressburg, pastors like Molnár in Pest, admired the Emperor and embraced reform, though the emotional logic of their passage to national consciousness would steer them into dangerous visionary channels, looking partly back to Hussite origins, partly forward to Panslav renewal.[25]

Much the same is true – *mutatis mutandis* – of the Orthodox communities. To some Serbs and Rumanians Josephinism came as a patriotic release.

> Joseph II, our ruler beloved,
> The world's sun and bestower of blessings . . .

Proud Minerva, the goddess of wisdom
Has enlightened thy spirit from childhood . . .
Ah, sweet season! Ah, Golden Age among us!
Among all men new affection is kindled! . . .
For our nation we shall all implore Him
And with tears to Him we shall petition . . .

That distinctive and personal statement was penned by Dositej
Obradović – indeed, it introduces his remarkable autobiography,
printed in 1783.[26] Obradović, as a radical seculariser, sitting *pro
tem*, in Leipzig, could indulge his rose-coloured muse in a way
that the ecclesiastical authorities at home, who had existing
political claims and new grievances (recall the burial issue!) were
less disposed to do. Most of his fellow-Serbs, even reformist ones,
like some spokesmen at the Temesvár congress of 1790, were
harder to persuade;[27] whereas the Rumanian intellectuals of
Transylvania, predominantly Uniate, often Vienna-trained, offer
better material for an argument about creative linkage between
state-service, which provided them with jobs and a *raison d'être*,
and their mission of popular vernacular awakening. Even if
relations with the Habsburgs proved in the longer term a
temporary alliance rather than a real bond of sympathy; even if
the view sometimes expressed that Rumanian national sentiment
was a construct of the eighteenth-century Viennese administration is seriously misplaced; yet the positive Austrian stimulus to
the Transylvanian School must be reckoned very considerable.[28]
The fullest evidence of positive associations between Josephinism and national awakening (and I am building up this skeletal
picture in reverse order of correlation) is to be found among
various Catholic Slavs of the Monarchy,[29] alike lacking any
adequate political structures into which they could pour the new
commitments of enlightened intellectuals. The Czech case raises
complicated issues, but the deep and lasting enthusiasm of
the *obrozenci*, or progressive patriots, for Joseph stands beyond
question. At times, as in the *Kniha Josefova* of the publicist
Kramerius (also the editor of the officially-approved vernacular
newspaper already mentioned) it turned into pure adulation.[30]
The 'Private Society' which grew up in the 1770s brought together
almost all leading intellectual luminaries of Bohemia, lay and
clerical. Whereas its founder was an outsider (Ignaz Born), and
royal patronage was only extended in 1790, its patriotic historical
and scientific proceedings occupied exactly the common ground

between *Aufklärung* ruler and subjects, that emotional dualism which Dobrovský articulated in his lecture of 1791 on the loyalty of Slavs to the House of Austria (*Ueber die Anhänglichkeit und Ergebenheit der slawischen Völker an das Erzhaus Oesterreich*); which Pelzl echoed when assuming the new chair of Czech at Prague University in 1793; and which Riegger likewise proclaimed in print the following year.[31]

The Croats were far less advanced, yet they owned a roughly equal debt to Josephinism. There is arguably a direct progression from Baltazar Krčelić, whom as historian we have already encountered justifying Maria Theresa's claim on Dalmatia, and who as court protégé fought for rational centralising administrative practice in Croatia, to his pupil, Nikola Skerlecz. Skerlecz, reforming official, cameralist, head of schools and High Sheriff at Zagreb, also issued the first modern defence of the Croatian tongue and worked especially for the economic development of his native regions. Thence the patriotic baton passed to his younger colleague, Maximilian Vrhovac, Joseph's nominee as Bishop of Zagreb, model of a cautiously reformist prelate, aesthete and bibliophile, having one foot in the Enlightenment and the other in the prehistory of the Illyrian movement.[32]

Nevertheless both Czechs and Croats (like Serbs and Rumanians) possessed some alternative focus for such aspirations, in historic claims and vested interests. It is striking how both Pelzl and Skerlecz – close contemporaries – passed from admiration to condemnation of Joseph during the 1780s; while neither Dobrovský nor Vrhovac – another pair of near contemporaries, born and deceased within two years of each other – felt any real spiritual sympathy for the Emperor.[33] Two more minor nationalities had nowhere else to turn. In later eighteenth-century Carniola, home of the Slovenes, the largest entrepreneur was an archetypal Josephinist figure, Sigmund Zois, Freiherr von Edelstein, a rational and practical scholar with particular interests (like Born) in chemistry. One of his brothers, while studying at the Theresianum, actually translated into Italian the Sonnenfels work we have been considering, as *Dell'Amore della Patria*. Yet the Freiherr von Edelstein was also Baron Žige Zois, patron and fellow-enthusiast of the first generation of admirers of the Slovene tongue, among them the Jansenist canon, Jurij Japelj, who – like Durych and Procházka in Bohemia – prepared a new Bible translation at the behest of the regime using sixteenth-century Protestant models. Zois's secretary, Kopitar, would become the founding father (along with the Dobrovský of the *Anhänglichkeit*) of later Austro-Slavism.[34]

Finally a glance at the Catholic Slovaks. In the 1760s and 1770s they threw up, in Maria Theresa's court librarian, A.F. Kollár, another Krčelić, argumentative and devious mixture of central and local patriot, defending Habsburg rights in Hungary and beyond with some consciousness of Slav solidarity.[35] Then in the 1780s came a squarely clerical phase, accompanying the foundation of the general seminary at Pressburg, with strong Slovak representation among both teachers and pupils, stressing the usefulness of the vernacular to priests and the need to overcome divergence, of dialect. A typical progressive cleric of the day, Anton Bernolák, led the way with a Slovak grammar, while his colleagues Bajza and Fándly embraced other aspects of vernacular-based reform, the first with a more radical literary critique of the Hungarian church establishment, the second with a series of tracts for peasant instruction (one of them, it may be recalled, about bees).[36] Altogether the Bernolák circle furnishes the purest example of national consciousness being furthered by both the detailed measures of Joseph II and the general ethos of the Austrian reform programme. How fitting, therefore, that the finest existing work on the history of Josephinism and language should be by a Slovak – and how characteristic that subsequent writers should have almost totally ignored it![37]

To conclude: the correlation between state-building and national awareness in the Habsburg lands was not always negative, even though – and sometimes precisely because – the impulse towards some kind of overall 'Austrian' identity proved so abortive. Indeed, the *Aufklärung's* important stimulus to linguistic claims and new forms of patriotism – especially where older forms of it were largely absent – seemed for many to be perfectly reconcilable within a Josephinist aegis. There might therefore be lessons for the interpretation of Joseph's policies if this kind of multinational comparative enquiry were properly pursued. Their greatest affirmative effect on nationality operated through the new accents they introduced into *ecclesiastical* loyalties, above all Catholic ones. That would help to underline the controversial priorities of Winter and Maass in stressing the religious objectives of the government programme (though not their narrowness of focus).[38] There might also be lessons for the interpretation of nationality. Identifying the roots of national movements in *Aufklärung* ideals of improvement and popular

guidance helps to sustain arguments about their material circumstances, but hardly Marxist ones, since the practical policies
were espoused by nationally-minded intellectuals less because of
existing 'bourgeois' development than from a desire to create it.[39]

Perhaps, to illustrate the overall argument, I can be forgiven
for modifying (and bifurcating) a famous comment by Józef
Piłsudski, who observed that he had travelled on the Socialist
tram, but only until 1918, when he got out at the station marked
'Polish independence'.[40] Joseph II, we might say, climbed aboard
the tram of *Aufklärung*, only to stop at the station marked
'*Staatsbürgertum*'. National communities in his Monarchy simultaneously climbed aboard a Josephinist tram, only to alight at
the station marked 'nationhood' (those having farthest to travel
being most moulded by the initial impetus), and creating in the
process, for the nineteenth century, an equal and opposite
distortion of Enlightenment. Let us recall our simple Slovene
apiarist, Janša. Is there not a bee-line (as it were) from him to
the age of full national organisation? After all, the cultural
headquarters of Austrian and Hungarian Slavs took their very
name of *Matica*[41] from the queen-bee, that essential symbol of
the practical groundwork of Josephinist renewal.

8. The Smaller German States

CHARLES INGRAO

I

THE small and middle-sized states of the Holy Roman Empire may very well be the last frontier of the debate over enlightened absolutism. They have generally been ignored by historians, who have all too often been content to regard them as insignificant postage-stamp states whose petty despots were more likely to expend their tiny revenues on comic opera armies and extravagant imitations of Versailles than on innovative domestic programmes. That has all changed. We now realise that many of the German regimes were as committed to introducing Enlightenment ideas and benevolent government as were the dynastic 'Big Three' of enlightened absolutism – Joseph II, Frederick the Great and Catherine the Great – themselves German-born monarchs who had sprung from the same cultural heritage.

Of late there has been increased interest in studying these smaller principalities – and with good reason. Like their counterparts in the Italian peninsula the German states offer an ideal opportunity to examine late-eighteenth-century government in an environment that was, in several respects, more conducive to the drafting and execution of domestic reform. Since they did not rule powerful states or command large armies the smaller princes were less likely to be consumed or distracted by the kinds of foreign policy and security considerations that so often predetermined the domestic programmes of the great powers. They were also less likely to incur the intense opposition of conservative corporate bodies, even though such institutions did exist in most principalities. This was partly because their territories were less dispersed and diverse, but also because the smaller princes had never pursued centralisation quite to the extent that it had been attempted by the great powers, thereby sparing them the otherwise inevitable clashes and legacy of mistrust that so typified crown–country relations elsewhere. Given their smaller size, the

221

8.1 Germany in the later eighteenth century

petty princes did not have to deal with as many intermediary levels of provincial estates or administrative offices that might defuse, dilute or divert their reform projects. Indeed, the modest size of their territories allowed them to devote more attention to the successful local application of individual initiatives. Finally, the lesser German princes were never too proud to learn from others. Whereas the great monarchies generally ignored developments in the smaller states, the German principalities continuously studied and readily copied their neighbours, both large and small. As a result they provided an ideal vehicle for launching numerous pilot projects, the results of which could be widely publicised by the hundreds of public affairs journals being published throughout the *Reich*.

It is this favourable constellation of circumstances and developments that has prompted several historians including the present writer to suggest that enlightened government functioned best within the smaller states of the Holy Roman Empire. Yet, as significant as its achievements were, it is equally important that we recognise those factors that shaped and ultimately limited its scope. One of the lessons to be learned from studying these principalities is that the potential for truly dramatic change was pre-empted from the outset by a number of entrenched values and institutions that predated the Enlightenment. Another is that many of the reforms decreed by the German princes that we have routinely termed 'enlightened' stemmed not so much from Enlightenment thought as from earlier ethical philosophies or systems of political economy. Indeed, before we examine enlightened absolutism in the German states – or anywhere else – we have to ask ourselves exactly what constitutes 'enlightened' policy. If by the word 'enlightened' we mean 'benevolent' in a broadly utilitarian sense, then the German regimes were models of enlightened government; if, however, it means 'stemming from Enlightenment ideas' then they were much less substantially so.

There can be little doubt that the leading echelons of German society generally embraced Enlightenment notions of natural law and human freedom, whether in the form of free trade, freedom of conscience, or the pursuit of knowledge through free inquiry. Moreover, educated Germans were supremely confident in the strength of human reason and the superiority of all things rational. They also endorsed the creation of wealth as a crucial element in the pursuit of human happiness. Yet, what the Germans called the *Aufklärung* operated in an environment totally different from France. Germany's rulers, whether they be in Vienna, Berlin or the host of smaller courts in between, had never discredited themselves or the system they represented quite to the extent that Louis XIV had in France. Hence, whereas the *philosophes* were determined to use the Enlightenment as a battering ram to destroy all that was wrong with the Ancien Régime, Germany's *Aufklärer*, publicists, and government officials generally concurred in advocating only limited, evolutionary change without seriously disrupting the status quo. Instead, Enlightenment ideas that operated as a destructive force west of the Rhine were readily integrated into the established matrix of ideas, values and institutions.

The *Aufklärung* had only a limited effect on the existing culture and patterns of behaviour of the German princely court. It played virtually no role, for example, in the princes' ongoing patronage of Germany's leading writers, artists and architects. To be sure, such philanthropy was laudable and of enduring value. In tiny Saxe-Weimar, Duke Carl August made a major contribution to modern German literature by retaining such literary giants as Herder, Wieland and Goethe, whom he employed as his chief minister. The prince-archbishops of Cologne and Salzburg were instrumental in promoting the early careers of Beethoven and Mozart, until both had landed more permanent positions in the capital of the Habsburgs. The Palatine and Bavarian elector Charles Theodore left his stamp on Mannheim, which still recalls his founding of the German National Theatre, and on Munich, where his chief minister, Count Rumford, took forty days to transform thirty-five acres of marshland into the city's famous English Garden. Landgrave Frederick II of Hesse-Cassel transformed his residence into one of Germany's most beautiful cities, until its near total destruction in World War II. That these rulers acted out of sense of obligation and not merely for personal pleasure is best evidenced by the efforts of Duke Ludwig Engelbert of Arenberg, who travelled to Great Britain, France and Italy in search of ancient and contemporary artwork, even though he was totally blind.

But did such laudable behaviour spring from the Enlightenment? Princes all over Europe had been sponsoring great artists and writers since the Renaissance. Indeed, such patronage was part of a composite court lifestyle that had come into vogue at the Versailles of Louis XIV and included numerous pursuits that were clearly more hedonistic than enlightened.

Moreover, the building of sumptuous palaces, the staging of elaborate court festivities, sporting events and hunting expeditions, and the maintenance of glamorous mistresses were expensive pastimes that easily overburdened the slender resources of many a small state. Such excesses did little for the prince's subjects, who had to support the court with their taxes and endure the damage inflicted by princely hunting parties with the loss of their crops. Many princes did cut back on these extravagances by assuming a modest lifestyle. No less typical, however, was the Duke of Württemberg whose many palaces drove him to bankruptcy, or the prince of Hohenlohe-Öhringen who continued his ruinous hunting expeditions because they were his 'only pleasure', or the landgrave of Hesse-Darmstadt who discontinued them only after twenty years of protests by his subjects.

The princes remained equally attached to the military life. Soldiering was, of course, the original, medieval calling of princes and nobles everywhere. Since then the religious and dynastic wars of the sixteenth and seventeenth centuries had done nothing to diminish its prestige as each state was compelled to defend its territories against one invading army after another. Then, in the mid-eighteenth century came Frederick the Great. It is difficult to overemphasise the tremendous influence that the Prussian king had on his fellow princes both as an advocate of Enlightenment values and as a practitioner of military aggression. His brilliant conquests at once infused new glamour to the profession of arms and opened numerous opportunities for a military command. Most German rulers either held commands in the Prussian, Austrian or Imperial armies, or had relatives who did. Thus, Carl August of Saxe-Weimar held a commission in the Prussian armies that invaded the Dutch Republic in 1787 and Revolutionary France five years later, while his cousin, Ernest Frederick III of Saxe-Hilburghausen was Frederick's unfortunate opponent at Rossbach. Louis IX of Hesse-Darmstadt willingly forsook the companionship of his loving wife and the comforts of her resplendent court in order to spend his entire twenty-two-year reign commanding a garrison at a desolate outpost on the French frontier. Nor were the princes content to serve the armies of other sovereigns. Virtually every state, no matter how small, maintained at least some troops, if only for ceremonial functions. In a great many instances the military's size corresponded to a principality's needs and resources. Yet more than a few princes were unable to restrain their passion for soldiers. Perhaps there is no better example than Prince William of Schaumburg-Lippe, who condemned military aggression in his memoirs, yet maintained a 1,200-man army in a country of only 20,000 people. Even the indisputably benevolent Carl August had to be restrained by his Estates' pleas for fiscal restraint when he increased Saxe-Weimar's forces from a mere 36 to 700 men in the first eight years of his reign.

To some extent militarism survived because society as a whole regarded warfare as a legitimate form of princely behaviour. King Frederick had, after all, been proclaimed 'Great' by contemporary Europeans not for the triumphs of his intellect but for the victories of his army. The same can be said for a widespread acceptance of certain elements of feudal society. Although all Germans were aware of the inequities of their *Ständegesellschaft* ('Society of Orders'), they did accept the legitimacy of noble privileges to the point that no one was prepared to eliminate them without providing adequate compensation. That would have been expropriation. Nor did even the most radical *Aufklärer* seek to create a truly egalitarian society. Like the Enlightenment elsewhere on the continent, the *Aufklärung* was sufficiently elitist that it assumed special station and treatment for society's upper crust. Moreover, in contrast to France, where monarchy had begotten tyranny and massive discontent, the *Aufklärer*, nobility and peasantry alike, shared their princes' confidence in the efficacy and legitimacy of the system of hereditary monarchy.

Finally, it is worth making the obvious point that Germany was a Christian society with deeply held religious beliefs. Enlightenment ideas could survive in Germany only when they could coexist with Christian thinking and values. The great *Aufklärer* Christian Wolff had, in fact, effected a marvellous synthesis of the two by placing the forces of nature and natural law under God's protection, and by making human reason an instrument for rationally studying the world He had created. Few Germans were, however, willing to go beyond Wolff by proclaiming deistic beliefs or attacking their principality's established church, as the *philosophes* had done in France.

Any programme of reform needed to be consistent or at least compatible with these values and institutions if it was to gain acceptance, not only among government officials, but even among the majority of educated Germans. This had already proven to be the case with cameralism, the existing system of government science and political economy that had evolved throughout Germany in the decades following the Thirty Years War. As the German regimes attempted to fund the armies and debts that the war had brought, they made a concerted effort to increase sources of state revenue through the scientific application of economic and administrative principles. They hoped not only to promote economic development by the conventional strategies then being employed in the France of Colbert; they also intended to organise and oversee both the people and sources of capital through a system of *Polizei*. In addition to raising taxable wealth, cameral science also stipulated how the increased state revenues could be handled most efficiently and effectively by the government fisc. Nor did cameralism stop there. Because they realised that healthy and happy subjects worked harder and more productively, the cameralists took a keen interest in promoting people's living standards, spiritual and psychological well-being, and overall happiness. It was with this comprehensive sense of responsibility in mind that rulers like the Duke of Saxe-Meiningen explicitly admonished local officials to 'further the welfare' of the downtrodden, and to be their 'friend and adviser', even to the point of helping to settle family or other personal problems.

By seeking to increase taxable wealth, cameralism addressed the state's fiscal needs in an era of ever growing armies and increasingly frequent dynastic wars. In its call for intensive government involvement in people's lives it also reflected not only the high degree of public confidence in royal authority but also contemporary Protestant appeals for enhanced social awareness and activism. In many respects it was also compatible with Enlightenment ideas that began infiltrating Germany at the beginning of the eighteenth century. Both advocated a rational, secular approach to securing the general welfare and happiness, which they largely defined in terms of material welfare. The two were, however, not quite identical. In its drive to enhance state revenue cameralism was likely to place fiscal integrity ahead of the popular welfare. Not unlike the modern social welfare state with which it has been compared, it was also bound to pursue mercantile and social policies that impinged on economic and individual freedom. Nevertheless, this was the system with which Enlightenment ideas now came into contact and with which it would have to conform, not only in government ordinances but in the minds of the men who drafted them.

It is impossible to present a single profile of the princes who ruled the Empire's 300-odd principalities. Certainly a great many of them thought of themselves as 'enlightened' and their age as one of Enlightenment. In vivid contrast to 'proprietary' monarchs like Louis XIV, they generally recognised as paramount their responsibility to serve the welfare of the state, which was broadly conceived to include the ruler's subjects as well as his dynasty. They were exposed to such notions at an early age, whether by university training or more often by childhood tutors, whose course of instruction typically combined Christian catechism and traditional princely pursuits like riding, dancing, swordplay and military history with philosophy, natural law and detailed cameral instruction in the nature of the principality's products and industries. That princes no longer ruled by divine right for their own benefit was made evident by their mentors, such as the tutor of the young Charles of Zweibrücken, who admonished his charge that 'though you are destined to appear on the world stage, you have only luck to thank. You are only a man and in this capacity you must fulfil your responsibilities toward God, toward your fellow man, and toward yourself.'

That this message usually found its mark is evident from the rulers' private and public correspondence. Remarks like William of Schaumburg-Lippe's observation that 'government should have as its objective the happiness of its subject peoples' were commonplace among their personal letters and directives. Like Frederick the Great and Catherine the Great, several rulers even corresponded with the brightest lights of the Age of Reason, including d'Alembert, Diderot, Holbach, and above all Voltaire, who counted no fewer than six German princelings among his numerous correspondents. Charles Frederick of Baden literally went further by frequently travelling to Paris in order to seek the advice of leading physiocrats like Mirabeau and du Pont de Nemours. He would ultimately share his insights in an anonymously written treatise on political economy.

Nor were Enlightenment-inspired ideas limited only to the Empire's lay princes. Notwithstanding the Roman Catholic hierarchy's determined opposition to most Enlightenment ideas, progressive government could be found among the Empire's two dozen prince-bishoprics and forty independent abbeys and ecclesiastical foundations that ruled about one-third of the German states' ten million people. Many ecclesiastical rulers were not career clergymen at all but were rather prominent nobles, imperial knights or cadet princes who were more inspired by politics and other worldly interests than by theological or clerical matters. Thus Archbishop-Elector Clemens Wenceslaus of Trier was a Saxon prince and former Austrian field marshal whose health had compelled him to forsake soldiering. Charles Francis von Velbrück of Liège not only lacked formal religious training but was a tireless proponent of the French *lumières* to boot. Though hardly an original thinker, Archbiship-Elector Maximilian Francis of Cologne was an Austrian archduke with an impressive command of Montesquieu and an appreciation of the Social Contract that obligated him to his subjects. Meanwhile, there were several less cerebral ecclesiastical rulers like Arch-bishop-Elector Emmerich Joseph of Mainz, who were content to live a comfortable life of hunting and entertaining just like their lay counterparts, while delegating everyday affairs to their ministers.

Whatever their predilections, the prince-bishops had the distinct advantage over dynastic princes in that they always came to the throne as mature adults bringing with them fresh ideas from a hybrid mix of families. This was especially the case in the period 1755–71, when fully three-quarters of the empire's bishoprics received new rulers, including all four of its relatively large archbishoprics. Furthermore, as celibate (though not necess-arily chaste) churchmen they invariably had no dynastic agenda and, therefore, needed neither large armies nor the taxes to maintain them. Small wonder that a popular refrain often heard in the church states was *Unterm Krummstab ist gut leben* – 'life is good under the bishop's crook'.

Unfortunately such examples of princely benevolence were not uniform. In Germany, as elsewhere, enlightened absolutism was sometimes a hit or miss proposition that depended on the luck of the royal succession. Although they were in the minority among their peers, Germany had its share of tyrants. Notwithstanding his tutor's childhood admonition, Charles II of Zweibrücken distinguished himelf as duke only through his personal greed and brutality. Despite his advocacy of various cameral reforms, the colourful bishop of Speyer was an outspoken enemy of the Enlightenment who once suggested that the entire University of Göttingen be closed down. (Fortunately for the professors, Göttingen was situated in Hanover.) In Bavaria, Elector Charles Theodore turned the clock back on his predecessor's enlightened reforms and devoted himself instead to showering honours and income on his bastard sons and court favourites, while maintaining a court of two thousand people that consumed a fifth of Bavaria's state revenue. In neighbouring Württemberg, Duke Charles Eugene actually did exclaim 'I am the state' as he launched a programme of palace building that only the Sun King could have appreciated. Though he too corresponded with Voltaire, it is fitting that the *philosophe*'s 162 letters dealt less with statecraft or philosophy than with his vain attempts to collect principal or interest on the 620,000 livres he had lent the spendthrift duke. Yet Voltaire fared much better than the prominent jurist Johann Jacob Moser and poet Daniel Schubart, both of whom he imprisoned for periods of five and ten years respectively following their criticism of his misrule.

Whereas the average German's chances of being ruled by a benevolent monarch were very much subject to the luck of the draw, he could be fairly certain of the quality of the princely administration. By the late-eighteenth century the German bureaucracies comprised a generally honest, hard-working group of university-trained professionals. Although their numbers included a high percentage of commoners at all levels, they were well-integrated into their state's power and social structure by a merit system that rewarded talent over birthright. Whereas the most accomplished non-noble officials were often ennobled, they were elevated as a whole to virtual nobility by special social status and privileges, as well as by generous financial incentives that usually included good salaries, adequate pensions, family survivors' benefits and the expectation of employment for qualified offspring. Indeed, the typical German state featured several *Beamtenfamilien* – family dynasties of civil servants that intermarried among themselves and, often, with the local nobility.

It was these men who were entrusted with the day-to-day application of government policy at the local level. They also enjoyed varying degrees of leeway in drafting and making policy at the very centre of the government, especially in the many instances where the ruler delegated authority but nonetheless appreciated the regime's responsibility to provide good government. German civil officialdom was never radical in its approach to reform. By integrating it into the ruling elite and tying its fortunes to the existing order, the German princes had guaranteed that their professional bureaucracy would value and protect the stability of the state and the Old Regime as a whole. It was, however, a driving force behind the kind of evolutionary change that constitutes enlightened absolutism in the German states. At the very least German officialdom applied the cameralist curricula of major universities like Göttingen by promoting economic development and accelerating the trend toward more secular and rational modes of thought. This contribution was crucial in states like Bavaria and Württemberg, where a truly conscientious ruler was lacking. It also proved pivotal to Hanover, whose elector was the absent King George III, but whose influential bureaucracy could effect cameral reforms that would have been unthinkable in Great Britain. More often than not civil officials were also converts to Enlightenment notions of education, freedom, religious toleration and social welfare that Christian Wolff had helped to introduce into Germany's universities. Several prominent ministers where close associates of Germany's leading *Aufklärer*, including Mainz's chief minister, Count Stadion, and Trier's Georg Michael de La Roche, whose philosopher grandson Clemens Brentano was named in honour of his elector, Clemens Wenceslaus. The acceptance of Enlightenment ideas by high officials is also illustrated by three of the Elector of Saxony's principal ministers, one of whom edited the works of Pierre Bayle, another who translated a physiocratic treatise on agriculture, and a third whose library housed the works of major Enlightenment *literati* such as Bayle, Bolingbroke and Locke, together with Christian religious tracts and several standard works on the science of cameralism. It is also evident in the widespread membership of civil officials in the reading clubs and masonic lodges that sprang up all over the *Reich* in the second half of the century. These German equivalents to the French salons became centres for the discussion of abstract Enlightenment ideals. The masonic lodges also sponsored a variety of philanthropic projects that reflected their best intentions of helping to improve society by reducing the sources of ignorance, intolerance, and various forms of social injustice. Although their Stuttgart chapter rejected the Duke of Württemberg's application, the freemasons readily admitted several princes including Duke George of Saxe-Meiningen, who demonstrated his disdain for less enlightened rulers by lambasting Charles II of Zweibrücken in one of the *Reich's* leading public affairs journals.

II

Armed with these good intentions Germany's princes and officials tried to translate this bevy of Enlightenment ideas, cameralist strategies and Christian values into a programme of practical reform. There is no such thing as a typical German state whose experiences were representative of the *Reich* as a whole. Not only was every prince different, but so was the mosaic of administrative, constitutional and economic structures that made every state truly unique. Some princes were absolute rulers while others, including most of the prince-bishoprics, enjoyed considerably less latitude *vis-à-vis* their corporate bodies. Several Protestant states, such as Hanover and Schleswig-Holstein, had absentee monarchs, while others, like Hesse-Cassel, Württemberg and Saxony were ruled by Roman Catholics. Nevertheless, it is possible to present a common profile of reform that was repeated over and over in dozens of otherwise different principalities.

The first order of business for most princes was to render their administrative system more rational and efficient. They generally eliminated superfluous posts and sinecures and also strove to end the common practice of accepting gratuities from petitioners by establishing firm and adequate salaries. In the name of greater precision the regimes tabulated and quantified all vital information in a veritable avalanche of statistical information that included the amount and condition of land, housing, crops, livestock, army recruits, feudal dues, and, of course, the population itself. In Hanover officials devoted twenty years to compiling a comprehensive topographical survey of the entire electorate. In several central German states even the names and functions of government officials were tabulated in published directories, or *Staatskalender*.

If the princes were drawn to administrative problems first, they usually devoted most of their time and money to economic development. As they sought to recover from the destruction and debts brought by the Seven Years War, they resorted to many of the same cameralist strategies that they had employed in the decades following the Thirty Years War. The regimes did everything they could to promote population growth in order to increase both worker production and demand for the goods they produced. Several governments simultaneously banned emigration, while enticing foreign settlers with various incentives. The duke of Arenberg even went so far as to offer one group of colonists free accommodation in one of his palaces. Some states offered marriage incentives, while certain Catholic countries like Bavaria and the archbishopric of Trier restricted the opportunities for young men to enter monasteries. The regimes also worked to attract and hoard all the specie they could, not because they subscribed to the bullionist philosophy of Colbert, but because cameralism preached the creative power of money placed in circulation. The net result was the same, however, as virtually every regime applied Colbertine restrictions on the export of raw materials and import of finished products, while encouraging the opposite.

The princes were also unabashed in their drive to create and protect a wide range of privileged luxury industries, just as Colbert had a century before. Porcelain was every prince's favourite, though few states were able to equal the high quality of Bavarian Nymphenburg or Saxon Meissen china. Various textile products were equally popular. Despite Germany's cool climate several principalities ranging from tiny Schaumburg-Lippe and Anhalt-Köthen to such northerly states as Cologne and Hesse-Cassel imported mulberry trees by the thousands in vain hopes of becoming silk producers. Even Charles Frederick of Baden was able to forsake his at times doctrinaire advocacy of physiocracy long enough to subsidise everything from steel and textile mills to the Black Forest's first cuckoo-clock factories.

This is not to say that the princes rejected out of hand the potential benefits of free trade. As a rule they concurred with those cameral scientists who decried the ruinous effects of government subsidies, monopolies, guilds, and tariffs. If they funded and protected so many business ventures it was because there was no private investment capital available to take their place. When and where circumstances permitted, various regimes did take a cautious step in the direction of *laissez-faire*. As the German states' most advanced industrial region, Saxony was in the position to replace monopolies and subsidised industries with free private competition in the years immediately following the Seven Years War. In neighbouring Franconia several states began to co-operate in reducing the prohibitive tariffs that had inhibited inter-state trade. Moreover, a large number of principalities sharply curtailed the guilds' ability to limit membership, restrict competition, and set wages and prices.

One of the features that distinguishes the economic initiatives of the late eighteenth century from those of the seventeenth is the attention they paid to agriculture. Although cameralism never placed quite the emphasis on agriculture that the physiocrats did, it did strike an even balance between the rural and urban

economies. Moreover, it received timely assistance from developments in Great Britain, where the dramatic improvements in
crop yields during the so-called agricultural revolution attracted
the immediate attention of several, principally northern and
Protestant German states. In the end many German agrarian
programmes were taken not from the abstract visions of the
French physiocrats but from the proven achievements of British
agrarian capitalists. Added to these impulses were the terrible
crop failures and famine of 1770–71, which gave new urgency to
the problems of the German farmer. Only after 1770, with the
infiltration of physiocratic ideas into states like Baden did French
ideas become a significant factor that merely intensified efforts
at agrarian reform.

British influences were most evident in the various strategies
that the princes employed to increase the amount of livestock
and farmland. Multiplying herd sizes was important to the
regimes because it would increase the all too slender supply of
both meat and dung for fertiliser. Since larger herds would need
more fodder, farmers were admonished to grow more and better
strains of clover and to feed their animals in stalls, where fodder
could be better rationed. The greater availability of dung made
it easier for the regimes to promote the replacement of the three-
field system with crop rotation. Indeed, the regimes employed
numerous tactics to increase the supply of farmland. Common
grazing land was parcelled out in several states, beginning in
1765 in Hanover, where 80 per cent of the arable land had
been community property. Several states drained marshes, most
notably Bavaria, which settled 600 families on 20,000 hectares
of land reclaimed from the Danube. Roman Catholic countries
like Bavaria and Mainz also followed the lead of Austria in
restricting mortmain, including the rights of monasteries to buy,
foreclose, or inherit private land.

The German regimes also followed the lead of the British in
encouraging potato cultivation. The hardy tuber was, however,
only one of several farm products that they promoted. In Trier,
for example, the government was instrumental in introducing
new strains of grapes that are today the pride of the Mosel wine
industry. In Hesse-Cassel, as in other states, newly-weds were
actually obligated to plant fruit trees within a short time after
their nuptials. These efforts were, in fact, part of an overall
strategy of boosting peasant investment and incentives that
stemmed chiefly from cameralist sources. A central feature was

government credit institutions that extended loans either for capital improvements or for emergency relief. Tiny Lippe-Detmold set up no fewer than three such funds for peasants, one that awarded cash gifts for disaster relief, another that granted loans of up to 3 per cent interest for poorer peasants to improve their plots, and a third that charged the general farming public 5 per cent. As an incentive some regimes like those in Würzburg and Erfurt showered industrious farmers with prizes or special privileges. Most governments also lengthened peasant leases to six or more years since tenants were more likely to improve plots once they were assured of long-term tenure.

It was primarily this desire to boost peasant incentives that encouraged several regimes to replace manorial obligations with free labour. Peasant dues and services varied widely not only from state to state but even within a given territory. But two things were almost universal among those regimes that did undertake reforms: their desire to boost productivity (and resulting tax receipts) by replacing these obligations with free labour, and their reluctance to end the manorial system altogether without first granting compensation to the feudal overlord, or *Grundherr*. As a result states like Baden, Hesse-Cassel and Hanover were usually content to institute these reforms only on the prince's own domains where there was no threat of noble opposition and where the increased productivity of free labour would most directly benefit crown revenues. But even they usually expected some kind of compensation for themselves, usually in the form of a cash payment paid by the peasant. Indeed, the commutation plan drafted by the Elector of Mainz called for the money equivalent of twenty years' dues. Meanwhile, elsewhere in Germany, most states were unwilling to go even this far and satisfied themselves with reducing, limiting, or regulating existing manorial obligations.

A final avenue of agrarian reform featured the regimes' attempts to protect the livelihood of the peasant, not only for humanitarian reasons but also because the state understood the peasant farmer's value to its tax base. This age-old policy of *Bauernschütz* took the eighteenth-century form of government legal assistance that ranged from the distribution of free, simply written handbooks in Osnabrück to the total assumption of lawyers' fees in Hesse-Cassel. It was also evident in several regimes' decision to outlaw the most destructive forms of hunting. Veterinary assistance and instruction was provided in a few states such as Würzburg, which offered free schooling for the sons of peasant farmers. Most states maintained inheritance laws (some dating from the sixteenth century) that forbade the suicidal subdivision of farmland beyond pre-defined subsistence levels. Moreover, following the famine of 1770–71 a number of princes introduced Prussia's system of grain magazines, albeit not for the military purposes originally envisaged by Frederick the Great.

Virtually every agrarian reform mentioned above had the dual objective of increasing productivity and preventing poverty, which the state saw as both a human tragedy and a fiscal problem. The same could be said for the regimes' welfare policies, which were designed not only to alleviate suffering but to minimise the numbers and maintenance costs of the indigent population. A great many ordinances were designed to forestall destitution by saving the peasant from bankruptcy. Rulers like Frederick II of Hesse-Cassel suspended taxes and forbade farm foreclosures during hard times, and even commuted long prison terms when they threatened to drive a convict's family into beggary. Numerous states went to great lengths in their sponsorship of fire insurance funds for both rural and urban areas. Every house in the bishopric of Eichstätt was numbered and appraised for registration in the country's compulsory *Brandkasse*. Though Hesse-Cassel's fund was voluntary, over 30,000 structures had been insured within its first four years of operation. Every state continued and many expanded existing sumptuary laws that banned those luxury goods that officials felt might ruin a family's finances or the work ethic that they felt went hand in hand with a simple life. Even proper health care assumed new prominence as an instrument for preventing poverty, since it was widely acknowledged that healthy people were not only happier, but more productive. Most regimes established medical commissions that oversaw the setting of standards of competence for midwives, doctors, surgeons, and apothecaries. In keeping with this commitment to preventive health care, Würzburg officials set minimum space and lighting requirements for schoolrooms, forbade parents from giving their children heavy work, and ordered parish priests to instruct all expectant mothers on proper baby care. In Hesse-Cassel, the home of Germany's first facility for unwed mothers, officials were even ordered to keep an eye on young women whom they suspected might conceal an unwanted pregnancy and later kill their babies.

Support was also made available for those who had already

succumbed to poverty. In most cases, however, such measures dated from the seventeenth century and represented the gradual, post-Reformation takeover of social services from the church, rather than Enlightenment ideas. One such measure that was applied virtually everywhere was export and price controls on staple products such as grain. The regimes were determined to keep bread prices within the reach of the urban poor and were generally afraid to experiment with free market forces. If prices were allowed to float, as they were in Münster, the government was likely to tie workers' wages to their rise and fall. The princes were especially active during shortages such as in the famine of 1770–71, when they embargoed food exports and subsidised the sale of imported grain. Indeed, it was not uncommon for a prince to pawn his silver service or crown jewels, or travel personally to major trade centres like Hamburg to purchase Russian or Polish grain for his people.

Whereas health care was upgraded and made available for the general public, it was generally provided free for the indigent. Several states maintained hospital facilities for them. Würzburg's *Juliusspital* even operated a health insurance plan supported by the city's guilds that provided free medical care to their generally poor apprentices and journeymen. Finally, there was a significant increase in the number of state-run orphanages and poor houses. No public welfare programmes attracted a closer personal interest from the princes themselves. It was not uncommon to find that the ruler himself had designed their uniforms, planned their meals, arranged their daily routines or, in the case of the prince of Hohenlohe-Neuenstein, actually housed them in a wing of his own palace. Once again, however, these efforts came not so much in response to Enlightenment impulses as it did to contemporary Christian views toward charity and cameralist efforts to rehabilitate people through a programme of regular work and self-discipline. Facilities such as Munich's famous 'Poor People's Institute' were essentially, in fact, workhouses where former beggars performed chores such as spinning and weaving in the expectation that a combination of time and work would reclaim them as useful members of society.

If the princes' administrative, economic and welfare initiatives drew more from cameralism than the Enlightenment, there were certain avenues of reform that were truly offspring of the Age of Reason. One was the practice of law, which became more efficient, professional and humane in keeping with the teaching of Christian Wolff. A few states attempted to streamline their law codes, although the resulting products were not necessarily models of Enlightenment jurisprudence. There was a move throughout the Empire to quicken the pace of justice, at least partly out of concern that litigants were spending too much of their productive time in court and too little working. To facilitate early settlements Hesse-Cassel fined lawyers who dragged out lawsuits, while Anhalt-Bernburg encouraged litigants to meet out of court – without their lawyers present. Elector Max Franz of Cologne actually established deadlines before which cases had to be settled. In the interests of greater honesty several princes raised judges' salaries and forbade them from accepting gratuities.

But the most noteworthy advances came in the treatment of accused and convicted criminals. Like the later teachings of Cesare Beccaria, Wolffian jurisprudence argued that the goal of prosecution was prevention rather than revenge, and excluded almost without exception the use of pre- or post-trial corporal and capital punishment. Several states introduced these restrictions beginning in the 1760s, after similar decrees had been enacted in Prussia. Yet, since many policy-makers still believed that corporal and capital punishment might deter some criminals, several states such as Saxe-Meiningen and Hesse-Cassel repealed them secretly by internal memoranda. Others, like Cologne's Max Franz, kept both on the books, but required the ruler's personal approval before such a sentence could be carried out. Even where executions persisted they were often sharply cut back and reserved for only the most heinous crimes. Meanwhile, as an alternative to execution, Hesse-Cassel simply conscripted many of its capital criminals forcibly into the army, while Arenberg sent them halfway around the world to the Dutch East Indies.

In terms of translating Enlightenment ideals into domestic programmes, education was surely the princes' field of greatest endeavour and accomplishment. One reason for the widespread promotion of education was its compatibility with existing forces. It conformed with the expectations of those pragmatists who expected it to make people more useful in promoting economic recovery from the destruction of the Seven Years War. It also reflected the princes' own agenda for secular, governmental control over all levels of society. Nevertheless, as a rule the German regimes also embraced the Enlightenment's absolute faith in education's effectiveness in promoting the happiness and welfare of an enlightened society. The prince of Hohenlohe-Öhringen recognised this confluence of objectives when he wrote at the conclusion of his reign that

> The foundation for a state comes not from masses of crude, uncivilised and ignorant men. Only a civilised, well-led and well-disposed people can give strength, wealth, power, [and] happiness to a country. To enlighten the people, give them sound morals, civic virtues, [and] clear religious understanding was always one of our first concerns.

One of the most frequently voiced criticisms of governmental educational programmes throughout the continent was that they favoured the elite social groups centred at court at the expense of the common people. It is undeniable that the thrust of public enlightenment in Germany was strongest at the court, if only because it made more sense to initiate projects near those elite social groups that were most likely to utilise them. In addition to the aforementioned private lodges and reading clubs, the princes founded a multitude of public libraries, museums, and academies of sciences, arts and letters in their capital cities. They and their ministers also manifested their fascination with science by funding numerous botanical gardens, menageries, anatomical laboratories, observatory telescopes, lightning rods, and manned balloon flights. This patronage was evident even at the court of the repressive Bavarian Elector Charles Theodore, whose chief minister Count Rumford and court chaplain each conducted experiments with such exotic forces as steam and electricity, and whose Academy of Sciences awarded first prize in a 1789 essay contest to the prince-abbot of St Emmeram for his scholarly treatise 'Concerning the Effectiveness of Lightning Rods on Thunderclouds'.

It would be unfair, however, to insinuate that the princes ignored those of their subjects who lived outside the capital. Virtually every German state published at least one public affairs journal for the country's professional classes (principally bureaucrats and merchants) that not only disseminated important legal and economic news but often promoted the values and books of the *Aufklärer*. It was with this same group in mind that the princes generally expanded the facilities of the Empire's numerous universities, while founding no fewer than seven additional institutions in the half century 1736–86. The curricula of most universities often gave special prominence to natural science and philosophy.

Nor were the common people left unattended. Several princes recast their public school systems from top to bottom. The quality of instruction was improved by giving teachers better pay and requiring them to complete additional coursework that was often taught by the local university faculty. A number of states such as Hanover and Baden launched ambitious school building programmes. Meanwhile, the schoolchildren themselves were subject for the first time to standardised schoolbooks, regular examinations, and compulsory attendance. There was, however, considerable variation in what they were taught. Most school systems emphasised practical skills such as the '3 Rs' and vocational training. Yet several states, most notably the bishopric of Fulda, decreed education laws that betrayed their confidence in the educability of the masses and their conviction that 'education gives a person worth [and] can enhance human happiness'. Charles Frederick of Baden actually dispatched his educators to Anhalt-Dessau's prestigious *Philanthropin*, where a faculty of distinguished pedagogues drawn from all over Germany promised to educate the 'whole man' according to the revolutionary teaching philosophy of Jean-Jacques Rousseau. Just as Rousseau had called for an end to severe discipline and strict memorisation of material, a great many states joined Fulda in seeking to promote greater human reasoning and understanding by employing the Socratic method that the Silesian educator Johann Ignaz Felbiger had introduced into the Austrian schools.

One casualty of educational reform had been the virtual monopoly that the church had heretofore enjoyed over the moulding of the public mind. Of course the regimes remained committed to Christianity as the indispensable instrument of public morality and salvation. Thus even Fulda's reformed school curriculum still reserved six of its 27 hours of weekly instruction for religion. Yet the school commissions and censorship boards that decided what people would read and hear were now in the hands of lay government officials whose commitment to rational and secular thought was unmistakable. One can sense this shift in the words of one educator who, upon reviewing the new qualifying examinations for Saxony's teachers, expressed satisfaction that there were 'none of the metaphysical statements, dogmatic beliefs, or piles of Biblical quotations' that had guided the teachers and students of past generations. This shift toward

secular values also manifested itself in the Saxon electors'
coronation oath, which was now rewritten to stress their obligation
to the public good, rather than to God. It was just as evident in
the tendency of German officials everywhere to attack religious
zealotry and superstition. Such popular religious practices as
witch-hunting and exorcisms were banned. Most Catholic states
also sharply curtailed elaborate religious ceremonies, processions,
and pilgrimages in favour of simple, more understandable religious services and liturgy delivered in German rather than Latin.

The reduction in religious processions and pilgrimages also
had an economic dimension because they competed with the
people's work time. Hence the bishop of Fulda banned all
overnight processions, while his counterpart in Augsburg even
prohibited them on Good Friday because they discouraged people
from working on the following day. It was with this same rationale
in mind that Würzburg's prince-bishop proscribed dancing and
drinking on the days preceding religious holidays. Indeed,
Protestant and Catholic governments alike significantly reduced
the number of religious holidays because they encouraged what
Elector Emmerich Joseph of Mainz referred to as 'sinful idleness'
that unduly reduced worker productivity.

This combination of secularisation and economic self-interest
also explains the readiness with which the princes accepted the
Enlightenment principle of religious toleration. Most states
increased the degree to which they tolerated religious minorities,
both because they deemed it just and because they were eager
to increase the productive work force. As Clemens Wenceslaus
explained in justifying Trier's Toleration Patent in 1783, 'On the
one hand, the elimination of all appearance of intolerance brings
credit to our holy religion; on the other, the settlement of wealthy
merchants and entrepreneurs serves the domestic economy,
employs lazy beggars, and brings foreign wealth into the country.'

Several states went out of their way to offer private worship,
schools and churches to religious minorities, especially if they
could establish that they were industrious and had marketable
skills. A few states went one step further by enrolling Jews in
their schools and universities, and by appointing Jewish or other
religious minorities to their faculties.

III

The long list of domestic initiatives pursued by the German
princes represents a compelling record of achievement and should

remove any doubts that the more enlightened regimes were committed to providing the best possible government to their subjects. Unlike the proprietary monarchs of the previous generation they focused their efforts on serving the state as a whole rather than on attaining selfish dynastic interests. Admittedly the bulk of their reforms owed as much or more to pre-existing ideas and values as they did to the Enlightenment itself. Yet, the princes' willingness to resort to a variety of systems need not call their sincerity into question. As later generations have discovered, abstract Enlightenment notions were not necessarily infallible or superior to the largely compatible but sometimes competing insights offered by cameralist or Christian thought.

It is difficult to quantify the extent to which these reforms benefited German society. In purely economic terms they generally helped increase agricultural output, improve the availability and distribution of food, and laid a firmer basis for subsequent industrialisation. More rational and responsible fiscal policies enabled a great many states to restore their finances, thereby enabling them to fund more and better domestic services. By providing greater public assistance, better justice, religious toleration, and limited protection from manorial exploitation, these domestic reforms also alleviated much of the suffering of the common people. Moreover, by making education more widely available they provided the key to even more rapid progress in the following century.

Admittedly, even the most enlightened princes were still limited in what they could accomplish. Quite aside from the restrictive role played by pre-existing values and patterns of behaviour, the German states were sometimes handicapped by their size. Their finite financial and human resources sometimes limited the availability of qualified teachers, doctors, or entrepreneurs, just as the Empire's seemingly infinite number of separate political jurisdictions foredoomed the opportunities for free trade and economic growth. Like their larger neighbours they also had to contend with conservative forces opposed to even moderate change. Although the princes generally enjoyed a better working relationship with their estates, they confronted the same widespread public traditionalism encountered by the most powerful European sovereigns. An enlightened pastor in Hohenlohe-Öhringen spoke for all European reformers in complaining that the 'townspeople and peasants do not always recognize the good intentions of the higher authorities'. They particularly resented changes in religious practice, such as in the reduction of religious holidays and ceremonies, the curtailment of Latin, or the toleration of minorities. But they also resisted such newfangled ideas as street lights, lightning rods, smallpox innoculations, crop rotation, potatoes and improved strains of existing crops.

Yet, if we are to judge the German regimes by their *intent* and by what progress they did make in achieving their goals, enlightened absolutism succeeded in Germany. Their success is perhaps best measured by the continued, widespread popular acceptance that they enjoyed following the outbreak of the French Revolution. In expressing their distaste for the Reign of Terror and the French occupation that followed, the German people made clear their preference for the moderate changes peacefully introduced by their own rulers. It was perhaps the greatest legacy of enlightened absolutism that this remained their preference well into the twentieth century.

9. The Danish Reformers

THOMAS MUNCK

ENLIGHTENED absolutism in Denmark followed patterns common to many other governments in Europe. The aim was not just greater efficiency in administration, maximisation of state revenues, liberalisation of internal trade and guild regulation, and development of economic resources, important though these naturally were to most European governments at the time. In keeping with the influence of the Enlightenment, Danish reformers also sought to improve educational standards, to encourage public debate by reducing censorship, and to achieve greater fairness before the law. In a European context, however, it was the Danish efforts to achieve effective agrarian and peasant reform that stand out, partly because a serious effort was made to deal both with land usage and with relations between peasant and landowner, but above all because the reforms carried out in the later 1780s – uniquely in later-eighteenth-century Europe – actually worked as intended, and survived the revolutionary period.

The period of enlightened absolutism lasted in Denmark from perhaps the 1740s or at least 1755 through to the 1790s. The process was far from continuous, however, and the brief stormy phase associated with the notorious Struensee from 1770 to January 1772 was not the most important. Because of the mental illness (schizophrenia) of the King, Christian VII (1766–1808), absolutism in Denmark could not continue to function in the way it had evolved since its establishment in 1660. After a period of growing unpredictability in the late 1760s, Struensee attempted to break through the complex bureaucratic system which had been inherited from Frederik V's reign (1746–66), but in doing so laid himself wide open to accusations of tyranny and usurpation. His fall brought a sharp reaction during which most of the reforms he had pushed through were revoked in favour of narrow-minded conventionality. It was only after another political upheaval in 1784, when crown prince Frederik (VI) became regent and formal head of a team of gifted and able advisers, that the most important aspects of government-initiated reform could be implemented.

I

Literate and government circles in Denmark had access to the European Enlightenment partly through some influential figures at the University of Copenhagen and at the aristocratic academy in Sorø, partly from mid-century through new periodicals and occasional publications, and ultimately through a hesitant but growing circulation of newspapers providing comments on and reviews of major developments abroad. Yet, despite distinctive elements of its own, and the popularisation initiated by the historian, philosopher and playwright Ludvig Holberg (d. 1754), the intellectual atmosphere even in Copenhagen was largely derivative. The effect of censorship on any discussion of domestic affairs was clearly apparent into the 1750s – perhaps until 1755, when, on the occasion of the king's birthday, a competition was announced for essays on any subject which 'may help to maintain the prosperity of the land'. Over the next years some of the wide-ranging submissions were published in the journal *Danmarks og Norges oeconomiske Magazin*, and the debate inevitably extended beyond the purely practical aspects of trade, industry and agriculture to more fundamental aspects of social inequality and the role of government.

At a more practical level, however, the privy council, the various government colleges (departments) and the inner court circles included many men of German or Schleswig–Holsteiner birth, familiar not just with the French and Anglo-Scottish Enlightenment but also in particular with German cameralism. German was to a large extent the administrative language, and many nobles owned land both in Denmark and south of the border; some naturally also pursued a career which included service both in German principalities and in Copenhagen. Notable amongst such Germans was the effective first minister of Frederik V, count Adam Gottlob Moltke, and the prominent diplomat and reformer baron Johan Hartvig Ernst Bernstorff. Moltke was from Mecklenburg, while Bernstorff's family held offices and estates in various areas, including Hanover, Lauenburg and Holstein. Both men, like other major landowners of their generation, and indeed some of the officials in charge of crown estates, were keen observers of English agrarian developments. By the 1760s Moltke was experimenting with *Koppelwirtschaft*, the north-German variant of convertible husbandry, while Bernstorff's nephew, Andreas Peter Bernstorff, later to play a crucial role in the reform administration of 1784, was familiar both with cameralism and with the western Enlightenment through his studies at various German institutions, in Geneva, in France and in England.

From the start, therefore, the central government was simply party, in matters of rural reform, to opinions held by some liberal landowners and increasingly discussed amongst the wider public. Already in 1746 local government officials had been asked to report on the depressed state of the rural economy, and Moltke had the same year suggested the establishment of a high-level permanent commission which in fact took shape only in 1757. It drafted legislation to facilitate the removal of demesne land from the restraints of common cultivation, and considered other aspects of enclosure. Some attention was also paid to the question of promoting consolidated peasant holdings and of improving peasant incentives through more secure tenure, but these matters were liable to touch seigneurial rights too closely and were passed by. Only with regard to its own land did the crown adopt a reformist stance, notably in the parcelling out of estates for sale as freehold direct to peasants themselves.

After the succession of Christian VII in 1766, however, the resulting power struggles at court re-opened debate on a number of deeper issues. Although the older and moderate Bernstorff remained a central figure in foreign and economic affairs for some years, the King's Swiss tutor Eli Salomon François Reverdil, amongst others, began seriously to question some of the fundamentals of seigneurial power over the peasantry. Claude Louis de Saint-Germain, a French general who had been Frederik V's military adviser, was particularly blunt in his criticism of the system of *stavnsbånd*. Less comprehensive and perhaps less degrading than east-European serfdom, the *stavnsbånd*, established by law in 1733 (perhaps on Brandenburg-Prussian precedent) to ensure manpower for the rural militia, required all males aged between 18 and 36 (by 1764 between 4 and 40) to remain on the estate of their birth and take up tenancies there when required, so as to be available for enrolment. Saint-Germain was critical of the system because he regarded it as inefficent from a military point of view: in fact bondage had proved convenient for landowners who were concerned to maximise available labour resources, and their right to select recruits could readily be abused for disciplinary purposes. Although not necessarily particularly oppressive in practice, the *stavnsbånd*, together with other punitive powers which landowners were entitled to use, could hardly be avoided in any 'enlightened' discussion of peasant conditions. In 1767 a new Agrarian Commission (*Landvæsenskommission*) was established with a wide brief, and it was reorganised the following spring as a permanent Board of Agriculture (*Generallandvæsenskollegium*) with a prominent position in the administrative system. The discussions in this institution clearly show that the aim was now not just to achieve technical improvement, but equally to try to reform peasant–seigneur relations more fundamentally, including forms of tenure, rents, labour services and other issues of vital importance for the rural population. But it was recognised, not least by Henrik Stampe (as *Generalprokurør*: the government's highest ranking legal adviser), that active policies on such issues could hardly fail to generate confrontation with landowning interests. No eighteenth-century government could undertake that lightly, and so it is not surprising that the legislation that emerged in 1769 was very tentative.

By then it had become more widely recognised amongst landowners that the old seigneurial system was too inflexible in a period of growth and rising prices. Complex seigneurial rights, regulations against the engrossing of peasant holdings in the demesne, and the shortage of grazing made the development of new land in the poorer parts of Jutland and elsewhere difficult, and landowners could not otherwise easily take advantage of rising prices as long as they were still in common cultivation with their tenants. Rents were also traditionally fixed, so seigneurs could normally only extract more income from their tenants through labour services. These were unregulated, but could best be exploited if the pattern of land usage was modernised. However, there was a major qualitative leap from, on the one hand, the promotion of enclosure and technical development to, on the other, external interference in the sensitive area of peasant obligations to his seigneur: the first need not lead to the second, and in the opinion of many should not. State intervention in peasant–seigneur relations beyond a certain level could in any case be regarded as contrary to the law. Some landowners in fact argued that a true *laissez-faire* approach in the countryside would be best, involving both an abolition of the *stavnsbånd* and the abandonment of all legislative protection on behalf of tenants. To landowners like Tyge Rothe, conscious of the trend in England towards commercial leasing of land to tenant-farmers, a freer contractual relationship between landowner and peasant could in the long run offer considerable advantages. More generally, however, the economic advantages of substantial change were far from obvious to many owners of lesser estates, who were concerned that order and discipline might easily collapse if traditional patterns were upset.

As recent research has shown,[1] peasant attitudes were indeed changing. Some tenants in Jutland and the duchies, encouraged by the rapidly improving economic climate of the 1750s and 1760s, had been persuaded to buy freehold rights over their holdings. On the Hørsholm estate north of Copenhagen the hereditary leases introduced from 1759 together with a commutation of all dues into an annual rent soon convinced the peasants there. The initiatives by A.P. Bernstorff in favour of hereditary peasant leaseholds or freeholds, together with a commutation of labour services on his uncle's estate just outside Copenhagen became a *cause célèbre*. Elsewhere crown tenants were also increasingly willing to contemplate reorganisation of holdings even if it meant enclosure and dispersal of the old village communities. Equally, there was interest in the application of new rotations to peasant holdings, which was of course possible with or without the full enclosure of tenancies. The Royal Society for Agricultural Science, founded in 1769 and soon counting amongst its honoured members a model reforming peasant, Hans Jensen Bjerregård, could thus preach to a sympathetic audience at different social levels. Its prize competitions and debates on the relatively uncontroversial issues of land usage and *udskiftning* (manorial and peasant enclosure out of common cultivation) ensured wide acceptance of current ideas on the subject. The gradual reorganisation of land into consolidated holdings which occurred all over the kingdom during the next decades and into the nineteenth century can be attributed partly to these efforts, partly also to government legislation (notably that of 1769 and 1776) and to the provision from 1781 of significant government grants to help cover the costs of moving farm buildings out of the village communes.

II

In the flurry of cabinet decrees that followed Johann Friedrich Struensee's rise to power in the late summer of 1770, agrarian reform was largely submerged in a much wider programme. Plans for abolition of the *stavnsbånd* were shelved, perhaps because of their controversial nature. Only the question of peasant labour services was actively pursued, prompted by the responses to an initiative taken by the preceding administration. An ordinance of May 1769, without making any recommendations, had encouraged landowners formally to agree with their tenants on the specific amount of services due and to report these to the central administration. It became apparent that in some parts of Denmark, especially its eastern half, labour services due from a peasant farm could amount to as much as 250 days per year, including 60 with horses and equipment. The cabinet on 20 February 1771, in its characteristic peremptory manner, decreed an immediate reduction to 144 days, including 48 with team. This ordinance not only explicitly aimed to protect the peasantry against abuse of seigneurial rights, but also represented a major potential encroachment on the hitherto sacrosanct entitlement of landowners to make the most of their property. Enforcement thus proved difficult, and the ordinance was allowed to lapse as soon as Struensee fell; in a decree of 12 August 1773 the government explicitly withdrew from formal regulation of peasant–seigneur relations.

From September 1770 Struensee and his few associates produced an avalanche of reformist decrees inspired by German cameralist thinking and especially by the French Enlightenment: well-intentioned and often overdue, but with little coherence or overall aim. More seriously, much of the legislation was precipitous, curt or based on insufficient knowledge of Danish conditions. This, combined with Struensee's arrogant failure to understand or speak Danish, was in any circumstance bound to generate criticism and outright hostility, not least as the Danes were becoming increasingly aware of their national identity and critical of the number of Germans in high offices. For more than a year, however, Struensee's dominance over the royal couple was sufficient for his cabinet to act, after suspending the old privy council, without any consideration of court faction or public opinion.

There were many ideas of value in the cabinet orders issued

in the name of the king by the new team. Following Prussian precedent, a new centralised finance department was established headed by Struensee's brother, Carl August, to give clearer indication of policy and overall resources in areas hitherto covered by three separate offices. Court pensions and honours were cut, and efforts were made to restrict the scale of fees and perquisites which officials charged to supplement their salaries. Promotion within the bureaucracy was to be more on the basis of proven service rather than patronage, and although Struensee in no way can be described as anti-aristocratic – acquiring the high-ranking title of count for himself in July 1771 – the effect was to alienate the old service nobility and isolate the cabinet. Efforts were also made to liberalise trade laws, guilds and tariffs along physiocratic lines, but there was little indication of wider economic policies. Provisions for poor relief were extended, again entirely in accordance with prevailing European notions of good government. But unlike reformers in western and southern Europe, or indeed in the Habsburg Lands, Struensee paid little attention to the state church beyond extending marginal toleration and securing a relaxation of some of the strictures on morality created during more than two centuries of Lutheran orthodoxy. He did, however, fall entirely into line in tackling the judicial system: the hierarchy of courts especially in Copenhagen itself was simplified, the judiciary was completely separated from the executive branch of government, some prison reform was attempted, and the use of torture limited.

Struensee also adopted one of the quintessential aims of the European Enlightenment when on 4 September 1770 he issued a cabinet order stating that 'it is our absolute conviction that it is as damaging to the pursuit of truth as it is hindering in the discovery of inherited prejudices and misconceptions when honest patriots are deterred from writing freely according to their convictions or from attacking abuses and prejudices'. Censorship was accordingly abolished, but like other reforming statesmen Struensee soon discovered two basic truths: firstly, that freedom of speech within a system of absolute monarchy was liable to turn against its creator, and, secondly, that once emancipated the press and public opinion were not easily brought back under control. Pamphlets making the most imaginative use of rumours of Struensee's now long-standing affair with the Queen were no doubt hardly what the minister had had in mind, and so a decree of 7 October 1771 ordered that all publications must bear the name of the author or the publisher, so that they could be subject to the laws of libel and of *lèse-majesté* in the usual way.

The Struensee reforms thus have many of the hallmarks of European enlightened absolutism: well-intentioned, sometimes in their details meeting a real need, but also underscoring the incongruity of trying to impose selective liberalisation by autocratic means. Poor harvests in 1770 and 1771, after a period of prosperity, brought some policies into unjustified disrepute. But the actual implementation of change was singularly chaotic even by the standards of the time, and little effort was made to explain underlying motives in a comprehensible way. When we remember that outside the court there was a deliberately pre-served ignorance of the true mental condition of the King, and that Struensee caused popular consternation and anxiety for instance by prescribing a regime for the crown prince taken directly from Rousseau's *Emile*, it becomes easier to understand the exhilaration which greeted the palace coup of January 1772.

III

The circle of conservative nobles who returned to power under the aegis of the king's stepmother and her protégé Ove Høegh Guldberg was hardly on firmer constitutional ground than Struensee had been; indeed despite initial intentions to restore the old forms of government, and the abandonment of most of his structural changes, they gradually relapsed into a style of government by cabinet decree similar to that of Struensee. But Guldberg, as first minister, gained popularity by appealing both to Danish nationalism and to broad conservative bewilderment or resentment against the recent experiences of enlightened autocracy. Characteristic of the new regime were the measures to re-affirm stolid Lutheran orthodoxy, and the law of 1776 which would henceforth debar foreigners from state service.

On the surface, the regime did not appear entirely one-sided: A.P. Bernstorff soon gained a seat in the privy council, and until 1780 followed in his uncle's footsteps as the person in charge of foreign policy and of relations with Schleswig-Holstein. The law on nationality of 1776, however, put him in an awkward position, and he was in any case increasingly isolated in an administration which had explicitly turned away even from the benevolent paternalism of earlier aristocratic reform. The emphasis was now placed on a consolidation of the traditional foundations of the state, and initiative was mostly reserved for state-sponsored projects like the canal connecting Kiel with the Elbe, completed in 1784, and various commercial and colonial developments seeking, in the West and East Indies, to exploit Danish neutrality in the war of American independence.

As noted earlier, the re-organisation of arable land into consolidated enclosed holdings continued all over the kingdom from the 1760s into the Napoleonic period. Such change did not necessarily affect peasant–seigneur relations, and if anything the Guldberg ministry sought to strengthen seigneurial control in the face of mounting criticism and growing uncertainty in the countryside. A new militia ordinance of 1774 increased recruitment and confirmed the *stavnsbånd*; the responsibility of landowners for the payment of taxes by their tenants also remained in force. Even the sale of crown land, which in the 1760s had been organised in such a way as to offer opportunities for the tenants themselves, was after the fall of Struensee geared much more towards the alienation of whole estates as fully operative manors complete with labour-yielding tenants – partly on principle, partly out of a need to ensure immediate cash and tax revenues rather than uncertain stakes in peasant mortgages. Economic conditions for agriculture were satisfactory in the later 1770s, especially 1777–79, but appear to have deteriorated significantly from 1780 through to around the middle of that decade. Although some historians have cast doubt on the traditionally very gloomy assessment of peasant conditions in this period,[2] there is some consensus that mounting concern in enlightened circles as well as grievances amongst some groups of the rural population itself could not be restrained for ever.

IV

The palace coup of 1784, transferring formal power to the 16-year-old crown prince Frederik immediately after his delayed confirmation in a manner worthy of contemporary opera, was remarkable both for the absence of violence and for the scale of real change which it brought at the centre. The queen mother disappeared into retirement, while Guldberg himself was given a post as provinicial governor which he fulfilled with efficiency and loyalty. Christian VII, persuaded to sign the ordinance of

14 April 1784 which invalidated Gulberg's cabinet system, remained an uncomprehending spectator for the remaining 24 years of his reign. His son, by making himself a required co-signatory with the King of all acts of state, in effect took over the extensive powers ascribed to the crown according to Danish absolutism. The coup had been planned in advance by a group including the prince's tutor, Johan von Bülow, and the Norwegian Theodor Georg Schlanbusch, probably with advice from *Generalprokurør* Stampe himself; and, at a more discreet distance, it had the cautious approval of count A.P. Bernstorff, the finance minister count Ernst Schimmelmann, the able but fairly conservative councillor, count Joachim Otto Schack-Rathlou, and count Christian Ditlev Reventlow. Titles notwithstanding, the aristocratic team formed from this group was not cut in the conventional mould of the 1760s. Bernstorff, 49 years old and from now on effectively first minister until his death in 1797, was becoming more firmly convinced of the need for substantial state-initiated reform of rural conditions in order to improve both economic performance and social stability. More outspoken in this respect, however, was Reventlow (1748–1827), who like his aristocratic contemporaries had travelled through much of Europe for his education and had a little first-hand experience of English economic and agrarian conditions. After inheriting his father's estates on Lolland he had started experimenting with peasant life tenure, commutation of labour services, more complex rotation-patterns and scientific forestry.[3] He was now, in addition to his other offices, placed in charge of the Exchequer, the department with primary responsibility for agrarian conditions.

As we would expect from the European experience, there was no unanimity in government circles about the essential priorities of reform. Divisions within the privy council itself, as well as the social and political conditioning of nobility which Reventlow never shook off, may explain why no general initiatives were taken for two years. A new sale of crown lands in northern Zealand after 1784, however, did give the government a chance to demonstrate its aims. The commission placed in charge of the operation was instructed to ensure not only a rapid completion of the peasant enclosure already in progress there, but also the best possible conditions for all tenants and cottagers on the estates, ideally through the granting of ownership rights if that could be achieved without too much damage to crown income. In fact the best practicable solution turned out to be hereditary tenure, more acceptable to the peasants because it did not saddle them with a heavy mortgage. By the summer of 1788 Reventlow could present the first batch of deeds to the tenants at Frederiksborg in a well-publicised ceremony laid on partly in order to bring pressure on opponents of change in Copenhagen. In his speech, Reventlow appealed to God 'to make these peasants an example ... for their brethren to follow, and to make the formerly unappreciated Estate of the peasants into a hard-working, happy and upright people, from whose prosperity all other Estates will blossom, and on whose loyalty and courage the king can rely as fully as on the most secure defence'.[4] By 1820 the re-organisation on the crown estates there was complete, and the peasantry had had their labour and other dues converted into annual payments, which, because of favourable economic conjunctures, involved no loss for the Exchequer.

By European standards the policies tried so far in Denmark did not represent anything unusual. In the Habsburg Lands, Maria Theresa's adviser von Raab had in the 1770s implemented hereditary tenure on crown estates, out of similar motives, and during the same period she and Joseph II had become increasingly conscious of the need also to limit abuses of seigneurial authority, notably in the local courts and amongst the local officials who were supposed to deal with peasant complaints. Joseph's well-intentioned legislation in 1781 to encourage peasants to buy their own holdings, to abolish serfdom formally, and to limit seigneurial rights to inflict corporal punishment, were naturally different in orientation because of substantial dissimilarities in practical conditions, but may nevertheless be regarded as springing from recognisably similar kinds of thinking to those gaining some ground in Denmark. Far more drastic were the implications of the Habsburg *Robotpatenten* of the 1770s, of which the only Danish equivalent was Struensee's ill-fated 1771 ordinance on labour services. But whereas Joseph came to grief over the even more radical tax-reforms of 1786–89, attempted at a time when he was heavily committed elsewhere, the Danish reformers were able to sustain a more limited but more carefully designed series of legislative interventions which, uniquely in Europe at the time, achieved a workable compromise between seigneurial and peasant interests without destroying the social and political fabric of the state.

V

It was a relatively minor routine enquiry which set reform in motion. An old bone of contention in the rural community had been the degree to which seigneurs controlled, as *de facto* executors, the inheritance-system from one tenant to the next. Such a practice could easily be abused, and in 1783 a *herredsfoged* (district judge) in Jutland suggested to the Exchequer that it would be fairer to have an independent third party (appointed by the local court) act as executor and value-assessor when a tenant died, so that the interests of heirs could also be taken into reasonable account. Reventlow eventually passed the report on to two legal advisers for the crown, the new *Generalprokurør* Ole Bang, and *Kammeradvokat* Christian Colbiørnsen, with the additional request that they examine the crucial issue of proper valuation of tenancies also at the beginning of a new tenure, to prevent a peasant being held responsible for losses or deterioration which he had not in fact caused. Both lawyers emphasised the urgent need of clarification in this area; Colbiørnsen added that 'conditions of tenure must therefore be fixed by law, and this law must, since it will have to protect a lesser Estate against a more powerful, be given such force and clarity that the latter by their power or the former by their weakness will not be able to shake it or hinder its serious implementation'.[5]

Reventlow did not pass on the comments of Bang and Colbiørnsen for more than a year, but in July 1786 he submitted a report referring to this matter and calling for a wider assessment of many aspects of the relationship between cottagers, *stavnsbundne* tenants, freeholders, landowners and the state itself, where, he said, many abuses had crept in. This was included in the privy council's instruction to the *Store Landbokommission* (Great Agrarian Commission) established on 26 August: it was ostensibly to clarify rather than interfere, but was also asked to examine 'anything that might serve to improve the condition of the peasantry', provided the rights of landowners to use their property in accordance with the law were not affected. The Commission was to consist of sixteen members, including a number of prominent government officials and four outside representatives of the landowing interest; it was to be chaired by Reventlow, and significantly the secretarial task was allocated to Colbiørnsen.

Christian Colbiørnsen (1749–1814) was a Norwegian, son of an officer in Romerike and younger brother of a very able lawyer and legal historian. A near-contemporary of Reventlow, he was by background and temperament far more emotional and radical in his outlook on contemporary social conditions. His ability and flair for argument had already been recognised in government circles, possibly making him a political target for conservative critics already in 1785. He was so enthusiastic an exponent of a strong interventionist state – of enlightened absolutism – that he never favoured any kind of devolution even for his own homeland, despite growing tensions there during the abortive Lofthuus peasant rebellion of 1786–87. His role within the new Commission has not been examined as thoroughly as it deserves,[6] but appears frequently to have been of critical importance, partly because of the influence he exerted through his secretarial role, partly because of his acute, sometimes belligerently caustic, interventions within the Commission and also in the wider public discussion and polemics that he amongst others deliberately encouraged. Reventlow had originally intended the proceedings of the Commission to be published, no doubt in the knowledge that particularly the Copenhagen intelligentsia (with little to lose) would help to put conservative seigneurial interests on the defensive. That this was not carried through to the end was partly out of recognition that provocation was undesirable in an already delicate situation. Colbiørnsen, however, soon became a controversial and outspoken representative of what was increasingly regarded as anti-aristocratic reform, gaining the deep hostility of many landowners, and enormous popularity in Copenhagen and amongst the rural population itself.

It is not the aim of this Chapter to discuss in detail the actual legislation proposed by the Commission and approved by the crown, but an outline is indispensable. Within a year, on 8 August 1787, the Commission's recommendations on the legal relationship between tenants and landowners were promulgated: each tenancy was now to be valued by a third party at the beginning and end of the period of tenure, and comparisons or damages settled accordingly. Security of tenure was confirmed, and evictions for default banned except after prior court hearing. In addition, the wooden horse and other forms of corporal punishment were no longer to be applied to farm tenants – though silence on the matter suggested that cottagers were not entitled to such basic protection. More generally, the peasant's obligation to serve his seigneur obediently was reiterated, and resistance, for example to seigneurial initiative towards enclosures, was prohibited.

On 20 June 1788 came the crucial ordinance irrevocably abolishing the *stavnsbånd* outright for those aged under 14 or over 36, with a limit of 12 years for the rest (so that all peasants would be emancipated at the latest in 1800). Opposition from the military may partly have been met in that conscription was fully maintained, but as a matter solely of state interest. The levy of troops was to be based on new district allocations according to a census of 1787 and would offer no scope for seigneurial interference. It was this legislation which provoked the unprecedented resignation of two members of the four-man privy council, Schack-Rathlou and Rosenkrantz, leaving only the crown prince and Bernstorff until replacements were found. Seigneurial resistance may also account for the fact that, after 58 working sessions over two years, the Commission was now placed in abeyance for a while, and Reventlow himself, despite his manifest political skills, to some extent side-tracked.

After a two-year gap came further ordinances from 1790 reiterating the prohibition against the enclosure of peasant holdings within a demesne: if no tenant could be found, the holding would have to be parcelled out amongst neighbours or to cottagers. Only then did the reformers tackle the most resented and controversial aspect of peasant–seigneur relations in Denmark: labour services. In 1791 a series of ordinances required all labour services to be made the object of voluntary but itemised contractual agreements between tenants and landowners, if necessary subject to crown arbitration if no agreement could be reached. Significantly, labour services were to be fixed at the current levels of 1790, which were in fact the highest ever; again, apparently at Reventlow's own insistence, the peasantry were enjoined to obedience. Over the next few years further elaboration proved necessary in respect of labour services and their possible commutation (1795 and 1799), just as further decrees were passed regarding the far less controversial enclosure; in essence, however, the basis had been created for positive protection of the tenant population on privately owned estates throughout the kingdom, and hence for the perpetuation of the type of relatively small-scale Danish farming which has survived until recent times.

This was by no means the sum of reforming initiatives that
sprang from the central administration after 1784. Reventlow
himself was prominent in other fields, including the establishment
of a credit bank in 1786 intended (despite inadequate resources)
to help peasants buy their own freehold and thereby get rid of
labour obligations. Apart from his continuing involvement in
overseas trade, he also remained interested in agrarian technical
improvements, land development and forestry, and was a driving
force behind effective initiatives to establish a better network of
roads. Colbiørnsen was promoted to the prestigious post of
Generalprokurør in 1788, but also became so deeply involved in
the Chancery's work on the administrative and legal implications
of the new legislation that he spent less time with the Commission
after 1790. The outbreak of the French Revolution did not bring
reforms to a halt: in addition to the regulation of labour services
by the Commission itself, work on peasant and manorial enclosure
continued, and options for the commutation of church tithes
were considered. An ordinance of 1792 gave notice that the slave
trade in all Danish possessions would be phased out over the
next ten years. In 1803 a new poor law was implemented to
replace those of 1708. A commission on schools was established
in 1789 which, in view of the atmosphere in Europe as a whole,
did not pursue more radical suggestions for secularised education,
but by 1814 a standard primary system in towns and country
was becoming effective. Not surprisingly the relative freedom of
expression of the 1780s was gradually lost, but to the law courts
rather than to the government: even the restrictive law of 1799
on censorship did not attempt quite the controls of pre-Struensee
days.

The major achievement, however, remained the agrarian
reforms, and we must attempt to explain why the Danish
government succeeded in its aims in this particular field, where
most other enlightened ministries failed to a greater or lesser
extent. One reason lies in the fact that while the peasantry were
helped, cottagers and smallholders decidedly were not. The
origins of the poor law of 1803 can in fact be traced back to a
committee of 1787, established in recognition of the fact that the
enclosure movement and other reforms was bound to damage
severely the interests of smallholders and cottagers. There were
more of them in Denmark than there were peasants on holdings

big enough for even meagre self-sufficiency: in eastern Denmark perhaps twice as many. They lost their common grazing rights and the protection of the village community, and were allowed to become the low-paid labour-supply not just for peasants but also for landowners. The government did nothing to protect them, beyond recommending the allocation of small plots of land for family supplies. By ordinance of 1807 seigneurs were allowed to impose unlimited labour services on any cottager on their estate, combined at their discretion with unregulated disciplinary corporal punishment. It seems as if the smallholders had to pay a compensatory price to landowners for what the peasants proper had gained.

Of undoubtedly greater symbolic and practical significance at the time, however, was the fact that the reforms were very much a compromise: seigneurs were by no means the obvious losers in real terms. Colbiørnsen was no Jacobin, and never encouraged subversion. Reventlow was and remained an aristocratic land-owner: as we have seen he was as concerned as other nobles about the growing unrest by 1789, when hopes of major improvement in rural conditions led many peasants if not to revolt then at least to refuse to fulfil their ever-increasing labour obligations. It is outside the scope of this article to consider peasant action,[7] but the recurrent insistence by the government that peasants remain obedient to their seigneur was a direct response. Most important of all, however, was the fact that the labour service ordinances of 1791 accepted seigneurial demands virtually without qualifica-tion: services were pinned at their highest level ever, and the Commission was quite explicit that if it came to arbitration it would not offer any leniency. If the abolition of the *stavnsbånd* and the regulation of peasant–seigneurial relations in law had imposed some limitations on the range of the powers exercised by landlords, they did not lose at all in terms of what they could legally extract from tenants. In 1792, in fact, the government went even further by permitting landowners to renegotiate the hitherto fixed rents (*landgilde*) for holdings that had been enclosed, thereby enabling landowners to start passing on to their tenants (admittedly on specific terms) the costs of enclosure itself. Landowners had also as far back as in 1784 been freed from responsibility for the taxes due from holdings sold off from an estate as freehold. Many Danish landowners, as mentioned, had long since found the traditional agrarian structures rather restrictive, and could genuinely welcome at least some change, especially at a time when there was a boom in Danish agricultural prices and productivity: seigneurs as well as peasants could reap the benefits. So indeed did the government itself in terms of increased tax revenues, an incentive to reform that may deserve more consideration than has hitherto been the case.[8]

The safeguarding of seigneurial interests, then, goes a long way towards explaining why the reforms survived (and here the contrast with the implications of Joseph II's late tax scheme are instructive). But despite the moderate, in some respects limited, nature of the changes, conservative landowning interests did not capitulate without a struggle. Their anxiety had been considerable ever since it had become apparent that the instruction to the 1786 Commission could be interpreted quite widely; the enthusiastic expectations of the Copenhagen intelligentsia (expressed notably in the press and in a flood of polemic pamphlets) and even of the rural population seemed to bear out their worst fears. Within the Commission itself Colbiørnsen's insistence on positive discriminatory state intervention in favour of the peasantry persuaded some members to argue for outright *laissez-faire* in peasant–seigneur relations: Adam Smith's work had become widely known in Denmark, and seemed to offer an even more radical solution than the policies proposed by Reventlow and Colbiørnsen. The able and experienced landowner Morten Qvistgaard, who attempted to resign from the Commission in December 1787, adopted this stand, as did Reventlow's own brother and even some of the Copenhagen *literati*. Other tactics were also adopted in the face of what some regarded as a threat of pervasive bureaucratic interference in seigneurial affairs. The reaction of two privy councillors to the ordinance of 1788 has already been noted, and conservative spokesmen tried to appeal to the Danish nationalism of the Guldberg days against what they could once more (with rather less justification) describe as alien ideas. Polemics, however, were liable to rebound. Qvistgaard had already published his opinions on the abolition of the *stavnsbånd*, only to find Colbiørnsen in print showing that Qvistgaard's text was plagiarised. For good measure, Colbiørnsen added that, according to Qvistgaard, 'the Danish peasant is to be tied to his birthplace until the day of his death: after that day the honourable councillor makes no additional claims on him – he has the privilege of letting himself be buried outside the estate, provided he can fully prove that he is thoroughly dead. . . .'

Appeals to the government had taken various forms, but

by 1790 anticipation of 'French contagion' and concern over crumbling discipline erupted into a major show of resistance which put the fundamental stand of the government to the test. A petition carrying the signatures of 103 landowners was presented to the crown prince in August 1790 by a delegation headed by Tønne Lüttichau. In anticipation of what was to come, the text asserted that the reforms to date were in breach of Danske Lov (the law code of 1683) and of the guarantees of property rights that had been fundamental to Danish absolutism. It was this, and the fear of further unrest amongst the peasantry, that made them ask for revision of the laws (except that terminating the *stavnsbånd*, to which they had resigned themselves). The crown prince, however, rejected any alteration in what had already been decreed, and allowed Colbiørnsen to prepare a full rebuttal. In the published version, Colbiørnsen turned from detailed arguments to a bitter accusation of treason, arguing that 'the king's government is being censured, the laws are described as unjust; the council of state which has considered them, the commissions which have drafted them, the officials whose duty it was to make proposals, are being accused of dishonesty and ignorance. . . . Is it the government of Nero, Caligula or Tiberius, rather than that of the seventh Christian, that is described here. . . ?'[9] The threat of treason charges were not lost on Lüttichau, especially when it was discovered that some of the 103 signatures were forgeries. In a storm of abusive pamphlets and public debate Lüttichau and Colbiørnsen ended up in April 1791 in a libel suit before the Supreme Court; in the circumstances, and with Colbiørnsen still holding the most prominent legal office in the state, Lüttichau was perhaps fortunate to escape with a heavy fine and exile.[10]

The petition movement of 1790 and its sequel mixed the truly conservative landowning cause and the case for total *laissez-faire* in such an unconvincing way that both were destroyed. Yet as noted, the actual material interests of landowners were largely safeguarded in the second stage of legislation in the 1790s, and in that respect the petitioners could feel that they had achieved something. The Danish crown, however, clearly had intervened in an unprecedented way in peasant–seigneur relations, and did subject landowning property rights to permanent bureaucratic control. That this could be done successfully is in large measure to the credit of the very able circle of ministers around the crown prince – and, if one remembers Joseph II again, thanks to

Bernstorff's skill in keeping Denmark out of war (except very briefly in 1788). The outcome also gives credence to the view that enlightened absolutism, suitably balanced and with public support, could achieve moderate change even in highly sensitive areas without provoking revolution.

10. Frederick the Great and Enlightened Absolutism

T. C. W. BLANNING

I

DESPITE periodic attempts to bury it, the issue of Frederick the Great's relationship with the Enlightenment keeps thrusting its way back to the surface. Over a decade after Betty Behrens suggested that enlightened absolutism was a concept best discarded, Frederick's most recent and most distinguished modern biographer, Theodor Schieder, identified the Enlightenment as crucial to a proper understanding of Frederick, his policies and his system.[1] Further acquaintance with Prussia in the eighteenth century appears to have led Miss Behrens herself to change her mind, for she has now written, in a recent study, that 'the ideology of the Prussian Enlightenment . . . was the ideology of the King himself'.[2]

It will be argued later in this chapter that the Enlightenment did have a discernible influence on Frederick's policies, and that this was indeed of great importance, especially for the subsequent history of Prussia and Germany. Enlightened absolutism will continue to confound attempts to deny its reality, because contemporaries believed that it existed and also for the equally good reason that the influence of the Enlightenment must be apparent to all eyes unrestricted by the blinkers of reductionist ideology. And within the pantheon of enlightened absolutists, pride of place will always go to Frederick the Great. It must be conceded, however, that potent criticisms have been levelled against the concept of 'enlightened absolutism', even in the Prussian context, and, moreover, against both parts of the composite: Frederick's regime, it has been argued, was neither enlightened nor absolutist.

Most persuasive has been the line of attack which concentrates on the continuity in Prussian policies and institutions. Everything done by Frederick, it is suggested, had been done already by his predecessors. His attachment to religious toleration – much publicised by himself and mainly responsible for his enlightened reputation among contemporaries – had been part and parcel of the Hohenzollern tradition for more than a century by the time Frederick came to the throne in 1740. It dated back at least to 1613 and to the Elector Johann Sigismund's conversion to Calvinism, a move which put the ruling house at odds with its predominantly Lutheran population and made toleration an essential condition for effective government. The need to attract immigrants to thinly populated and unalluring Brandenburg, especially urgent in the aftermath of the devastating Thirty Years War, provided a further motive for each and every ruler: Frederick's great-grandfather had welcomed Huguenots from France, his grandfather had welcomed Pietists from Saxony and his father had welcomed Protestants from Salzburg. Yet not even the most elastic definition of that pliable word 'enlightened' could allow its application to Frederick William the Great Elector, Frederick I or Frederick William I.

Less striking but also apparent was the continuity in judicial reform, the other major 'enlightened' achievement of Frederick. Here the continuity was personified by Samuel Freiherr von Cocceji (1679–1755), usually associated with the reforms of Frederick but very much a man of the previous reign. Like Frederick William I, Cocceji believed that all law had its origin in the will of God, not a notion usually associated with the Enlightenment.[3] Also a staunch social conservative, he maintained the aristocratic essence of Prussia's judicial system and even led noble opposition to a mild measure of agrarian reform proposed by Frederick in 1751.[4]

Indeed, it can be argued persuasively that differences between Frederick and his father were of style rather than substance. At first sight, there could hardly be a more obtrusive contrast than that between the flute-playing, philosophical aesthete and the brutal, psychopathic *dévot*. On one and the same day, Frederick William I had a senior official found guilty of dereliction of duty garrotted in front of his office, with all personnel obliged to watch, and then received the Salzburg refugees, with the entire court obliged to get down on their knees to pray with them.[5] Yet although suffering more than most at the hands of his tyrannical father (whose dementia may have been due to the fashionable royal disease of porphyria), Frederick recognised that it was Frederick William I who was the true creator of Prussia's great-power status. In his *History of My Own Times* he paid to his father tribute which was as sincere as it was generous; for example: 'The fame to which the late king aspired, a fame more just than that of conquerors, was to render his country happy; to discipline his army; and to administer his finances with the wisest order, and oeconomy. War he avoided, that he might not be disturbed in the pursuit of plans so excellent. By these means, he travelled silently on towards grandeur, without awakening the envy of monarchs.'[6]

It is some measure of Frederick's subordination of his emotions to reason that he was able to see past the treatment he had suffered to the true nature of his father's achievement. His admiration he advertised most eloquently by imitation. The all-important administrative system he left fundamentally unchanged. Frederick William I had achieved so much in this all-important sphere that there was little left for Frederick to do (an advantage which distinguished him sharply from – say – Catherine the Great or Joseph II). The army was more than doubled in size, but the recruiting, training, discipline, weaponry and tactics of its most important arm – the infantry – remained those of Frederick William I.

Similar consistency was to be found elsewhere. In his promotion of trade and industry, Frederick was an arch-conservative, retaining and developing all the classic features of seventeenth-century mercantilism: treasure-hoarding, protectionism, monopolies, *dirigisme* and autarky. The only changes introduced to the fiscal system were the increase in the number of commodities liable to the excise and the privatisation of its collection (by the detested French-run *Régie*). He was notoriously conservative in his attitude to popular education, insisting that peasants should be taught enough to make them useful – but not so much as to give them ideas above their station. The school regulations published during his reign betray a strong line of continuity dating back a century and more, as do the equally old initiatives to found village schools.[7] The attempts that the centre did make to improve primary education almost invariably foundered on resistance in the localities.[8] Until deep into the next century, private initiatives were more important than those of the state.[9] The obligation for all children to attend school, which was to make Prussia the most literate of all the great powers by the next century, was first imposed not by Frederick but by his father.[10] It was also Frederick William I who first expressed the wish to abolish serfdom, in 1719.[11]

In short, all the policies pursued by Frederick turn out to have been initiated by his predecessors, none of whom – by any stretch of the imagination – can be deemed 'enlightened'. Moreover, in two important respects, he was *more* conservative, not to say reactionary. The first was his attitude to the nobility. Frederick William I had still been anxious about the political ambitions of the Junkers, preferring docile commoners for many senior positions in the civil service. He had also launched a vigorous and successful campaign to expand his land-holdings at the expense of noble land-owners suspected of misappropriating portions of the royal domain. Frederick put an immediate stop to that; in one of his first instructions to the General Directory, he ordered that:

> The officials shall be forbidden on pain of death to harass the nobility and to resurrect old claims against them. All a noble landowner has to do is to prove that he was actually in possession of a piece of land before 1740. And if any disputes arise between noble landowners and officials of the royal domains, then not only shall the General Directory make sure that justice is done, but it shall also make sure that it is I rather than the nobles who suffers any injustice. For what for me is a small loss can be a great gain for the nobles, whose sons it is who defend this country and whose quality is such that they deserve to be protected in every way.[12]

Frederick was loyal to this programme, both in principle and in practice, throughout his reign. In his 'Political Testament' of 1752, which contained his innermost thoughts on government and were not written with an eye to publication, he elevated support for the nobility to the status of a general political maxim: 'a sovereign should regard it as his duty to protect the nobility, who form the finest jewel in his crown and the lustre of his army.' Other social groups might be wealthier, but none surpassed the nobles in valour or loyalty.[13] In practice, this meant an end to the special favour shown to commoners by his father. Even if the aristocratic composition of Frederick's civil service has been exaggerated,[14] he did favour nobles for senior positions, both in the General Directory at the centre and in the 'War and Domains Chambers' in the provinces, and he did restore to the Junkers the right to nominate candidates for the crucial position of Landrat.[15] It goes without saying that the great majority of army officers were nobles – nine out of ten by the end of Frederick's reign.[16]

Not only did Frederick confirm the public institutions of the Prussian state as instruments of outdoor relief for his nobles, he also did all he could to preserve their private wealth. With the price of grain and the value of land rising sharply, if erratically, it was a time of great opportunity, but also high risks, for the Junkers. For the prodigal, unenterprising or unlucky, danger threatened in the shape of the increasingly numerous, wealthy and acquisitive commoners, who wanted nothing more than a Junker estate – a '*Rittergut*' – with which to acquire social prestige. Nothing illustrates Frederick's social conservatism better than his letter to Cocceji of 29 December 1750, in which he noted with alarm that estates belonging to old noble families were passing into the hands of *nouveaux riches*. He ordered that in future no noble estates could be alienated from the class without his express permission.[17] The prohibition was renewed in 1762.[18] (This attempt to turn back the economic tide was a predictable failure: by the end of the century, more than ten per cent of Junker estates had passed into the hands of commoners.[19]) With the same object in view, Frederick encouraged noble families to entail their landed property, thus making it invulnerable to genetic accident – but with an equal lack of success.[20]

More concrete was the assistance given during the dark days following the Seven Years War, after Junker families had been decimated by the exceptionally high casualty rate suffered by the officer corps and after Junker estates had been devastated by the French, Swedish, Austrian and Russian invaders. Frederick intervened to impose a moratorium on bankruptcy proceedings for two years, to make cash grants for the repair and restocking of farms, and to establish rural credit institutions (*Landschaften*) for the supply of cheap and easy mortgages.[21]

Closely allied to Frederick's social conservatism was his militarism, understood here to mean the paramount status assigned to military needs and military values. His father had been no mean militarist himself, characteristically describing the bloody battle of Malplaquet as the happiest day of his life,[22] but it was Frederick who gave Prussia the reputation of being 'the most militaristic state in Europe'.[23] By the end of his reign it was possible for one well-placed observer to comment that 'the Prussian monarchy is not a country which has an army, but an army which has a country, in which, as it were, it is just billeted'.[24] There is plenty of statistical information to support this impression; for example: in the course of the eighteenth century the proportion of the Prussian population in the armed forces rose from one-twentyfifth to one-sixth; even after the Seven Years War military expenditure accounted for 70 per cent of the budget; by 1786 Prussia was the thirteenth largest state in Europe in terms of population, the tenth in terms of area – but the fourth (or perhaps even the third) in terms of the size of its army; soldiers and their dependants accounted for a quarter of Berlin's inhabitants, 20 per cent of those of Halle, Brandenburg, Stettin and Frankfurt-an-der-Oder and 10 per cent of those of Königsberg and Breslau.[25]

Frederick knew full well that he was a militarist – and was proud of it. In his Political Testament of 1752 he wrote: 'I have been brought up in the army since I was a child. My cradle was surrounded by weapons. I have served in all ranks from captain upwards.'[26] In fact he had been given his first command at the tender age of six, when a company of 130 cadets had been formed for him to drill – an exercise which, like everything else ordered by Frederick William I, was to be taken seriously.[27] Little Frederick loathed most of his education, most notably the large doses of Pietist Christianity forced down his throat, but he appreciated his juvenile military exercises to the extent of recommending them for his own successor: 'it is of the greatest importance that he should be given an inclination for military matters. To this end, he must be told by everyone at every opportunity that any man of good breeding who is not a soldier is a miserable wretch.'[28]

Frederick took great pains to set a personal example of devotion to military matters. He was never seen out of military uniform, he lavished attention on the army, he demonstratively shared all the privations suffered by his subordinates when on campaign, he made his brothers serve in the army (and broke his brother August Wilhelm with sickening brutality when he failed to come up to expectations) and made it clear that he esteemed military values above all others. So he decreed that an ensign who had served on campaign outranked a royal councillor.[29] Virtually the only way to the top in Frederick's Prussia was through military service. As he told the minister for Silesia, Ernst von Schlabren-dorff, in 1763:

> Let me make it plain once and for all that I will not sell titles and still less noble estates for money, to the debasement of the nobility. Noble status can only be gained by the sword, by bravery and by other outstanding behaviour and services. I will tolerate as vassals only those who are at all times capable of rendering me useful service in the army, and those who because of exceptionally good conduct and exceptional service I choose to raise into the estate of the nobility.[30]

Military pretensions unsupported by military success makes only for derision – one thinks of poor Joseph II, for example – but Frederick's undeniably charismatic personality and undeniably heroic feats in the face of apparently impossible odds added the vital ingredient of glamour to make militarism a potent, popular and enduring force. It was not just the Junker officers who were seduced by the siren call of martial glory. A particularly acute French observer, the comte de Guibert, wrote after a visit to Prussia a year after Frederick's death that: 'the ordinary people in Prussia, even among the lowest classes, are imbued with the military spirit, speak with respect of their army, know the names of their generals, refer to their feats of arms and the times they have distinguished themselves'.[31] Recalling his childhood in Berlin, the romantic poet Ludwig Tieck wrote in his memoirs:

> The King appeared at military parades and reviews as the great warlord, who had defied successfully a coalition of all of the rest of Europe, and at the head of his troops, who had won so many battles. When there were military exercises or manoeuvres outside one of Berlin's city-gates, perhaps the Hallesche or the Prenzlauer, then the citizens of Berlin streamed out in their hordes to watch. My father [a master-carpenter] also used to take his children out to these popular festivals. Among the pressing crowds of people, the rush of artillery-trains and the marching soldiers, we were prepared to put up with the dust and the heat for hours on end, just to catch sight of our old Fritz surrounded by his dazzling retinue of celebrated generals.[32]

But the militarisation of Prussia went beyond the generation of popular enthusiasm; it entered into the very bones of the state and society. This was social militarisation, the remoulding of society to fit military needs, so that the military system and the social system became identified and every part of society had a dual civilian/military role.[33] The Prussian noble landowner was also an officer, so that the protection afforded to the former was essentially the protection of the latter (*'Adelsschutz'* was *'Offiziersschutz'*).[34] The Prussian peasant was also a soldier, so that the protection afforded to the former was essentially the protection of the latter (*'Bauernschutz'* was *'Soldatenschutz'*).[35] Nor was the urban population free from the all-pervasive influence of social militarisation. If the townspeople were mostly exempt from conscription, their opportunities for material, social and cultural advance were also artificially constricted by the paramount demands of the military.[36]

The same obsession with military might which led Frederick to inflict over-exertion on Prussia's social system also led to narrow limits being fixed for the expression of opinion. Philosophical and theological issues could be discussed with relative freedom, but anything impinging on politics was strictly taboo. At the beginning of his reign, Frederick had ordered that the Berlin press should be granted 'an unlimited freedom', telling a sceptical official that 'if they are to be interesting, newspapers should not be obstructed'.[37] It was of little significance that that order was simply not obeyed, for it was not long before Frederick himself changed his mind. By 1751 Lessing could report from Berlin to his father that the newspapers there were sterile and tedious, on account of 'the strict censorship'.[38]

It was also Lessing, the greatest literary figure of the German Enlightenment, who delivered the definitive indictment of Frederick's censorship. To understand fully this passage, part of which is often quoted, it is necessary to know the context: the poet Klopstock had drawn up a plan for an academy at Vienna, to promote the arts and sciences, and had submitted it to the Austrian court. If it had been realised, Lessing would have been entrusted with the supervision of the Viennese theatres. In a letter to Lessing in July 1769, Friedrich Nicolai – Berlin bookseller, prolific author and warm admirer of Frederick – had poured cold water on the plan, pointing out that a work by Moses Mendelssohn, like Nicolai a prominent member of the Berlin Enlightenment, had just been confiscated by the Austrian censors. Lessing replied:

Vienna may be what it is, but there is more to be hoped for
there than in your Frenchified Berlin. If Mendelssohn's book
has been confiscated in Vienna, then it was only because it
was published in Berlin and it could not be conceived that
anyone in Berlin would write in favour of the immortality of
the soul. And don't talk to me about your Berlin freedom to
think and write. It's just the freedom to market as many insults
about religion as one likes. An honest man should be ashamed
to avail himself of it. Just let someone in Berlin try to write
about other matters as freely as Sonnenfels has done in Vienna.
Let him try to tell the truth about the rabble of courtiers as
Sonnenfels has done. Let him try to make a stand for the rights
of the subject and against exploitation and despotism, as is
being done in Denmark and in France – then you will soon
see which country is the most slavish in Europe.[39]

As if these criticisms of the enlightened status of Frederick's
Prussia were not enough, there is one more which needs to be
outlined, of an even more fundamental kind. Marxist historians,
in particular, have argued that the concept of enlightened
absolutism is flawed by a contradiction between the social
character of its two component parts. Absolutism was a political
system which was essentially aristocratic, for it marked the
transition from 'the concealed dictatorship of the nobility to the
open dictatorship of a representative of the nobility in the interests
of the whole noble class'.[40] The Enlightenment, on the other
hand, was the ideology of the emergent bourgeoisie, whose
historical mission it was to destroy absolutism and the feudal
order which underpinned it.[41] So what looks like enlightened
absolutism was an attempt to resolve the growing tension between
feudalism and capitalism, an attempt which was doomed to
failure because a system which represented the interests of the
nobility could not adopt an ideology belonging to a class which
was necessarily hostile.[42]

The class character of old regime Prussia, it is argued, is
revealed as soon as one scratches below the surface of Frederick's
apparent autocracy. The socially neutral absolutist state, standing above class interests, was a myth: everything Frederick did
was in the long- or short-term interest of the Junkers. As soon
as he tried to do anything of which they disapproved, the
limits of his power were exposed immediately. The Estates, the
provincial organisations of the nobles, were still powerful and

still able to ensure that the arm of the state reached only as far as they wished.[43] A particularly good illustration of the true location of power in Prussia was supplied by Frederick's attempt to abolish serfdom in Pomerania after the Seven Years War. After the Junkers there had rejected his plans emphatically, using the cogent argument that it was only serfdom which allowed them to serve in the army in such numbers, Frederick backed down.[44] At the centre, the reign saw a constant struggle between royal autocracy and the noble-dominated bureaucracy for supremacy. Frederick may have delayed the latter's victory – but he could not prevent it.[45]

II

In the face of this composite image of a mercantilist, aristocratic, conservative, militarist despotism, it is difficult to see how the Enlightenment might gain admission to Frederick's Prussia. Yet it does belong there, even if entry can be gained only at the cost of some important qualifications. In the first place, it is clear that Frederick was both familiar with and deeply attached to the major thinkers of the Enlightenment. During his exceptionally difficult adolescence he took flight from paternal brutality to the urbane and sophisticated world of progressive French and English culture, assembling a clandestine library in a back room of Ambrosius Haude's bookshop of more than 3,000 volumes, including the major works of Descartes, Bayle, Locke and Voltaire.[46] He had rejected revealed religion by the time he was 13, was signing himself '*le philosophe*' at 16 and joined a masonic lodge in 1738.[47] He was also thoroughly acquainted with the work of the most influential German enlightened philosopher, Christian Wolff, whose work he had translated into French, so as to be able to understand it better.[48] Just one week before he became king, Frederick wrote to Wolff to thank him for a copy of his latest work on natural law:

Every man who thinks and loves truth must take an interest in your book; every man of integrity and every good citizen must regard it as a jewel, which your generosity has given to the world and which your acumen has discovered. I am all the more moved by it because you have dedicated it to me. Philosophers should be the teachers of the world and the teachers of princes. They must think logically and we must act logically. They must teach the world by their powers of judgment, we must teach the world by our example. They must discover, and we must translate their discoveries into practice. I have been reading and studying your works for a long time now and am convinced that all who read them must esteem their author. That can be denied to you by no one.[49]

This early and thorough immersion in the Enlightenment left Frederick dyed-in-the-wool. Having absorbed certain basic tenets about the world in general and politics in particular, he developed little. Under the influence of Voltaire, he abandoned the elaborate metaphysics of the kind favoured by Wolff but otherwise he remained loyal to the culture of his youth. He also took philosophy out of the study and on to a wider stage, proving to be certainly the most celebrated and probably the most effective populariser of the Enlightenment in Germany, if not Europe. Every educated European knew about Frederick's well-publicised friendship with Voltaire, who corresponded regularly and came to stay on several occasions, most notably from 1750 to 1753.[50] Although that particular sojourn ended in bitter mutual recrimination, the close relationship between the most famous writer and the most charismatic ruler of the century could not help but propagate the image of *le roi philosophe*.

Frederick encouraged the association by his prolific writings on the theory and practice of politics. For a German prince to be able to write coherent prose was remarkable in itself; when the prose proved to be lively, witty and incisive, the impact was sensational. While Frederick's treatises may have lacked originality and profundity, they crackled with an enviable combination of intelligence and malicious humour. Whether it was the *Anti-Machiavel* of 1739 or the *Oeuvres du philosophe de Sanssouci* of 1750, or the *Essai sur les formes de gouvernement et sur les devoirs des souverains* of 1777, or even the notorious *De la littérature allemande* of 1780, Frederick's publications kept him continuously in the front-rank of Europe's intelligentsia.

He had won their attention and approval at the very start of his reign by his demonstrative recall to the University of Halle of Christian Wolff, who had been banished from Prussia by Frederick William I in 1723 after accusations of impiety. In Frederick's own self-congratulatory but memorable phrase, the triumphant return of the most important philosopher of the

German Enlightenment to the scene of his former persecution represented a 'conquest in the land of truth'.[51] His own repeated and well-publicised use of the word '*philosophe*' to describe himself added to his reputation as the most enlightened ruler of his age. So did his virtual refoundation of the Berlin Academy in the early 1740s and the attraction to his capital of some of the leading figures of the French Enlightenment – among others, Voltaire, Maupertuis, d'Argens, Raynal, La Mettrie, d'Alembert and Mirabeau.[52] The visitors found much to dislike in its austere, rigorous, militarised atmosphere. Voltaire spoke for them all when he remarked to d'Alembert: 'The king has greatly embellished Sparta, but he has transported Athens only into his study'.[53] Yet even the most critical could appreciate that, for all his faults, Frederick had set an example of enlightened kingship which had effected a sea-change in the way in which all Germans were governed:

Today, if one excludes a very small number of principalities subjected to imbecilic tyrants, one can discuss in Germany, at least in theory, every question relating to theology, philosophy, economics and politics; books for which one would have been burned before the reign of the propagator of the Enlightenment, are now printed and sold publicly. Profound contempt is now the fate of anyone wishing to use force to repress or punish freedom of thought. The princes and the men of letters now restrain each other: and if that is not the best state of affairs, it is at least a thousand times more preferable than that which lasted for centuries. There we have an immeasurable advantage which the rest of Europe, as well as Germany, has reaped from the example set by Frederick.[54]

Inside Prussia, Frederick's commitment to the Enlightenment was revealed most clearly in his political thought, which was one area in which 1740 clearly did mark a watershed in Prussian history. There could hardly have been a sharper contrast between father and son. To describe Frederick William I's utterances on the subject as 'political *thought*' is to abuse the language, as his political testament of 1722 demonstrates. A grotesque, semi-literate but oddly impressive farrago of piety, banality and common sense, it is as far removed from the cool secular rationalism of Frederick as it is possible to imagine. Characteristically, Frederick William began with a prayer: he had always placed his trust in God, certain of salvation, and repented of his sins. His successor must take no mistresses (or rather whores as they should properly be termed); he must lead a godly life and set an example to country and army; he must not indulge in excessive eating or drinking; he must tolerate no theatre, opera, ballet, masquerades or public dancing in his dominions, for they were all the work of the Devil; the house of Hohenzollern had always shunned such things – and that was why God had smiled on it; he must fear God and never start an unjust war, but must also stand up for Hohenzollern rights; and so on, and so forth.[55]

Through these crude, misspelt, ungrammatical effusions there beats the imperious pulse of a man who knew what he wanted and how to command it. The following observations on the nature of political obligation give a fair indication of the rudimentary nature of Frederick William's political thought: 'So long as God gives me breath, I shall assert my rule like a despot', 'My subjects must dance to my tune or the Devil take me: I'll treat them as rebels and have them hanged and roasted like the Tsar does' and – most eloquent of all – 'After all, we are lord and king and can do what we like'.[56] In the event, Frederick William's behaviour was a curious mixture of over-exertion at home and cautious passivity abroad. So he forged the weapons his son was to use but never dared to take them campaigning himself.[57] His theoretical omnipotence was limited by three powerful constraints: respect for the Holy Roman Empire, concern for the Hohenzollern family and, above all, fear of his terrible Calvinist God.

Imperial law, dynasticism and Christianity meant nothing to his son, who despised all three. Where Frederick William's thinking was prescriptive, particularist and pious, Frederick's was rational, universal and secular. To legitimate his authority, he postulated a social contract, by which the inhabitants of a state of nature delegated to a sovereign sufficient authority to maintain external security and internal order. By the middle of the eighteenth century, such a concept was hardly original, but its advocacy by a ruler of a backward and hitherto minor state certainly was. Frederick's clearest exposition was given in the course of a discussion of religious toleration in his *Essay on the forms of government and the duties of sovereigns* of 1777:

One can compel by force some poor wretch to utter a certain form of words, yet he will deny to it his inner consent; thus the persecutor has gained nothing. But if one goes back to the origins of society, it is completely clear that the sovereign has no right to dictate the way in which the citizens will think. Would not one have to be demented to suppose that men said to one of their number: we are raising you above us because we like being slaves, and so we are giving you the power to direct our thoughts as you like? On the contrary, what they said was: we need you to maintain the laws which we wish to obey, to govern us wisely, to defend us; for the rest, we require that you respect our liberty. Once this agreement had been made, it could not be altered.[58]

This is, of course, an authoritarian interpretation of the social contract – the grant of sovereignty is irrevocable and unconditional, the subjects have no right of resistance – but it does impose on the ruler the obligation to serve the interests of the whole. Frederick made that explicit in a passage following that just quoted:

Here we have, in general, the duties which a prince should carry out. So that he never neglects them, he should often recall to mind that he is a man just like the least of his subjects; if he is the first judge, the first general, the first minister of society, it is not so that he can indulge himself, but so that he can fulfil the duties involved. He is only the first servant of the state, obliged to act with honesty, wisdom and with a complete lack of self-interest, as if at every moment he might be called upon to render an account of his stewardship to his fellow-citizens.[59]

'The first servant of the state' was a phrase repeated by Frederick time and again: it became the *Leitmotiv* of his political system. In a letter to Voltaire written the year before he came to the throne, he used a striking simile to identify the role of the ruler: he was like a heart in the human body, receiving blood from all members and then pumping it out again to the furthest extremities of the body politic; he received loyalty and obedience, he sent out security, prosperity and everything else which could further the welfare of the community.[60] It was this sense of responsibility to the whole which rescued Frederick from any charges of despotism. The essence of despotic power was held to be its arbitrary, capricious character, its dependence on the whim

of its wielder. That was what led contemporaries to castigate the
Russian and Ottoman Empires as 'oriental despotisms'. In
Prussia, on the other hand, the absolute power of the king was
limited by the obligations imposed by the social contract,
especially by the rule of law. Although Frederick did indeed
abuse his absolute power on occasions – did act despotically, in
other words[61] – he was loyal to his enlightened political thought
with sufficient consistency to enable Prussians to reject with
indignation the charge that they lived in a despotism.[62]

Frederick's early ingestion of the clear (if thin) draughts of the
Enlightenment may not have made him an original or profound
thinker, but it did help him to abandon the constraints which
operated so powerfully on his father. He never cared anything
for his family, demanding that the interests of the Hohenzollern
dynasty be subordinated to the interests of the Prussian state.[63]
He had only contempt for the Holy Roman Empire, despising
its 'antiquated, fantastical constitution' and dismissing its central
body – the *Reichstag* – as 'but a kind of phantom. . . . The envoy
which a sovereign sends thither resembles a yard-dog who bays
at the moon'.[64] But he reserved his most withering scorn for his
father's most potent sanction, dismissing Christianity as 'an old
metaphysical fiction, stuffed with fables, contradictions and
absurdities. It was spawned in the fevered imagination of the
Orientals, and then spread to our Europe, where some fanatics
espoused it, where some intriguers pretended to be convinced by
it and where some imbeciles actually believed it.'[65]

No one can say – and Frederick himself probably did not know
himself – just why he took the fateful step to invade Silesia in
December 1740. As he recognised, a simple desire to make a
reputation probably had a good deal to do with it.[66] It is at least
possible, however, that the liberation of his thinking from the
traditional constraints which kept his father inactive helped him
to recognise the opportunity and to overcome any scruples about
making the most of it. Consequently, the sharp division which
is usually drawn between enlightened domestic policy and
unenlightened foreign policy is perhaps overdone. Frederick
himself saw no necessary contradiction between using the power
of the state to expand its territory, enhance its prosperity and
improve its security and using the power of the state to improve
its domestic arrangements – and neither did many enlightened
contemporaries.[67] It needs to be remembered that what late-
twentieth-century man regards as enlightened and what Freder-

ick's own contemporaries understood by the word do not overlap completely.

It is also important to bear in mind the constraints imposed on Frederick by his state's geopolitical position. Strung out across Europe, from the Rhine to the Vistula, and divided into disparate parcels of territory, much of it acquired only very recently, Prussia was a country of frontiers and enclaves for whom danger threatened with the same insistence as opportunity knocked. If we know, with the advantage of hindsight, that the Swedish and Polish empires were in a state of terminal decline, Frederick's knowledge of recent history warned of possible revival. If we know that Austro-Prussian dualism was to end with defeat for the Habsburgs, Frederick's direct experience told him that the opposite result was the more likely. If we know that the Russian empire was a colossus with feet of clay, Frederick could see only the size. In this fluid world of constantly shifting frontiers, his paramount concern had to be external security. As the Thirty Years War had taught with merciless severity, to try to opt out of great-power politics in Eastern Europe was to invite despoliation. To analyse this situation with reason unfettered by prejudice was to employ the methodology of the Enlightenment, even if the lessons the process taught – tactical aggression and strategic militarism – were apparently at odds with enlightened principles. It was the enlightened Prussians who survived the eighteenth century as a great power; it was the unenlightened Poles who were partitioned off the map of Europe.

This primacy of foreign policy also dictated the limits of domestic reform. Frederick constructed his political *theory* on a *tabula rasa*, but his political *practice* had to be implemented in a world made intractable by centuries of custom. In principle, for example, Frederick was utterly opposed to serfdom: 'of all conditions, the most wretched and the one which revolts the rest of mankind. It is certain that no man was born to be the slave of a fellow human-being; such an abuse is rightly detested.'[68] But he also recognised that the whole agricultural and social system of this state was constructed on the basis of serfdom and that consequently the destruction of the latter would shake the former to its foundations.[69] This consideration prompted him to move gingerly, bending and retreating whenever Junker opposition became serious.[70] Yet a good deal was achieved in the course of his reign. Even if the protection accorded to the peasants was really protection accorded to potential soldiers,[71] protection it was. The Prussian peasants enjoyed a legal status superior to that of their counterparts in neighbouring countries, while their material status may well have been superior to their equivalents in France.[72] The dangers inherent in trying to move too fast too soon in the direction of full emancipation was demonstrated by the fate of Joseph II in the Habsburg Lands in the late 1780s.

Frederick's self-consciously rational approach to foreign affairs and the domestic implications of the primacy of foreign policy was also deployed in military matters. Despite all the attention he lavished on the army, he was very careful never to allow the military sphere to take on a life of its own. Frederick used his army as an instrument of policy; he began a war with a precise, limited objective and when he had achieved it, he stopped (or rather he tried to stop – his enemies were not always co-operative). Very sensibly, he did not allow military success to inflate his original objectives, with the result that his successes were more modest but also more enduring than those of Napoleon, who fatally disregarded the golden rule of limited aims. In short, Frederick provided the perfect illustration of Clausewitz's maxim that war is nothing but the continuation of policy by other means – which is hardly surprising, when one remembers that Clausewitz based much of his theorising on his study of Frederick's reign. Consequently, as Theodor Schieder has observed: 'if by "militarism" one understands the application of a military way of thinking to the political sector, then Frederick can only be classified as a militarist with reservations, for his military actions were always determined by his political objectives.'[73] What Schieder does not point out is, of course, that this qualification does not draw the sting from the undeniable social militarisation of institutions and values in Prussia during the course of his reign.[74] Nevertheless, between 1740 and 1786 Prussia was at war for a shorter period of time than any other major power.

Inside Prussia, Frederick's attachment to the Enlightenment was revealed most clearly in his policy of religious toleration. Certainly this was part of the Hohenzollern heritage; certainly he was also motivated by his simple contempt for Christianity; but equally certainly he was also influenced by the thinkers of the Enlightenment, for whom toleration was *the* central axiom.[75] It just will not do to argue that because toleration was useful, it therefore had nothing to do with the Enlightenment – the most influential advocates of religious toleration (Voltaire and

Beccaria, for example) employed just such utilitarian arguments. Less than a month after coming to the throne, he had issued two unequivocal instructions on the subject: 'All religions must be tolerated and the sole concern of officials is to ensure that one denomination does not interfere with another, for here everyone can seek salvation in the manner that seems best to him' and 'all religions are just as good as each other, so long as the people who practise them are honest, and even if Turks and heathens came and wanted to populate this country, then we would build mosques and temples for them'.[76] It was just because his belief in toleration was based on principle that he took it so much further than his predecessors, expanding it to embrace pagans, deists, agnostics, atheists – even Jesuits. It was Frederick William I who banished Christian Wolff for supposed heterodoxy; it was Frederick who brought him back.

In present-day Western societies, religious liberty has come to be so taken for granted, and religion itself has declined so much in importance, that Frederick's achievement in this sphere has often been underrated. It should be remembered that it was still possible in France, in 1766, for a youth to be sentenced to have his right hand cut off, his tongue cut out and to be burned at the stake for desecrating a wayside crucifix and for uttering some childish blasphemies.[77] Other flesh-creeping stories of religious persecution could be gathered from one end of Europe to the other. Frederick's Prussia was the most tolerant state in Europe – bar none. Moreover, it was recognised as such by, among others, the most influential German intellectual of the day, Immanuel Kant. As a professor at the Prussian university of Königsberg, Kant was well placed to assess this aspect of his sovereign's policy. In the most accessible of his publications, the short article 'What is Enlightenment?', published in a Berlin periodical in 1784, Kant wrote of Frederick:

A prince who does not regard it as beneath him to say that he considers it his duty, in religious matters, not to prescribe anything to his people, but to allow them complete freedom, a prince who thus even declines to accept the presumptuous title of *tolerant*, is himself enlightened. He deserves to be praised by a grateful present and posterity as the man who first liberated mankind from immaturity (as far as government is concerned), and who left all men free to use their own reason in all matters of conscience.[78]

The influence of the Enlightenment was also clear and important in the all-important area of law reform. Especially in countries with long traditions of representative government, law reform is naturally viewed as peripheral to the political process. In the absolute monarchies of continental Europe, however, the law represented the only protection available to individuals in disputes with the state or with fellow-citizens. Here the law was the constitution. That is why contemporary Germans, for example, attached such importance to the creation of a *Rechtsstaat*, a state in which the rule of law had been established. That is why Frederick's innovations in this sphere were given such intense – and approving – attention by progressive opinion.

Frederick began his reign by abolishing torture and reducing drastically the number of capital crimes.[79] Thirty-seven years later, he wrote to Voltaire that his enlightened policy of seeking to prevent rather than to punish had led to the virtual elimination of capital punishment: on average, only fourteen to fifteen death sentences were pronounced each year.[80] Once again, the capricious brutality of the penal systems of most European countries should be borne in mind when assessing the novelty and significance of this achievement.

More important, if less spectacular, than the changes in the criminal code was the reform of civil law. Frederick William I may have recognised the need, but very little was actually done during his reign.[81] If Frederick inherited the man for the job, Samuel von Cocceji, it was he who laid down the principles (at a crucial conference held at Potsdam on 15 September 1746) and – most important of all – it was he who gave Cocceji the necessary support when he ran into opposition from vested interests.[82] By the time Cocceji died, in 1755, he had 'laid the foundations of a national system of justice that proved increasingly capable of removing many of those evils of which the French complained but were unable to remedy'.[83]

The Seven Years War and its difficult aftermath then brought the process of reform to a halt. It was revived by the notorious 'Miller Arnold' affair of 1779–80, when the King intervened to correct what he saw as an injustice inflicted on a miller by a Junker landlord.[84] In a cabinet order of 14 April 1780, Frederick instructed the Chancellor von Carmer to begin the preparation of a legal code for Prussia which would be comprehensible to the common man.[85] Although it was not published until 1794, the General Code (*Allgemeines Landrecht*) was very much an achievement of Frederick's reign.[86] For all its social conservatism, it was also deeply imbued with the Prussian Enlightenment, through Carmer's two most important collaborators: Carl Gottlieb Suarez and Ernst Ferdinand Klein.[87]

Prussia was no *Rechtsstaat* by 1794, let alone by 1786, even if Frederick stated in his will, written in 1769, that he had established the rule of justice and the laws.[88] He may also have stated his intention never to interfere in the courts and to allow the laws alone to speak,[89] but he proved unable to resist the temptation to abuse his power – most spectacularly in the Miller Arnold affair. One of the central axioms of enlightened jurisprudence – the separation of justice from the regular administration – was never fully achieved.[90] The Junkers continued to influence judicial appointments, through the Estates, and continued to enjoy a privileged position.[91] Equality before the law was not even an objective of the General Code, let alone achieved by it.

These necessary qualifications must not be allowed to obscure the sense of movement in the right direction which Frederick's reforms created. At a time when the legal procedures of – say – France and Great Britain were characterised by confusion, expense, lethargy, arbitrariness and brutality, the Prussian system was distinguished by its relative speed, cheapness, uniformity, impartiality and humanity. It is quite false to suppose that the King's justice stopped at the gates of the Junker's estate. The Junker did indeed enjoy seigneurial jurisdiction, but his court had to be run by properly qualified judges, it could not try criminal cases and in civil cases there was a right of appeal to the royal courts.[92] There is a mass of evidence to show that the peasants did avail themselves of the opportunity to invoke the protection of the law against their lords – and did so successfully.[93]

For contemporary Prussians – indeed, for contemporary Europeans – this was dramatised by the Miller Arnold affair. This may have been a 'judicial catastrophe', in the sense that Frederick interfered with the course of justice, but it certainly gave the lie to the notion that the Prussian state was merely an instrument for promoting the interests of the nobility. Among the senior officials dismissed by Frederick as a result of the affair were the Grand Chancellor von Fürst, the Landrat von Gersdorff and the President of the *Regierung* Count Finck von Finckenstein.[94] Frederick also announced that he was going to make an example of the unjust judges, 'because they must learn the lesson that the humblest peasant, yes and what is more even the beggar, is just as much a human being as His Majesty himself . . . and whether it is a prince accusing a peasant, or the other way round, the prince is equal to the peasant before the law'.[95] Paradoxically, the affair also demonstrated the independence of the judiciary, for the Berlin *Kammergericht* passed judgment uninhibited by royal interference and when the minister of justice, von Zedlitz, refused to discipline them for it, Frederick did not dare to carry out his threat to dismiss him.[96] So the upshot of the affair was that while angry bureaucrats visited the dismissed chancellor to present their condolences, the townspeople and peasants demonstrated in support of the King.[97] As a Berlin clergyman observed, the affair had 'transfigured' Frederick in the eyes of the common man.[98]

The common man did not admire Frederick because he believed him to be enlightened, he did so because he believed Frederick was 'the champion of the peasant against the landlord and of the weak against the strong'.[99] It was the intelligentsia of Prussia who found his enlightened qualities so attractive. Contrary to the article of faith advanced by Marxist historians,[100] the Enlightenment in Prussia was not a bourgeois phenomenon, neither in terms of those who wrote the works nor in terms of those who read them. Both producers and consumers were ideologically united – but socially mixed, including within their ranks nobles, clergymen and many kinds of commoners. Indeed, members of the capitalist bourgeoisie were conspicuous only by their absence. The statistical evidence on this point is clear: Table 10.1, for example, gives figures relating to the social and/or occupational background of the 165 authors from Berlin listed in Nicolai's *Allgemeine Deutsche Bibliothek* as having published during the course of 1784. Those figures are confirmed by an analysis of contributors to the *Berlinische Monatsschrift*, the most important journal of the Enlightenment in Berlin.

The most striking characteristic of these statistics (and of many others available like them) is that most members of the Prussian intelligentsia were employed by the state. They enjoyed a common education, a common training and a common commitment to the ideals of the Englightenment.[101] While relations with their king were sometimes strained, they were bound to his state by both interest and conviction, seeing in it the best chance of progress and modernisation.[102] That is why any examination of enlightened absolutism should not deal just with the person of the monarch

TABLE 10.1 *Contributors to the* Allgemeine Deutsche
Bibliothek, *1784*

	%
Nobles	15.8
Clergy	19.4
Officials	30.3
Schoolteachers	10.9
Academics	18.2
Independent writers	1.2[103]

TABLE 10.2 *Contributors to the* Berlinische Monatsschrift

	%
Nobles	15
Professors and schoolteachers	26.7
Officials	20
Clergy	16.7
Army officers	3.3
Merchants or bankers	1.7
Booksellers	0.7
Craftsmen	0.3[104]

but should be extended to take in the whole bureaucracy. All
aspiring Prussian civil servants were obliged to undergo a rigorous
course of theoretical and practical training. It was a process
which included a degree from a Prussian university and it was
there, especially at Halle and Königsberg, that the Prussian
version of the Enlightenment was absorbed by so many. By the
end of the century, it is true, a growing number were moving
on from the increasingly old-fashioned seeming *dirigisme* of
enlightened absolutism to the *laissez-faire* philosophy propagated
by Adam Smith, who came to be venerated in Prussia more than
in any other country (including his own).[105] Nevertheless, during
Frederick's reign the enlightened ethos of his state attracted
progressives not only from Prussia but from all over Germany.
The most talented of these *Wahlpreussen* (Prussians-by-choice)
turned out to be Baron Stein, who came to serve and to 'form
himself' under the man he chose to call 'Frederick the Unique'
in 1780.[106]

III

The unenlightened characteristics of Frederick's Prussia were both manifold and manifest. To the eyes of a critical contemporary, such as Lessing, it might well have appeared 'the most slavish country in Europe'.[107] Moses Mendelssohn, on the other hand, who lived all his life in Prussia, formed quite a different impression:

> I live in a state in which one of the wisest sovereigns who has ever ruled mankind has made the arts and sciences to blossom and a sensible freedom of thought so widespread that their effect has reached down to the humblest inhabitant of his dominions. Under his glorious rule I have found both opportunity and inspiration to reflect on my own destiny and that of my fellow-citizens and to present observations to the best of my ability on the fate of mankind and providence.[108]

Lessing believed that freedom of speech in Prussia was confined to the right to be rude about religion (in itself not a privilege to be sneezed at, it might be added). The British traveller John Moore formed quite a different impression:

> Nothing suprised me more, when I first came to Berlin, than the freedom with which many people speak of the measures of government, and the conduct of the King. I have heard political topics, and others which I should have thought still more ticklish, discussed here with as little ceremony as at a London coffee-house. The same freedom appears in the booksellers' shops, where literary productions of all kinds are sold openly. The pamphlet lately published on the division of Poland, wherein the King is very roughly treated, is to be had without difficulty, as well as other performances, which attack some of the most conspicuous characters with all the bitterness of satire. A government, supported by an army of 180,000 men, may safely disregard the criticisms of a few speculative politicians, and the pen of the satirist.[109]

That last observation was a forerunner of Kant's celebrated observation in 'What is Enlightenment?', in support of his argument that 'a lesser degree of civil freedom gives intellectual freedom enough room to expand to its fullest extent': 'only a ruler who is himself enlightened and has no fear of phantoms, yet who likewise has at hand a well-disciplined and numerous army to guarantee public security, may say what no republic would dare to say: *Argue as much as you like and about whatever you like, but obey!*'[110] These illiberal but enlightened sentiments could also be cited in support of the earlier warning against assuming that eighteenth-century definitions of enlightenment are the same as those of today. Kant also supplies the best possible example of an enlightened Prussian who disapproved fundamentally of many aspects of Frederick's state – notably its serfdom, autocracy and militarism – but who still believed that it was advanced, progressive and heading in the right direction:

> If it is now asked whether we at present live in an *enlightened* age, the answer is: No, but we do live in an age of *enlightenment*. As things are at present, we still have a long way to go before men as a whole can be in a position (or can even be put into a position) of using their own understanding confidently and well in religious matters, without outside guidance. But we do have distinct indications that the way is now being cleared for them to work freely in this direction, and that the obstacles to universal enlightenment, to man's emergence from his self-incurred immaturity, are gradually becoming fewer. In this respect our age is the age of enlightenment, the century of *Frederick*.[111]

If Frederick's Prussia was considered by the greatest philosopher of the age to be experiencing enlightenment, then it is difficult to see how the concept of enlightened absolutism can be denied reality, even if only at the level of contemporary impressions. In fact, as we have seen, despite all the qualifications, a good case can be made for establishing the actual influence of the Enlightenment on Frederick's policies, especially in the crucial areas of religious toleration and law reform. For the subsequent history of Prussia, however, what was most important was the image Frederick's state acquired of being in the vanguard of progress and responsive to changing needs. It was this which prompted enlightened Prussians to observe patronisingly that the French revolutionaries were trying to follow the trail blazed by Frederick and, in the next century, for the liberal Friedrich Dahlmann to liken the Prussian state to 'the magic spear, which heals as well as wounds'.[112]

11. Catherine the Great

ISABEL DE MADARIAGA

BELIEVERS in the theory of enlightened despotism thought that Russia was the ideal country in which to apply their ideas. Since it had no ancient institutions which would need to be rooted up, it was a blank page on which the *philosophe* might inscribe what he wished.[1] Nothing could be less true. When Catherine seized the throne in 1762, Russia was an absolute monarchy, placed at the despotic end of the spectrum which extended through the Prussia of Frederick II to the France of Louis XV.[2] There were no institutional limitations on the power of the ruler, who was even entitled to name his, or more often her, successor. There were no constituted bodies or 'estates', no 'intermediate powers' of the kind that existed elsewhere in Europe. As head of the executive the sovereign exercised authority through a series of functional colleges headed by boards under presidents, whose work was co-ordinated by an appointed administrative Senate of some twenty or thirty people. This misleadingly-named body had no legislative powers, which were lodged entirely in the ruler. Local administration had been allowed to collapse into an under-financed chaos when Peter I's top-heavy organisation was dismantled in 1727. Vast *guberniyi*, divided into provinces, then into districts, were placed under governors and *voyevodas*, almost the only officials who received a salary. The rest were supposed to live off the people whose lives they administered, and whose lawsuits they adjudicated.

If the organs of government were too few and too slack, society itself was very tightly regimented. Russia had remained much as Peter I (Peter the Great) had left it, with a very small nobility of some 50,000 males, some 200,000 registered townspeople out of approximately 1.1 million towndwellers, and about 10 million male peasants. It had a population second only to that of France, and the lowest population density in Europe.

Bondage of one sort or another pervaded Russian society. Until 1762, the noble was bound to serve the state in a military, or civil capacity, as long as he was called upon, usually some 25 years or more. His service life and social status were regulated by the Table of Ranks, instituted by Peter I, a code of precedence in military, civil and court life, modelled on codes in use in Western countries such as Prussia, Denmark, and even England, but applied with far more rigidity. All commissioned officers in the fourteenth rank became hereditary nobles (unless they were already nobles by birth), while civil employees acquired hereditary nobility, as distinct from personal nobility, with the eighth rank. Since the ownership of estates with serfs was limited to hereditary nobles, there was tremendous pressure for promotion among non-nobles, while noble officers tried to climb the ladder leading to rank and wealth. However, the self-image of the noble had undergone a remarkable change for the better in the brief reign of Catherine's deposed husband, Peter III, when, in the manifesto of February 1762, the nobility was freed from compulsory service.

The townsman was registered in his town; he was responsible, together with his fellow townsmen, for the collection of the poll-tax and for the performance of a number of unpaid duties on behalf of the state. The peasantry was, in 1762, divided into three broad groups: private serfs, church peasants and state peasants. The private serfs formed about 56 per cent of the peasantry. They were bound to their villages for tax purposes and for military conscription, but landowners were entitled to remove them and re-register them in other estates, to sell them individually or as families, with or without land, and to take them into domestic service. By the middle of the eighteenth century their position had been assimilated to that of earlier Muscovite slavery, indeed almost to slavery in the Western sense. Their masters were legally bound to provide for them in times of famine, and in old age, and were responsible for the payment of the poll-tax on their behalf. They exercised jurisdiction over them for all petty offences, but major crimes such as rioting, brigandage, or murder went before the state courts. The situation of the church peasants was slightly better since they could not be sold. State peasants paid the poll-tax and an additional sum in quitrent to the state, intended to counterbalance what the serfs paid to their masters. They could be conscripted to work on state building or engineering projects, in foundries, mines or other industries. An enterprising state peasant could however move to a town, make enough money, and eventually change his status from peasant to towndweller, even to registered townsman. All these social groups, except the nobles, were liable to conscription in the armed forces. These were the deep-rooted social institutions that any reforming ruler would have to confront. But in this society, so strictly divided into different orders, there were no established corporate structures, no noble assemblies, no urban corporations, which could be entrusted with a share in internal administration. The country was at that stage, already superseded elsewhere, when the government needed to draw upon the co-operation of members of the various estates in the tasks of administration, and in order to recruit them effectively, had first to organise and give legal form and rights to the necessary corporate institutions.

In the years before she seized the throne, Catherine had occupied herself with systematic reading of political literature, ranging from Plato to Bayle's *Dictionary*, from *Henri IV* by Hardouin de Péréfixe (a ruler whom she greatly admired) to the *Encyclopédie*. Throughout her life she continued in a completely pragmatic and empirical way to borrow ideas from authors who struck her as modern, sensible and rational. She worked systematically and very hard, with a small number of chosen advisers, and in the case of major administrative reforms, she tried them out on a small scale before introducing them throughout the empire. Most of her senior servants remained with her for decades, strengthening the impression of stability of her government, in spite of occasional changes of direction. She was also very attentive to manifestations of public opinion.

Catherine set out her political theory in the famous *Nakaz* or Instruction, which she wrote for the Legislative Commission, summoned in 1767 to draft a Code of Laws for Russia. Both have been widely misunderstood, at the time and later. The Instruction has been viewed as a Code of Laws, or even as a 'constitution'[3] which was never promulgated, the Commission as a Parliament which failed. As with many of Catherine's policies, there were precedents in previous reigns. But previous legislative commissions had been small *ad hoc* separate consultations with deputies from the nobility and the towns alone. What was novel about Catherine's experiment is that, only five years after her *coup d'état*, she felt sufficiently confident to embark upon the kind of large-scale consultation advocated by Diderot in the *Encyclopédie* (article *représentants*), by calling together representatives of all the free estates of the realm, government bodies, and non-Russian peoples, to examine the chaotic state of Russian law, and to draft a new code. Over 500 deputies were elected, each bringing with him an 'instruction' from his electors setting out local grievances, and the whole enterprise was managed with considerable pomp and publicity. Such a public consultation had not been held for more than a hundred years in Russia, and was not to be made again until the meeting of the first Duma in 1906.

It has been pointed out by Catherine's critics that the serfs were not represented. This is of course correct, but it has to be seen in the context of contemporary theories of representation. In Ancien Régime institutions it was not individual opinion, but corporate interests which were represented, and in no country at that time were peasants represented in diets or estates except in Sweden. And in Sweden it was only the peasants living on crown land, the equivalent of the Russian state peasants, who elected deputies since the interests of peasants living on noble land were held to be represented by the noble landowners. Thus, in Russia, serfs were held to be represented by their landowners, while state peasants sent deputies to the Commission. The Commission thus did not reflect individual opinion, let alone party opinion. Any deputy who attempted to set himself up as the advocate of the private interests of his electors, to run a 'surgery', was promptly called to order.

The clearest evidence of the influence of both Enlightenment and cameralist though on Catherine is provided by the Instruction she wrote in 1765–68 to guide the Commission in its labours. In it she analysed the government and society of Russia as she thought it ought to be, and expounded the general direction she wished to follow. Her principal source was Montesquieu's *L'Esprit des lois*, from which some 294 out of 526 articles were almost literally copied. She has been accused not only of plagiarism (to which she freely admitted), but of distorting Montesquieu's views, in that she applied to Russia (which he regarded as a despotism) the maxims which he applied to a 'moderate' monarchy; that she located the limitations on absolutism, the *pouvoirs intermédiaires*, in the bureaucracy and not in social groups such as the nobility, and that she rejected the doctrine of the 'separation of powers'. This judgment is partly true. Considering the low regard in which despotism was held, Catherine naturally rejected its application to Russia, and preferred to argue that the ruler of Russia must be absolute in view of the size of the country. She justified her argument by postulating the existence in Russia of fundamental laws and the acceptance of a degree of self-limitation in the ruler's use of absolute power. Catherine does not, it is true, locate the *pouvoirs intermédiaires* in social forces such as the nobility or the town corporations, and in this she can be said to deviate from her mentor – but we must remember that neither the nobility nor the towns had as yet any corporate structure in Russia. However in considering the lawcourts as intermediate bodies she does follow Montesquieu's principles. The courts were part of the bureaucracy in Russia, but then they were part of the state apparatus in France – a fact not appreciated by many commentators[4] – where the Parlements were not independent elected bodies, but emanations of the king in his capacity of supreme judge. Finally, though Catherine does not even discuss the separation of powers in her Instruction, one must remember that Montesquieu does not discuss this concept except in relation to one specific form of monarchy, which has political liberty as the main object of its constitution, namely the English monarchy. But she did endorse the separation of functions, i.e. the principle that the sovereign should not act as a judge, and that the judiciary should be separate from the executive.

Though Catherine had no sympathy for the theory of the social contract on which Beccaria's analysis of society was based, she was influenced by his utilitarianism, and by his humanitarianism which coincided with her own personal inclinations. His ideas are reproduced throughout the chapters of the Instruction dealing with the judicial process, torture, the nature of evidence and the nature and object of punishments, and Catherine put before the Russian public a condemnation of cruelty which had never been heard in Russia before, let alone from the throne. While the influence of Montesquieu predominated in the chapters dealing with social organisation, cameralist ideas were echoed in the sections of the Instruction dealing with public order and the economy.

The Instruction was not a code of law, nor did it have any legal validity. But it exercised a considerable influence on the intellectual climate of Russia in the eighteenth century and even in the nineteenth, since it opened the way to public discussion of many issues which had never previously been openly debated. Though its circulation seems originally to have been limited to officials, the Instruction was nevertheless freely available to the public from the beginning, since it was advertised as for sale in the Senate bookshop for 50 kopeks as early as 1768. Though its natural law principles, which paid scant regard to Russian positive law, led to its having but slight influence on subsequent Russian legislation, the Instruction had a beneficient influence on Russian practice, particularly in the field of penal law, when it was frequently quoted in favour of moderation and the reduction of penalties.

As a codifying body, the Legislative Commission proved both unsuitable and unsuccessful, though the at times acrimonious debates, and the materials collected by it and through it, were to prove invaluable in subsequent legislation. It was prorogued, not dissolved, on the outbreak of war with the Ottoman Porte in autumn 1768, since it was impossible for Catherine, with the resources in manpower and money available to her, to embark on reforming programmes while Russia was at war. It is very probable that the Pugachev revolt which raged from September 1773 to August 1774 explains the failure to recall it, though we have at present no direct evidence of Catherine's thoughts upon the subject. But the need to restructure Russian local administration had revealed itself to be so great that Catherine could not wait for the leisurely process of codification, and adopted a different method, namely the preparation and promulgation of laws by means of consultation with chosen advisers.

In 1764 Catherine had already put through a small scale reorganisation of the civil administration, introducing salaries for all officials, and pensions after twenty-five years of service, inaugurating an intensive anti-corruption drive, and strengthening the network of procurators whose function it was to supervise the legality of the decisions of the executive, and who were responsible directly to the Procurator General, in turn responsible directly to her. At the same time the collection of the poll-tax was removed from the military and handed over to civilian officials.

The Statute of Local Administration of 1775, many of whose provisions lasted until 1864, some until 1917, was not simply the continuation of policies adumbrated in previous reigns. It represents a quite novel amalgamation of earlier Russian attempts to solve the problem with the doctrines of Catherine's Instruction, particularly those derived from Montesquieu, and a substantial dose of cameralism introduced by her Baltic German advisers. One of them, Count Jacob Sievers, also reinforced the English influence on Catherine. He had lived in England, greatly admired the British political system, and no doubt helped the Empress to thread her way through its mysteries as expounded by Blackstone, whose *Commentaries on the Laws of England* she was reading at the time in a French translation.[5]

The reform of 1775 was based on a number of specific principles, namely the multiplication of administrative centres in the provinces, and the corresponding transfer of functions from the centre to the localities; the separation of functions previously combined in one instance; the establishment of specific institutions for the free estates of the empire, and some degree of corporate organisation of these free estates to enable them to participate in the election of representatives forming part of local administrative institutions.

By the division of the vast *guberniyi* of Russia into smaller units, subdivided into districts of some 30,000 inhabitants, the various judicial instances were multiplied and brought much closer to the people, and a number of settlements of varying size, some extremely small, were raised to the status of towns. The separation of functions took quite elaborate forms: general administration, including what would today be regarded as welfare and was then grouped under the general heading of police, was entrusted to the governor, assisted by a board of appointed councillors, and a Board of Social Welfare, on which sat elected representatives of the various free estates. Financial administration including the collection of taxes and the administration of the state peasants became the responsibility of the deputy governor.

The judiciary was completely reorganised, and it is here that the 'estate' principle, and the participation of elected assessors is particularly striking. At the apex of the system in each *guberniya* there were a civil and a criminal court, from which appeal lay to the Senate (the governor was specifically forbidden to act as a judge). Three separate sets of courts were established below, in three tiers, for the nobles, the townspeople and the state peasants respectively, to which these social groups elected a number of assessors. The only court of first instance for all classes was the so-called Conscience Court. Here again we have an institution which has no roots at all in the Russian past, and was taken over entirely from foreign models seen through a Catherinian prism.[6] For the Conscience Court was to act as an equity court, a concept hitherto unknown in Russia, and which, it must be said, remained unclear both to the founder of the whole system and to the judges who operated it. The court was supposed to temper the harshness of the law, but it was given additional functions, namely dealing with cases concerning minors, sexual offenders and witchcraft, etc. But it was also to this court that an entirely new function was entrusted: anyone detained for more than three days without charge could appeal to the Conscience Court to be released on bail, except in case of serious crimes such as *lèse majesté*, or murder. This was a complete departure from Russian practice, clearly borrowed from the English example of habeas corpus, though not from any specific English court. The extent to which this right was actually made use of in Russia remains shrouded in mystery.[7]

At the lowest level, the policing of the countryside was entrusted to a 'lower land court', presided over by a land commissar, with two noble assessors elected by the local nobles, or where there were few nobles, appointed by the crown, and with elected peasant assessors. The Russian Statute clearly describes this institution as a court, but states that the land commissar is not empowered to act as a judge on his own. Nevertheless historians have often remarked that this police organ was wrongly called a court. The confusion may arise because of failure to appreciate that here again there are traces of English influence, notably that of the procedure of indictment by justices of the peace. The English J.P. was not an agent of the executive; he operated by judicial means, i.e. by prosecuting offenders for failing to carry out specific duties (e.g. repairing bridges or clearing manure off roads), and since the land commissar was charged with many of the duties carried out by Justices of the Peace, it is possible that Catherine endeavoured to establish an institution with in-built guarantees against arbitrary misuse of powers such as the English system might seem to provide in her eyes through her reading of Blackstone.[8]

The participation of representatives of social groups in local government institutions, except in the administration of finance which remained firmly in government hands, was provided for by election at the lower levels, by appointment at the higher levels. A rudimentary corporate structure of the nobles and the towns was laid down in the Statute, to organise the elections: the nobles of a district met to elect a marshal, who became the lynchpin of the subsequent establishment of local noble assemblies. The towns elected a town chief, who in turn became the centre for the future development of urban corporate organisation. In these institutions we have pale shadows of Montesquieu's '*pouvoirs intermédiaires*'.

The Statute of 1775 was introduced only gradually throughout the Empire. Even so there were serious difficulties in implementing a policy which was in so many ways designed for a more advanced society than Russia. It did not take account of the enormous variation in type of population (serf or state peasant); of the presence of resident noble landowners; of the size of towns and the presence of enough educated merchants and townsmen to fill elective posts; of economic potential, communications, racial homogeneity, and the need for a considerable infrastructure of surveyors, doctors, teachers, clerks, etc. Nevertheless the new institutions did bring life to provincial towns, where energetic governors proceeded to erect new stone buildings, set up schools, hospitals, inoculation centres, to open libraries, patronise the theatre, give balls and enliven social and intellectual life, and encourage manifestations of local initiative in the field of social welfare.

But in spite of the increased services available to the local inhabitants, both urban and rural, and the presence of some conscientious and honourable officials, there is little doubt that Russian local administration continued to be inefficient, corrupt, and often brutal. In spite of new attitudes of mind inculcated into the elite,[9] the mental outlook of centuries could not be changed overnight. The habits of servility to superiors and bullying of inferiors were deeply ingrained, the notion of public service as a vocation and not a burden or a means of enriching oneself still limited to the most enlightened. Even more difficult, because so alien to the Russian tradition, was the attempt to introduce the concept of legality in the judicial process. The total absence of legal training (the first professor of Russian law was appointed only in 1768) helps to account for the failure to grasp the first principles of the rule of law among even governors, whose experience was often entirely military. The efforts of the government to provide better trained officials only began to have results in the nineteenth century.

Judging by contemporary accounts, the brigandage and disorder endemic in some outlying areas was now brought under some degree of control. It is almost invariably assumed by historians that the land commissar, the lowest rural official elected by nobles to police the countryside, always acted in the interests of the nobility. Some were undoubtedly local despots, but what is known of their behaviour is almost entirely based on memoir literature, not on archival sources. Strange as it may seem, there has been scarcely any modern study, based on administrative or judicial records, of the actual functioning of any of the institutions set up in 1775. Some work has been done on a few of the Conscience Courts, but nothing is known about the work of the local government boards; there is no study of the work of any single Board of Social Welfare.[10] Thus the historian can analyse the policy but must suspend judgment on its implementation.

If the Statute of 1775 reflected the ideas of Montesquieu in the elaboration of separate institutions for the different estates, the Police Ordinance of 1782 was closely modelled on cameralist legislation. It incorporated all the aspects of paternalistic social engineering associated with the seventeenth-century concept of 'police', and included detailed rules on how the police was to perform its function of moral guidance, borrowed probably from the *Traité de la Police* of Nicolas de la Mare, which in turn resumed the experience of Louis XIV's great lieutenant of police, La Reynie.[11] Police jurisdiction was limited to trivial offences for which they were authorised to inflict minor summary punishments. All other offences had to go before the courts. In accordance with the practice of *Polizeiwissenschaft* elsewhere, the police was charged with the regulation of fire fighting, public hygiene, street lighting, weights and measures, public decency, the control of public assemblies, and the arrival and departure of strangers.

If the new system of local administration was to work at all, it had to draw on the voluntary participation of elected representatives of the free estates. Catherine therefore considered it necessary in the two Charters of 1785, to the Nobility and to the Towns, to enumerate and enact the basic civic, property, status and corporate rights of these estates, and to define their membership. From now on, nobles could not be deprived of rank, honour, property or life without trial by their peers. The noble's immunity from corporal punishment and personal taxation was confirmed.[12] Noble land was freed from any state restrictions on its exploitation, and the noble's right to buy land with serfs was expressly stated, as was his right to set up manufacturing enterprises, to leave Russian service, to travel and to serve abroad. The Charter extended the corporate rights granted in 1775, authorising the nobles in each *guberniya* to meet every three years to elect their marshal, who acted as the head of the local corporate nobility. A register of the nobles was introduced on a *guberniya* basis: the nobles were divided into six groups according to their origin, though all equal in rights, and the register was to be kept in the office of the local noble assembly. To some extent the decision as to who was a noble was thus placed in the hands of the nobles themselves. The corporate nobility was also allowed to 'make representations' to the Senate or directly to the Empress on local needs. It was this right which was used by the noble assemblies, for example in Tver, in the period leading up to the emancipation of the serfs in 1861.

The Charter to the Towns granted the same kind of personal property and civil rights to the townspeople, though corporal punishment was banned only for the wealthiest groups, namely rich merchants and entrepreneurs, and for a new category of 'distinguished citizens', mainly professional people. The Charter of 1785 also provided the towns with an elaborate system of local administration, in which elected officials took part, an extension to the towns of the provisions of the Statute of 1775.

Catherine's doctrinaire approach is illustrated by the unpromulgated Charter to the State Peasants which dated roughly from the same time. The Charter divides the peasants into six groups according to their origin in almost exactly the same terms as those used for the other social classes. It sets out their personal, property and civil rights, and the upper stratum of the peasantry is also granted immunity from corporal punishment. A number of provisions were made for the election of elders and the organisation of communal self-government. The sheer absurdity of the attempt to force the Russian peasantry into the same kind of social mould as the nobles and the townsmen may explain why the Charter was never promulgated. It is also possible that Catherine was persuaded to abandon a scheme that might easily lead to unrest among the serf population. Or the outbreak of the second Turkish war in August 1787 may have led her to shelve this, as so many other plans.

The decentralisation of administrative functions by transferring them to provincial centres led to the phasing out of a number of central government colleges, particularly of a commercial or economic type, such as the Colleges of Mining or Manufactures, or the College of Estates. We know that in the 1780s Catherine was thinking of a fundamental reconstruction of the central government of Russia, for which the Charters to the free estates were in some sort a preliminary. She was working on the reform of the Senate, and she even drafted a plan for an imperial council, and a high court of justice with some legislative functions, to which elected assessors from the nobility, the townspeople and the free peasants would be elected. These drafts were to some extent anticipations of the projects which emerged among the advisers of Alexander II in the late 1870s. In this case we know also that Catherine laid these projects aside on the outbreak of war in 1787.[13] There was not enough qualified manpower, nor did the ageing Empress have enough energy to proceed with elaborate plans for reform in the middle of a war, and from 1787 to Catherine's death Russia was almost never at peace. The French Revolution also probably made her abandon ideas of thoroughgoing reform. Yet, if we take the Statute of 1775 and the two Charters of 1785, as well as the Police Ordinance of 1782, it is clear that many of the demands put forward in the Legislative Commission were met. The laws represented that codification which the Commission itself had failed to produce, and introduced major social and administrative changes into the structure of Russian government and society.

What remains paradoxical is that Catherine II was attempting to create and give legal form to a society of orders (*Ständegesellschaft*) at a time when this type of society was being increasingly undermined by other, more radical, centralising reformers, such as Joseph II. This was not just a question of more or less conservative opinion, but a reflection of the state of the social and economic development of Russia. The country was simply not in a position to entrust internal administration to a class of literate officials with a tradition of public service, such as had been produced by the universities of Germany for their princes. Hence administrative tasks had to be delegated to the only social group qualified by years of military experience to undertake them. By attaching rank and salary to the various posts, Catherine was hoping to attract the poorer nobles who had left service to continue to serve in the provinces. It is a matter of controversy among historians whether in delegating service in this way the state – if it existed[14] – was abdicating its rights into the hands of a power hungry and greedy class of nobles, or whether an autonomous state maintained its ultimate control over the resources of the country, allocating the largest portion to that class, the nobles, to which it had entrusted administrative and military duties.

One of several reasons why Russia had not been able to evolve a class of competent non-noble government servants was the

educational backwardness of the whole country. Peter I's attempt
to introduce secular education had met with but limited success;
Moscow university was only founded in 1755, and senior Russian
public servants were totally untrained in law or public administration except for those (and the number was increasing) who
pursued mainly government-subsidised studies in German,
French, Dutch, English and Scottish universities.

Concern with education was central to enlightenment thought,
whether in absolutist Prussia and Austria, or in the Polish gentry
republic. Catherine, however, was no admirer of Rousseau's
educational theories; on the whole she preferred Locke, though
she disagreed with him over the issue of corporal punishment,
which she totally rejected. Her early experiments centred around
the somewhat fanciful notions of her adviser, I. I. Betskoy, to
create 'a new race of men' (and women) by sedulously shielding
children from all contamination by the coarseness and brutality
of the surrounding world. It was a programme suitable for an
elite, but not for a national school system, which Catherine had
adumbrated in the Statute of 1775. The moving spirit behind
her subsequent educational reform was Joseph II, who in the
course of his visit to Russia in 1780, informed her of the policy
pursued in the Habsburg lands after the expulsion of the Jesuits.
Joseph sent her an Austrian adviser, F. Jankovich de Mirievo,
an Orthodox Serbian, expert in the methods devised by the
educational reformer, Abbot Felbiger, who had applied them
first in Prussian Silesia and then in the Habsburg Monarchy (see
above, pp. 175–6). Catherine's Statute of National Education
of 1786 set up a comprehensive system of high schools and junior
schools, free, secular, co-educational, open to all classes including
serfs if they had the permission of their masters. A centrally-
devised curriculum was imposed, teaching aids and textbooks
were produced including elaborate regulations on the duties of
teachers and a textbook setting out the moral and political
principles to be inculcated into the pupils, designed to produce
dutiful and obedient citizens.

There was inevitably a considerable discrepancy between an
ambitious plan and a more modest achievement. The Boards of
Social Welfare, charged with setting up and running the schools
had neither the funds nor the staff. There was an enormous
shortage of qualified teachers, and though they were not ill-paid
compared to other public servants, and though they were given
positions in the Table of Ranks, with the possibility of rising to

hereditary nobility after twenty-two years service (the ladder by means of which Lenin's father rose to noble rank) the teachers were too uncouth, and the members of the Boards of Social Welfare too inexperienced to be widely successful. Moreover the education provided was still too sophisticated to respond to the basic needs of Russian townspeople, who did not demand more than the three Rs and the ability to read devotional works.

Nowhere was Catherine more truly a daughter of the Enlightenment than in her patronage of arts and letters. St Petersburg became one of the most beautiful European cities, once the banks of the Neva were clothed in granite and the canals spanned by elegant bridges. Less visible to the general public was Catherine's private sponsorship of a wide programme of translations into Russian which she subsidised to the tune of 5000 roubles a year. In fifteen years the Translation Society formed by her was responsible for the publication of 112 translations, thus giving employment to many noble and non-noble intellectuals.[15] The works selected included the best that France could offer, selections from the *Encyclopédie*, Montesquieu's *L'Esprit des Lois*, works by Corneille, Racine, Voltaire, Mably (translated by Alexander Radishchev), and many of Rousseau's writings, though not the *Social Contract*.

The numbers and the titles published far outpaced the capacity of the reading public to absorb them, and the problem of marketing translations became crucial. It must be remembered that most cultured nobles and non-nobles could read the classics of the French and German Enlightenment in the original languages. German had not yet been ousted by French in St Petersburg – if it ever was – and there may have been a larger market for German books than for French. Indeed the works of J. G. Herder, and those of Immanuel Kant, were published in German in Riga, where Herder became head of the gymnasium before he left for Weimar, and the complete political works of Rousseau were published in Riga in German translation in 1779–82.

Catherine's critics have found it difficult to explain why in 1783 she authorised the establishment of printing presses, hitherto a state monopoly, by anyone, subject to registration with the police. The most recent scholar in the field suggests that Catherine believed that the independent presses 'posed no serious political or ideological threat to the government' and that it was the desire to let private enterprise rather than the state bear the losses on publishing which proved the most significant consideration.[16] Until 1783 the government presses were all self-censoring; after 1783, there was still no central censorship, though works had to be submitted to the local chiefs of police to ensure that they contained nothing offensive to order, religion or public decency. Until the French Revolution the Russian censorship was laxer than ever before or since, with the exception of a few years at the beginning of the reign of Alexander I. How lax it was is illustrated by the police chief's approval of Radishchev's *Journey from St Petersburg to Moscow* in 1790.

Only recently has the hitherto widely-accepted account of Catherine's relationship with the publisher N. I. Novikov been revised by Soviet and non-Soviet scholars. The traditional view presupposes a duel between the elderly (she was 39 years old!) and reactionary Catherine and the 24-year-old radical Novikov, conducted in 1768 in the pages of *All Sorts of Things*, edited by Catherine's secretary, and *The Drone*, edited by Novikov. Yet it was not known at the time that the Empress took a personal part in *All Sorts of Things*. All the journals which appeared in the 1760s indulged in backbiting and satire against each other, as well as with *All Sorts of Things*.[17] It used to be stated that Catherine closed down Novikov's *The Drone* because of his attacks on her. There is no evidence to support this, whereas there is evidence that she supplied him with 200 roubles to assist him with publishing his next periodical *The Painter*, in which he published even more biting satires on uncouth nobles and impoverished serf villages. The subsequent rift between the Empress and Novikov, culminating in the latter's arrest and incarceration in Schlüsselburg in 1792, arose as a result of Novikov's participation in the Rosicrucian Masonic movement, directed from Prussia, which was alleged to have attempted to recruit the Grand Duke Paul, at a time of great tension between Russia and Prussia. There is also no evidence that Catherine resented the attempt of Novikov and other Moscow masons to undertake social welfare independently of the state, though she was perturbed at the huge sums of money they seemed to dispose of.

By 1790 Catherine woke up to the dangers implicit in unrestricted publication, not only of masonic obscurantism but of revolutionary ideas. Until 1789, and well into 1790, Russians had been surprisingly well-informed about events in the United States and France. Novikov had published, in a supplement to the *Moscow Gazette*, a series of articles praising George Washington as the 'founder of a republic which will probably be the refuge of freedom, exiled from Europe by luxury and depravity'. The French 'Declaration of the Rights of Man' of 1789 was published in No 74 of the *St Petersburg Gazette* in that same year. There were graphic descriptions of peasants burning chateaux in France, of the invasion of the French royal family's apartments in Versailles, reports of the debates in the National Assembly. French Revolutionary pamphlets, leaflets and periodicals were on sale in Russian bookshops.

The first indication of Catherine's dismay at what was being published came with Radishchev's *Journey from St Petersburg to Moscow*. It is very probable that until then she had felt able to allow tirades against foreign, or irrelevant classical, tyrants, but she drew the line at attacks on herself or her government. This explains the eventual exile of Radishchev to Siberia (in conditions infinitely more comfortable than the modern GULAG, since he lived with his sister-in-law and his servants, and was allowed to receive both books and money).

By 1790 even the advertisements of books for sale had to be submitted to the police,[18] and in 1791, the Academy of Science was given directives on what foreign news to publish in order to eliminate items about Russia, and to present news from France in a hostile light. Finally, in September 1796 Catherine withdrew the freedom she had once granted, and set up formal censorship of imported and locally printed books. One cannot help wondering what political events dictated this extremely severe reaction in 1796. It may not have been events in France, but the situation in Poland, where the second and third partitions had led to a veritable war against Russia which had undertones of revolutionary fervour about it.

Catherine II has on the whole been harshly treated by historians, particularly in the late nineteenth and twentieth centuries. Even non-Marxist historians have been conditioned by Marxist interpretations. Many have totally rejected her claim to be counted among the enlightened despots.[19] The somewhat disparaging portrayal of her personality and her reign now current is based on a number of issues which have been differently viewed at different times. These include the murder of her husband, her numerous lovers, the charge that she did nothing to put the 'liberal' maxims of her Instruction into effect, and that they were designed by a consummate hypocrite to throw dust in the eyes of the *philosophes* in Paris; that in her reign serfdom was both intensified and extended; that she brought unnecessary wars on Russia and that she partitioned Poland. Let us take them in turn.

At no time can one overlook the murder of Peter III, which casts a shadow on Catherine's reign. She did not order it, but she must have been aware that it was inevitable. She benefited from it, and was therefore an accessory after the event. The murder of a husband has always been perceived by society as a more heinous crime[20] than the murder of a son (Peter I) or of a father (Alexander I), or of the lover of one's wife (George I's Hanoverian father had it done for him). Coupled with Catherine's licentious private life, which offended Victorian susceptibilities far more than those of her own time – as Alexander Herzen put it in 1859, 'the history of Catherine cannot be read to ladies'[21] – it led historians to conclude that she was incapable of any virtuous action, let alone enlightened reform. The judgment is still repeated, even though its moral basis has been totally eroded. The permissive twentieth century may perhaps take a more charitable view of Catherine's private life. She was not in fact promiscuous, and the procession of her lovers was very orderly.

The charge that Catherine tried systematically to deceive the *philosophes* can be easily dismissed. Probably the most cultured person ever to sit on the Russian throne, and a genuine blue-stocking, she belonged by right to the cosmopolitan circle of intellectuals who subscribed to the *Correspondance Littéraire et Politique*, edited by her friend Melchior von Grimm. Her own temperament shines out in her letters to Grimm and to Voltaire – gay, witty, pragmatic, totally secular in tone. We do not know what she said to Diderot, but the notes he prepared for his discussions with her show how free and wide-ranging they were. If Diderot was disappointed in Russia, though not, it seems with Catherine herself, it was because he saw it before she had undertaken any of her great reforms, and because his visit burst the bubble of 'le mirage russe', the illusion that Russia was a blank sheet on which a *philosophe* could trace new institutions with the stroke of a pen. Diderot never commented on the Statute of 1775, and he died before the promulgation of the Charters of 1785 or the Statute on National Schools of 1786. He was in fact very ill-informed about Russia.

Similarly the charge that Catherine failed to put the maxims of her Instruction into effect belongs to a period when her major internal reforms were viewed from the perspective of liberal, populist or socialist historians, who never expected any good to come from the tsar. It also derives from the assumption that the Instruction represents a 'liberal' period in Catherine's reign. Yet there is nothing liberal in the political sense in the Instruction though many of its provisions were in fact implemented in a variety of ways and little by little. The attempt to introduce the rule of law in the courts, organised on estate principles; the abolition of torture in law and to a great extent in fact; the distinction introduced between detention pending trial and imprisonment as a punishment; the introduction of release on bail, these are but a few of the specific recommendations of the Instruction which were put into effect. The Charters of 1785 also represent the translation into law of some of the basic ideas expressed in the Instruction. What Russian liberal historians missed was any attempt to establish a permanent representative institution, or to limit absolute power by means of a constitution. Catherine never for one moment contemplated any dilution of her political power, and never deceived anyone in this respect. The first attempt to force her, as a woman with but a tenuous claim to the throne, to govern in consultation with a council of high officials, appointed by her but with the right to countersign laws, was brushed aside by December 1762. But to nineteenth-century liberal historians, the Enlightenment was identified with constitutionalism. This is a highly anachronistic approach, particularly if we bear in mind that the Legislative Commission met twenty years before the American constitutional convention.

It is necessary to refine the precise meaning of the concept of Enlightenment in the context of Catherinian reform. Like her contemporaries, she believed that it was desirable to maximise and hold on to state power in order to achieve reforms necessary for the welfare of the people. Many of the policies she considered useful had their origins in German cameralism. But by the middle of the eighteenth century, they were expressed in the 'rhetoric' of the eighteenth-century Enlightenment.[22] The style of government, the language in which policies were explained, reflected the values of the Enlightenment. The classical paternalism of the 'police' state was disguised in the language of 'progress'. This is particularly noticeable in the series of edicts of the years 1775–86 which laid down the basic parameters of a uniform adminis-tration and uniform rights throughout the empire, with scant regard to historical tradition. The language of the Enlightenment is also used to support the domination of the state over the church coupled with religious toleration in practice, if not in law. Catherine asserted her right to regulate Catholic, Protestant and Moslem religious organisation, including the appointment of Catholic bishops, while the Jews were not only granted religious toleration but more extensive civil rights than anywhere in Europe at the time. The educational programme was totally secular, and priests were not allowed to teach in state schools. The regulation of health and hygiene and the provision of inoculation centres and hospitals belongs more to the world of cameralism, as does the whole undercurrent of the control of national resources, improvement of communications, the land survey, efforts to improve agriculture (including the introduction of the potato with instructions on how to cook it), the freeing of manufacture from state shackles, but the language is still that of the Enlightenment.

The most damning indictment of Catherine is that she did nothing to improve the lot of the serfs. This has increasingly become the sole criterion for judging her claim to be an enlightened despot. Again, it is a criticism which only began to be made in the second half of the nineteenth century, when the peasant question dominated Russian political thought. It was given substance and weight by the massive and scholarly study of V. I. Semevsky published in 1903.[23] It was one sentence used by Semevsky which set the general tone: under Catherine, serfdom had become more intensive, 'in depth and in extent', not his overall judgment of her policies.

Nowhere was state initiative in fact likely to be more dangerous for the government than in matters concerning the peasantry. Serfdom was fundamental to the stability of the state, and the most important tax, the poll-tax, depended for its assessment and collection on the registration of the tax-payers where they lived. But at the beginning of her reign Catherine was able to implement a policy, already adumbrated by her predecessors, namely the secularisation of the church and monastic lands, thus winning the praises of Voltaire, and turning about one million serfs into state peasants.

Serf riots tended to mark the beginning of a new reign, since there was always the hope that the new ruler would take over private estates and turn them into state lands. There had indeed been a larger number than usual of riots on serf estates in the years 1765–68. It is difficult to pinpoint the causes, but in all probability serf awareness of the manifesto freeing the nobles from compulsory service deprived serfdom of legitimacy in the eyes of the serfs since it should logically have been followed by their own emancipation. Serfs were also well aware of the calling of the Legislative Commission to which the state peasants, many of whom lived in the same districts as the serfs, were sending deputies. It was also in these years that Catherine sponsored some experimental projects on court estates, with a view to trying out new methods of land tenure, and an essay competition which launched public discussion of the peasant question.

Unrest declined in the first two years of the Turkish War, but the cumulative impact of increased conscription, higher taxation, rising prices, sporadic food shortages, the plague which raged in Moscow and South Russia in 1771–72, the resentment bred in some of the borderland Cossack hosts by 'militarisation' and the heavy-handed repression of a local but quite serious Cossack mutiny led to the great rising which goes by the name of its leader as the Pugachev Revolt. Described as a peasant war, above all in Soviet historiography, it was in fact Cossack inspired, organised and led. It drew for support on the workers and peasants attached to industrial enterprises, on disgruntled Cossacks and state peasants, on Old Believers and on non-Russian tribes such as the Bashkirs. When the revolt moved into areas in which serf estates existed or predominated, the serfs of course joined in. It was typical of the 'monarchist illusions' of the leadership that Pugachev bound his followers to him by asserting that he was the legitimate tsar, Peter III, who had survived his wife's attempt on his life, and set up a court and a government modelled on that he intended to overthrow, down to the very titles of Catherine's courtiers.

In spite of the large area over which it eventually moved, the revolt of Pugachev never acquired enough consistency to threaten the heart of Russia, and its undisciplined hordes could not stand up to professional soldiers. It was over by August 1774, and was followed by a brief but savage period of repression. In more ways than one it was a tragedy for the peasantry. Their somewhat inchoate hopes for a Cossack Utopia were defeated. And the government saw to it that no such rising should ever occur again by disbanding the wild Zaporozhian Cossack host of the Dnieper, and binding the Don Cossacks, who had not taken part, even more closely to its service. But the revolt also put paid to any attempt by the government to regulate relations between serfs and their masters on a nationwide basis. Fear of an outbreak of the same kind led Catherine to prefer to consolidate the administration of the countryside in the hands of the class both qualified and desirous of performing such duties.

More than eighty years after he made it, the time has surely come to re-examine Semevsky's verdict quoted above. For instance, the law enabling landowners to exile serfs to Siberia for settlement antedated Catherine, and it extended the same power to state peasants, church serfs and townspeople. The peasants were the only social group to preserve this power until the twentieth century. Landowners, it is true, were empowered to send serfs to hard labour in the Admiralty (not in Siberia as frequently stated) without any form of state trial, but Russian legislation was so confused that Alexander I thought this power had been withdrawn with the Statute of 1775. The most severe law, of August 1767, which prohibited complaints against land-owners, except in cases of treason, on penalty of the *knut* and hard labour in the mines of Nerchinsk, repeated a provision of the Code of 1649 still in use, and seems to have been a temporary measure which was never repealed. The 'extension of serfdom' to the Ukraine is a misnomer. In the interests of uniformity, as advocated by Catherine, the poll-tax was extended both to the Baltic Provinces and to the Ukraine in 1783. This involved no change in the status of the peasants in Livonia, who were already fully enserfed. But in the Ukraine it implied the loss of the right to move from one estate to another, subject to the landowners' permission, which the peasant on privately-owned land still enjoyed in theory. Whether this involved the extension to the Ukrainian landowners of the full rights of Russian landowners (sale of serfs without land, removal from one estate to another, etc.) has not been adequately explored.

Catherine has also been criticised for making grants of settled estates, i.e. land with peasants, totalling some 800,000 people of both sexes, as rewards to servitors and favourites, thus turning 'free' state peasants into serfs. Semevsky himself, however, pointed out over eighty years ago that more than three-quarters of Catherine's land grants were in fact made from estates in territories annexed from Poland in the three partitions. It was thus Polish crown peasants and church and private serfs who were granted away, in accordance with the method, traditional in a country poor in specie like Russia, of rewarding faithful service. Catherine made very few grants from Russian state peasants, and she had a further reason for distributing Polish lands, namely the desirability of introducing Russian landowners and hence officials into non-Russian lands.

The assertion frequently made that Catherine's land grants were made in order to win the support of the nobility does not hold water if one examines the dates on which the bulk of the land grants were made. Until 1772, that is to say during the early period of Catherine's reign, when she was most in need of support, some 132,000 peasants of both sexes were granted. The remaining grants all date from after 1773, when Polish lands first became available, 260,000 being granted in what was to prove the *last* year of her reign. Finally one should note that Catherine limited the number of people allowed to own serfs, and stopped all previously-existing means of enserfing free men and women (including self enserfment) which had in the past led to many abuses.

Though Catherine certainly pursued traditional Russian aims in foreign policy, attempting to complete the work Peter had left unfinished, namely the acquisition of a foothold on the Black Sea, her extremely high-handed manner of conducting foreign policy provoked both the first and the second Turkish wars at a high cost in blood and treasure. War was taken for granted in eighteenth-century Europe as a means of acquiring territory and hence power, and in this respect Catherine does not compare unfavourably with her fellow monarchs. But her role in the destruction of Poland is particularly unsavoury. It is not so much that she took the initiative in the first partition of Poland, for the first hint came from Frederick II, as well as the ultimate pressure. It is rather that the whole of her policy towards Poland, from the election of Stanislas Poniatowski onwards, created a situation which rendered partition possible and the extinction of Poland inevitable after the Polish revolution and the constitution of 3 May 1791. Though Catherine shares the guilt for the partitions with the rulers of Prussia and Austria, yet she adopted bullying tactics almost unequalled at the time.

If one is to sum up, it is time that Catherine as a ruler should be re-evaluated. She was a rational human being; she worked hard, with well-chosen advisers, whom she trusted and who stayed in their posts for decades, not one of whom was ever sent into exile even when he was dismissed. She did not go about in snuff-stained coats like Frederick the Great, or shabby uniforms like Joseph II, and her court outdid theirs in brilliance and extravagance. In private life she preferred simplicity; she lit her own fire and made her own coffee in the morning, because she disliked being waited on, and her private parties at the Hermitage were remarkable for the ease and informality which prevailed. But in public and on state occasions she still believed in using the magic of monarchy, regarding the court as a necessary mechanism of government, a means of holding the allegiance of those who hoped to win power, fortune and favour, and of impressing the illiterate multitudes. She travelled extensively in her wide domains, attempting to acquaint herself with conditions.

Perhaps her main service to Russia was that she created a framework for government and society, more civilised, more tolerant, more free than ever before or after. But she did not find it possible to improve the lot of the private serfs, and it is this contrast between the freer and more orderly life of the upper classes, and even to some extent of the state peasants, and the unreformed state of private serfdom, which produces the impression that the situation of the serfs worsened during her reign. Yet they evidently benefited too from the increase in agricultural production, the growing domestic market, the development of craft industries, and even sometimes from facilities for education and promotion.

The difficulty about making a final assessment, or embarking on a political or sociological interpretation of Catherine's reign, is that though we know a great deal about the laws she enacted, we know so little about how they actually worked in practice. Was torture really no longer used? Was corporal punishment never inflicted in state schools? What sort of justice did the new courts dispense? Were the guarantees of property, life and honour observed? Were the rich and the powerful able to bend the law in their favour? Was Catherine merely the willing tool of a rapacious nobility, or did she, as she believed, preside over an autonomous state? In view of the present state of research we must suspend judgment for the time being. Experience suggests that practice lagged far behind theory, and many of Catherine's reforms were distorted in her reign already, and certainly in that of her son Paul, so that their ultimate import is hard to grasp. Nevertheless, in what she attempted to do, and in her style of government, she belongs clearly in the tradition of those who governed in what they believed to be the interests of the people and who go by the name of enlightened despots.

Bibliography

(Prominence is given to works in English, though foreign-language books and articles are cited where they are of particular importance or where no titles in English can be given.)

INTRODUCTION: THE PROBLEM OF ENLIGHTENED ABSOLUTISM
1. SOCIAL FORCES AND ENLIGHTENED POLICIES

The best introductions to these decades are the comprehensive and authoritative M. S. Anderson, *Europe in the eighteenth century 1713–1783* (3rd edn, London, 1987) and the notably lively William Doyle, *The Old European Order 1660-1800* (Oxford, 1978), while D. McKay and H. M. Scott, *The Rise of the Great Powers 1648–1815* (London, 1983) is an up-to-date survey of international rivalry. Norman Hampson, *The Enlightenment* ('The Pelican History of European Thought', vol. 4; Harmondsworth, 1968) is clear, readable and accessible, and is the best introduction to eighteenth-century thought, though Alfred Cobban, *In Search of Humanity* (London, 1960), remains a distinctive and enjoyable account of the Enlightenment. An important but rather more difficult discussion is Peter Gay, *The Enlightenment: an Interpretation* (2 vols, New York, 1966–69), while Ernst Cassirer, *The Philosophy of the Enlightenment* (1932; English transl., Princeton, 1951) was and, to a considerable extent, remains widely influential. There are good introductions to French thought by J. H. Brumfitt, *The French Enlightenment* (London, 1972) and Maurice Cranston, *Philosophers and Pamphleteers: Political Theorists of the Enlightenment* (Oxford, 1986); while the admirable 'Past Masters' series, published by Oxford University Press, contains several concise guides to major eighteenth-century thinkers, notably Peter France, *Diderot* (1983); Roger Scruton, *Kant* (1982); Judith N. Shklar, *Montesquieu* (1987) and Peter Burke, *Vico* (1985). For Voltaire, see Haydn Mason, *Voltaire: a biography* (London, 1981) and Peter Gay, *Voltaire's Politics* (Princeton, 1959); for Rousseau, James Miller, *Rousseau: dreamer of democracy* (New Haven, Conn., 1984); for Condorcet, K. M. Baker, *Condorcet: from natural philosophy to social mathematics* (Chicago, 1975). The important collection of essays edited by Roy Porter and Mikuláš Teich, *The Enlightenment in National Context* (Cambridge, 1981) reflects the present state of scholarship; see also three significant articles on the historiography of the Enlightenment: Robert Darnton, 'In search of the Enlightenment: recent attempts to create a social history of ideas', *Journal of Modern History*, 43 (1971); Lester G. Crocker, 'The Enlightenment: problems of interpretation' and John Lough, 'Reflections on "Enlightenment" and "Lumières" ', both in *L'età dei lumi: Studi storici sul settecento Europeo in onore di Franco Venturi* (2 vols, Naples, 1985): the second of these is reprinted in the *British Journal for Eighteenth-century Studies*, 8 (1985). For titles on the Italian Enlightenment, see the Bibliography to Chapter 2, below, p. 317; for works on German Cameralism, see that to Chapter 8, below, p. 333. Particular subjects and themes of importance are examined by: Franco Venturi, 'History and reform in the middle of the eighteenth century', in *The Diversity of History: Essays in Honour of Sir Herbert Butterfield*, ed. J. H. Elliott and H. G. Koenigsberger (London, 1970); Harry C. Payne, *The Philosophes and the People* (New Haven, Conn., 1976); Charles G. Stricklen, Jr, 'The *philosophe's* political mission: the creation of an idea, 1750–1789', *Studies on Voltaire and the Eighteenth Century*, 86 (1971), a valuable exploration of a crucial subject; D. D. Bien, *The Calas Affair* (Princeton, 1960); and John McManners, *Death and the Enlightenment* (Oxford, 1981), a brilliant study and one with wide implications; while the remarkable contemporary influence of Beccaria's *On Crimes and Punishments* (1764) is apparent from Franco Venturi's admirable edition: Cesare Beccaria, *Dei delitti e delle pene [con una raccolta di lettere e documenti relativi alla nascita dell' opera e alla sua fortuna nell' Europa del Settecento]* (Turin, 1965) with its notable introduction by the editor; there is an abbreviated French edition (Paris, 1965). Legal reform is also the focus of John H. Langbein, *Torture and the Law of Proof* (Chicago, 1977) and the collection of extracts edited by James Heath, *Eighteenth-Century Penal Theory* (Oxford, 1963). The real if limited impact of the Enlightenment on international relations is examined in contrasting ways by F. H. Hinsley, *Power and the Pursuit of Peace* (Cambridge, 1963) and F. Gilbert, 'The "New Diplomacy" of the eighteenth century', *World Politics*, 4 (1951), reprinted in the same author's *History: choice and commitment* (Cambridge, Mass., 1977); while particular statesmen influenced by the Enlightenment are examined by W. J. McGill, 'Kaunitz: the personality of political algebra', *Topic*, 34 (1978); O. T. Murphy, 'Charles Gravier de Vergennes: profile of an old regime diplomat', *Political Science Quarterly*, 83 (1968) and Lawrence J. Baack, 'State service in the eighteenth century – the Bernstorffs in Hanover and Denmark', *International History Review*, 1 (1979).

Brief introductions to the historiography of enlightened absolutism are provided by M. S. Anderson, *Historians and Eighteenth-Century Europe 1715–1789* (Oxford, 1979), pp. 119–31, and Charles Ingrao, 'The problem of "Enlightened Absolutism" and the German States', *Journal of Modern History*, 58: Supplement (1986). *Enlightened Absolutism (1760–1790): a documentary sourcebook*, ed. A. Lentin (Newcastle-upon-Tyne, 1985) is a well-chosen collection of short extracts from contemporary correspondence, and Stuart Andrews (ed.) *Enlightened Despotism* (London, 1967) an anthology of historians' views and eighteenth-century texts. Two brief discussions of value, though both are becoming dated, are John G. Gagliardo, *Enlightened Despotism* (London, 1968) and Henry E. Strakosch, *The Problem of Enlightened Absolutism* (London, 1970). Leonard Krieger, *An Essay on the Theory of Enlightened Despotism* (Chicago, 1975) is a difficult and at times impenetrable discussion, but it remains the only sustained attempt to grapple with a fundamental problem. Despite its title, François Bluche, *Le despotisme éclairé* (Paris, 1968) is primarily a comparative study (and a good one) of eighteenth-century monarchy. K. O. Freiherr von Aretin (ed.) *Der aufgeklärte Absolutismus* (Cologne, 1974) is an important collection of articles; the papers published in *Annales historiques de la Révolution Française*, 51 (1979) are far inferior, though they do include a convenient statement of Albert Soboul's views. Marxist approaches to the problem of enlightened absolutism can be followed in two volumes of proceedings from the 'Colloque de Mátrafüred': *Les Lumières en Hongrie, en Europe centrale et en Europe orientale*, eds, B. Köpeczi, E. Bene and I. Kovács (Budapest, 1977) and *L'absolutisme éclairé*, eds B. Köpeczi, A. Soboul, É. H. Balázs and D. Kosáry (Budapest, 1985) and in the more impressive writings of the distinguished East German historian Ingrid Mittenzwei: see in particular 'Über das Problem des aufgeklärten Absolutismus', *Zeitschrift für Geschichtswissenschaft*, 18 (1970) and 'Theorie und Praxis des aufgeklärten Absolutismus in Brandenburg–Preussen', *Jahrbuch für Geschichte*, 6 (1972). Some account of her approach can be had from Chapter 4 of Andreas Dorpalen, *German History in Marxist Perspective: the East German Approach* (London, 1985). A brief statement of Marc Raeff's views is to be found in 'The well-ordered police state and the development of modernity in seventeeth- and eighteenth-century Europe: an attempt at a comparative approach', *American Historical Review*, 80 (1975); the full-scale exposition is his *The Well-Ordered Police State: social and institutional change through law in the Germanies and Russia 1600–1800* (New Haven, Conn., 1983). The possibilities and the problems presented by this approach are made clear by the symposium in *Slavic Review* 41 (1982) (see in particular the contribution by Isabel de Madariaga) and by James Van Horn Melton, 'Absolutism and "Modernity" in Early Modern Central Europe', *German Studies Review*, 8 (1985). The best brief guide to the forces within states which resisted and inhibited reform is Dietrich Gerhard, 'Regionalism and corporate order as a basic theme of European history' in R. Hatton and M. S. Anderson (eds) *Studies in Diplomatic History: essays in memory of David Bayne Horn* (London, 1970).

Alfred Cobban, *A History of Modern France, vol. I: 1715–1799* (London, 1957) and C. B. A. Behrens, *The Ancien Régime* (London, 1967) are established introductions to eighteenth-century France, though both are now rather dated. The extent of Miss Behrens' second thoughts can be seen from her more recent *Society, Government and the Enlightenment: the experiences of eighteenth-century France and Prussia* (London, 1985). The best introduction is now *The French Revolution and the Creation of Modern Political Culture*. Vol I: *The Political Culture of the Old Regime*, ed. K. M. Baker (Oxford, 1987), a notably lively and up-to-date collection of essays in English and French. Detailed studies of importance include: Gerald J. Cavanaugh, 'Turgot: the Rejection of Enlightened Despotism', *French Historical Studies*, 6 (1969–70); Douglas Dakin, *Turgot and the Ancien Régime in France* (London, 1939); Steven L. Kaplan, *Bread, Politics and Political Economy in the Reign of Louis XV* (2 vols, The Hague, 1976), a massive study that has become a classic; Durand Echeverria, *The Maupeou Revolution: a study in the history of libertarianism – France, 1770–1774* (Baton Rouge, 1985); Robert D. Harris, *Necker: reform statesman of the ancien régime* (Berkeley and Los Angeles, 1979); Jean Egret, *Necker: ministre de Louis XVI 1776–1790* (Paris, 1975); two surveys of the reforms of Calonne and Brienne: Jean Egret, *The French Pre-Revolution 1787–88* (English trans, Chicago, 1977), Chapters 1–3, and Albert Goodwin, 'Calonne, the Assembly of French notables of 1787 and the origins of the "Révolte Nobiliaire"', *English Historical Review*, 61 (1946); and Harvey Chisick *The Limits of Reform in the Enlightenment: attitudes towards the education of the lower classes in eighteenth-century France* (Princeton, 1981). Studies which specifically examine enlightened absolutism in France include Louis Trénard, 'L'abolutisme éclairé: le cas français', *Annales historiques de la Révolution Française*, 51 (1979); a seminal study by Thadd E. Hall, 'Thought and practice of enlightened government in French Corsica', *American Historical Review*, 74 (1969); and two important articles by Maurice Bordes, 'Les intendants éclairés de la fin de l'Ancien Régime', *Revue d'histoire économique et sociale*, 39 (1961), and 'Un intendant éclairé de la fin de l'ancien régime: Claude-Françiois Bertrand de Boucheporn', *Annales du Midi*, 74 (1962).

There are some good recent treatments of the reign of Gustav III: see in particular Stewart Oakley, 'Gustavus III of Sweden', *Studies in History and Politics*, IV (1985) (Special Issue: *Essays in European History in honour of Ragnhild Hatton*) and H. Arnold Barton, 'Gustav III and the Enlightenment', *Eighteenth-Century Studies*, 6 (1972); a valuable account of his reign can also be had from H. Arnold Barton, *Scandinavia in the Revolutionary Era 1760–1815* (Minneapolis, 1986). The best non-Scandinavian biography is Claude Nordmann, *Gustave III: un démocrate couronné* (Lille, 1986). There is a detailed examination of the King's *coup* by Michael Roberts, '19 August 1772: an ambivalent revolution', in the first volume of *L'età dei lumi: studi storici sul settecento europeo in onore di Franco Venturi*, and a valuable introduction by the same author to *The Age of Liberty: Sweden 1719–1772* (Cambridge, 1986), which provides essential background.

2. THE ITALIAN REFORMERS

There are two recent and very useful general treatments: D. Carpanetto and G. Ricuperati, *Italy in the Age of Reason, 1685–1789* (London and New York, 1987) and the earlier chapters of S. J. Woolf, *A History of Italy, 1700-1860: the social constraints of political change* (London, 1979). The former gives rather more attention to intellectual life, though it may at times be somewhat difficult reading for the student with no background knowledge of the subject. In Italian there is the great work of F. Venturi, *Settecento riformatore* (5 vols, Milan, 1969–87), of which Vols I, II and V deal with Italy. This is a masterpiece of scholarship but extremely detailed and a study of ideas rather than of events. An older and more manageable single-volume account is F. Valsecchi, *L'Italia nel Settecento. Dal 1714 al 1788* (Milan, 1959). The most useful short account in English of Italian intellectual life is probably O. Chadwick, 'The Italian Enlightenment', in R. Porter and M. Teich (eds), *The Enlightenment in National Context* (Cambridge, 1981), while the essays in F. Venturi, *Italy and the Enlightenment* (London, 1972) throw light on particular aspects of the subject. The problem of the church is well treated in O. Chadwick, *The Popes and European Revolution* (Oxford, 1981) and for a short but important part of the century in F. Venturi, 'Church and reform in Enlightenment Italy: the sixties of the eighteenth century', *Journal of Modern History*, 48 (1976), 215–32. C. A. Bolton, *Church Reform in Eighteenth-Century Italy* (The Hague, 1969) is an interesting discussion of the work of Scipione di Ricci and the Synod of Pistoia. The essential biography of the Grand Duke Leopold is A. Wandruszka, *Leopold II: Erzhertzog von Österreich, Grosshertzog von Toskana, König von Ungarn und Böhmen, Römischer Kaiser*, 2 vols (Vienna, 1964–65): the first volume covers his work in Tuscany. There is also a useful short account of this in H. Holldack, 'Die Reformpolitik Leopolds von Toskana', *Historische Zeitschrift*, 165 (1941) 23–46, and some relevant information in H. Burr Litchfield, *Emergence of a Bureaucracy: the Florentine patricians, 1530–1790* (Princeton, 1986). E. W. Cochrane, *Florence in the Forgotten Centuries, 1527–1800: a history of Florence and the Florentines in the age of the grand dukes* (Chicago and London, 1973) is the only significant general work in English. On the duchy of Milan there are the useful though poorly written study of D. M. Klang, *Tax Reform in Eighteenth-Century Lombardy* (Boulder, Colo., East European Monographs, 1977) and D. A. Limoli, 'Pietro Verri: a Lombard reformer under Enlightened Absolutism and the French Revolution', *Journal of Central European Affairs*, 18 (1958), 254–80. The most recent Italian account of the duchy's history is that in D. Sella and C. Capra, *Il Ducato di Milano dal 1535 al 1796* (Turin, 1984). On Naples there is the readable but superficial H. M. Acton, *The Bourbons of Naples, 1734–1860*, rev. ed. (London, 1961), while early attempts at reform are covered in M. Schipa, *Il Regno di Napoli al tempo di Carlo di Borbone*, 2 vols (Naples, 1923). P. Chorley, *Oil, Silk and*

Enlightenment: economic problems in XVIIIth-Century Naples (Naples, 1965) is an interesting detailed study whose scope is wider than the title suggests; and D. Mack Smith, *A History of Sicily*, ii, *Modern Sicily after 1713* (London, 1968), is the only good English discussion of the island and its problems. The progressive but hardly enlightened government of Savoy-Piedmont is covered in G. Symcox, *Victor Amadeus II: Absolutism in the Savoyard state, 1675–1730* (London, 1983), and in great detail in G. Quazza, *Le Riforme in Piemonte nella prima metà del Settecento* (Modena, 1957).

3. POMBAL: THE PARADOX OF ENLIGHTENMENT AND DESPOTISM

The Pombaline period, despite the importance attributed to Pombal by historians, remains very much understudied. The principal biographies dating from the nineteenth century are more doctrinal tracts in the ongoing battle between liberals and traditionalists than the products of serious archival research. The body of this literature – both for Pombal and against him – is immense. The National Library of Lisbon published a bibliography of over 3,000 items on the bicentennial of Pombal's death in 1982 (*Marquês de Pombal: Catálogo Bibliográfico e Iconográfico*, ed. António Barreto, Lisbon, Biblioteca Nacional, 1982.) In English the best account is still that of John Smith published in the 1840s. It is highly favourable to Pombal. A more recent popular biography was written by a British diplomat Marcus Cheke in the 1930s (*Dictator of Portugal: Life of the Marquis of Pombal*, London, 1938.) The 1982 bicentennial of Pombal's death provoked a series of publications (and republications) of mixed value. The Jesuits produced a special edition of the magazine *Broterio* (2 vols, 1982) and the Masons, *Pombal Revisitado* ed. Maria Helena Carvalho dos Santos (2 vols, Lisbon, 1982). The most useful collection, however, is that contained in the *Revista de História das Ideias: O Marquês de Pombal e os seu Tempo*, ed. Luis Reis Torgal e Isabel Vasques, 2 vols (Lisbon, 1982). There was also an excellent catalogue, *Exposição, Lisboa e o Marquês de Pombal*, 3 vols, (Museu da Cidade, Lisbon, 1982) which contains many illustrations. The best general introduction to Pombal in Portuguese remains José Lucio de Azevedo, *O marquês de Pombal e a sua época* (2nd edn. Lisbon, 1922), though the volume by Verissimo Serrão, *Pombal* (Lisbon, 1982) is helpful. The most comprehensive overview of the theoretical background to Pombal's legislation is Francisco José Calazans Falcon, *A época Pombalina: política económica e monarquia illustrada* (São Paulo, 1982).

Brazilian historians have made a substantial contribution to our knowledge of the imperial dimensions of Pombal's concerns which help clarify many elements of his actions. The eighteenth century was after all one of Portugal's great and most creative colonial ages, and it deserves its just place alongside the history of the adventures and misadventures of the Portuguese experience overseas in Asia during the Renaissance and in Africa in more recent times. The European dimension, both diplomatic and economic, has also received some attention, especially the latter, as has the history of ideas and role of the Enlightenment in Portugal.

Portuguese historians have been on the whole unconcerned with the Pombal's placed among the reformers of his age. The most recent contributions to Pombaline scholarship in fact go so far as to deny Pombal any originality at all, something even his contemporaries recognized. This national self-effacement may be in the nature of Portuguese historiography which is more concerned with what might be called the vertical rather than the horizontal dimension – that is Pombal is almost always seen in terms of the projection of his activity into the disputes of the nineteenth century rather than his projection within the world of the eighteenth century. Pombal, however, despite this remains one of the more interesting rulers of the period, and the body of his own writing is surprisingly rich, which means the historian can discover more about his thoughts and motivations than is the case with many of his contemporaries. Among the major contributions to Pombaline studies are those of Jorge Borges de Macedo (*A Situação económica no tempo de Pombal*, Oporto, 1951) and Vitorino Magalhães Godinho (*Prix et Monnaies au Portugal 1750–1850*, Paris, 1955), who have transformed our view of the economic context – the conjuncture – within which Pombal acted. There are very few analyses of economic and social structures. Albert Silbert, *Le Portugal Méditerranéen à la fin de l'Ancien Régime, XVIIIᵉ – Début du XIXᵉ siècle* (2 vols, Paris, 1966) is useful, but concentrates mainly on the nineteenth century and is concerned only with Central and Southern Portugal. The Anglo-Portuguese commercial and diplomatic connections are well covered in H. E. S. Fisher *The Portugal Trade* (London, 1971), and David Francis, *Portugal 1715–1808* (London, 1985); Virgilio Noya Pinto, *O Ouro Brasileiro e o Comercio Anglo-Português* (São Paulo, 1979) has provided important data on colonial economic history. José Augusto França has delineated the social and cultural dimensions of one of the most lasting of Pombal's inspirations, the reconstruction of Lisbon in *Une Ville des Lumières: la Lisbonne de Pombal* (Paris, 1965). Romulo de Carvalho and Antonio Alberto Banha de Andrade have helped reshape views of the intellectual environment. Especially useful are the collected essays of Banha de Andrade, *Contributos para a História da Mentalidade Pedagogica Portuguesa* (Lisbon, 1981) and Rómulo de Carvalho, *História da Fundação do Colégio Real dos nobres de Lisboa (1761–1772)* (Coimbra, 1959). Also useful are the three substantial articles by J. S. da Silva Dias in *Cultura* (Lisbon) vol. 1 (1982) 45–114, vol. 2 (1983) 185–318, and vol. 3 (1984) 27–151. For the military reform there is a good coverage in Christina Banaschik-Ehl, *Scharnhorsts Lehrer: Graf Wilhelm von Schaumburg-Lippe in Portugal: Die Heeresreform 1761–1777 – 3 Studien zur Militärgeschichte, Militärwissenschaft und Konfliktsforschung*

(Osnabrück, 1974). Marcos Carneiro de Mendonça has provided a series of fundamental documentary collections including the important correspondence of Pombal and his brother, *A Amazonia na era Pombalina* (3 vols, Rio de Janeiro, 1963). Susan Schneider provided a fascinating analysis of the Douro during Pombal's rule in her *O Marquês de Pombal e o Vinho do Porto* (Lisbon, 1980). In the case of the administrative reforms overseas, Dauril Alden wrote an account of the viceroyalty of the marquês of Lavradio, one of Pombal's more impressive collaborators, *Royal Government in Colonial Brazil* (Berkeley and Los Angeles, 1969) and Heloise Liberalli Bellotto has provided an excellent complement to this with her study of the activities of the Morgado de Mateus as governor of the captaincy of São Paulo in Brazil, *Autoridade e Conflito no Brasil Colonial: o Governo do Morgado de Mateus em São Paulo* (São Paulo, 1979). The monopoly companies have been investigated by several historians, among them José Ribeiro Junior on the Pernambuco company, *Colonização e Monopólio no Nordeste Brasileiro: a Companhia Geral de Pernambuco e Paraíba 1759-1780* (São Paulo, 1976), and Manuel Nunes Dias on the company of Pará and Maranhão, *A Companhia Geraldo Grão Pará e Maranhão 1755-1778* (2 vols, Pará 1970). On the role of the church we now have the excellent monograph by Samuel J. Miller, *Portugal and Rome c. 1748-1830: an aspect of the Catholic Enlightenment* (Rome, 1978). This chapter draws on the author's previous work, especially Kenneth Maxwell, 'Pombal and the Nationalization of the Luso-Brazilian Economy', *Hispanic American Historical Review* 40 (1968) and *idem, Conflicts and Conspiracies: Brazil and Portugal 1750-1808* (Cambridge, 1973). The focus of the latter is mainly on colonial policy.

Great lacunae remain. We still know very little about the man himself, his family, and even less about the court and the enigmatic Dom José I, without whose acquiescence (or passivity) Pombal would have achieved nothing. We need to know more about the Portuguese actions in Africa and Asia in this period despite the work of Fritz Hoppe, *A África Portuguesa no Tempo do Marquês de Pombal 1750-1777* (Lisbon, 1970) and António Carreira *As Companhias Pombalinas* (Lisbon, 1983). There is paucity of serious studies of the international influences or connections during the Pombaline period, especially in as far as these had a direct role in Pombal's measures. These were, however, surprisingly extensive and a potential topic for some interesting comparative work. Some preliminary indications of these interactions can be seen in *Pombal Revisitado*, vol. 1, pp. 287-306 (Lisbon, 1982), and in Romulo de Carvalho's remarkable *Historia do Gabinete de Fisica da Universidade de Coimbra* (Coimbra, 1978). The reform of the University of Coimbra was of particular interest to Spanish reformers, as can be seen in the correspondence between Cenáculo and the Count of Aguilar in Seville and Campomanes in Madrid. Antonio Mestre, *Ilustración y reform de la Iglesia: pensamiento político religiosa de Don Gregorio Mayans y Siscar (1699-1781)* (Valencia, 1968), apéndice documental no. 9, pp. 495-97. Also, Claude-Herni

Freches, *Voltaire, Malagrida et Pombal* (Paris, 1969). An excellent account of the remarkable career of 'Antonio Ribeiro Sanches, élève de Boerhaave, et son importance pour la Russie', by David Willemse, is to be found in *Janus: Revue internationale de l'histoire des sciences, et de la médecine, de la pharmacie et de la technique*, vol. VI, suppléments (Leiden, 1966). We have little available on the bureaucracy and the state apparatus, however, despite the fact that Pombal's action was to make the state all-encompassing and monopolistic in its power. The vast body of Pombaline legislation is rarely examined for what it says about the motivations and justification for the reforms codified. We know little in detail about the households of the aristocracy. It is also clear that we need a new look at literary production during his period both high and low. The whole propaganda effort of the Portuguese in this period for instance cries out for attention.

4. CHARLES III OF SPAIN

The number of works in English on Caroline Spain is not great. However the pioneering study of later-eighteenth-century Spain by Richard Herr, *The Eighteenth-Century Revolution in Spain* (Princeton, 1958), remains perhaps the best place to begin in any language. Despite the revisions historians have made of several of his conclusions, Herr's work is fundamental. A perceptive and subtle brief analysis of Spain's condition and of the reform movement may be found in the first two chapters of Laura Rodríguez, *Reforma e ilustración en la España del siglo XVIII. Pedro Rodríguez Campomanes* (Madrid, 1975). Two outstanding general syntheses covering the entire century are A. Domínguez Ortiz, *Sociedad y estado en el siglo XVIII español* (Barcelona, 1976) and Gonzalo Anes Alvarez, *El antiguo régimen. Los Borbones* (2nd edn, Madrid, 1976), while Vicente Palacio Atard, in *Los españoles de la ilustración* (Madrid, 1964) presents a collection of his provocative essays, primarily on changing attitudes, and dealing with neglected issues like the role of women in elite culture.

Economic development of the period has been analysed in three excellent studies: Pierre Vilar's classic *La Catalogne dans l'Espagne moderne* (3 vols, Paris, 1962) one of the fundamental starting points of recent Spanish historical studies; Antonio García-Baquero González, *Cádiz y el Atlántico (1717–1778)* (2 vols, Seville, 1976) and the collection of essays on regional developments, economic as well as cultural and political, edited by Roberto Fernández, *España en el siglo XVIII. Homenaje a Pierre Vilar* (Barcelona, 1985). The editor's introduction provides an outstanding analysis of recent research and of the evolution of Spain during the period. Joseph Townsend looked at Spain with a strong economic interest in his *A Journey Through Spain in the Years 1786 and 1787* (2nd edn, 3 vols, London, 1792).

Among the important studies of the life and reign of Charles III are two essays by men who knew him – Francisco Cabarrús in his *Elogio de Carlos III Rey de España y de las Indias* (Madrid, 1789) raises the King and his accomplishments as does Carlos Gutiérrez de los Ríos, conde de Fernán Núñez in *Vida del Rey d. Carlos III de España* (Madrid, 1898, 2 vols.). The latter, written by a cultivated aristocrat and courtier provides a detailed picture of Charles's palace life and of his conversations. In English, Anthony H. Hull's *Charles III and the Revival of Spain* (Washington, D. C., 1980) provides a helpful synthesis despite the author's failure to take full advantage of recent research. Harold Acton, in the early chapters of *The Bourbons of Naples (1734–1825)* (London, 1956) provides a gossipy but informative view of Charles' Neapolitan reign, and William Coxe, *Memoirs of the Kings of Spain of the House of Bourbon* (5 vols, London, 1815) remains an important source. The fullest study of the reign is Manuel Danvila Collado's monumental *Reinado de Carlos III* (6 vols, Madrid, 1893–96), still a starting point for serious study of the period and helpful for Charles' years in Naples. Antonio Ferrer del Río, *Historia del Reinado de Carlos III en España* (4 vols, Madrid, 1856) also remains important. Political history of the age has invited little significant research in recent years, and detailed analysis of Caroline politics is lacking. Among the few important studies are those by Rafael Olaechea and José A. Ferrer Benimeli, *El Conde de Aranda. Mito y realidad de un político aragonés* (2 vols, Zaragoza, 1978), an unusually fine revisionist study based on important new research; and Juan Hernández Franco, *La gestión política y el pensamiento reformista del conde de Floridablanca* (Murica, 1984). Also revealing is another work by Rafael Olaechea – 'El anticolegialismo del gobierno de Carlos III', in *Cuadernos de Investigacíon. Geografía e Historia*, II: 2 (1976), while Janine Fayard's 'Los ministros del Consejo Real de Castilla (1746–1788)', *Cuadernos de Investigación Histórica*, 6 (1982), demonstrates the impact of Charles's determination to break colegial power in the crown's foremost deliberative body. Vicente Rodríguez Casado's *La Política y los políticos en el reinado de Carlos III* (Madrid, 1962) is helpful although the author insists on the middle-class nature of Caroline reform.

The Enlightenment and the relationship of reform to it have been widely studied. Among the important older works is the pioneering *La España ilustrada de la segunda mitad del siglo XVIII* by Jean Sarrailh (Mexico, 1957), originally published in French, as well as Richard Herr's work cited above. More recent studies include: Antonio Elorza, *La ideología liberal en la ilustración española* (Madrid, 1970), which analyses the radical transformation of Spanish thought in the 1780s; Francisco López, 'La historia de las ideas en el siglo XVIII: concepciones antiguas y revisiones necesarias', *Boletín del Centro de Estudios del Siglo XVIII* (BOCES XVIII) no. 3 (1975) a revisionist study based on recent reading; Antonio Mestre, *El mundo intelectual de Mayáns* (Valencia, 1978), a collection of revealing essays about reformist ideas and politics in

Valenica and Madrid; Marcelin Defourneaux, *Inquisición y censura de libros en la España del siglo XVIII* (Madrid, 1973) which studies the intellectual atmosphere in which the Inquisition operated; and Paula de Demerson, *María Francisca de Sales Portocarrero (Condesa de Montijo). Una figura de la ilustracíon* (Madrid, 1975), a biographical study of a leading woman courtier and intellectual. A helpful work in English on Feijoo and his intellectual world is *Benito Jerónimo Feijoo* by I. L. McClelland (New York, 1969).

Among leading recent studies of the uprisings of 1766 are the articles by Laura Rodríguez, 'The Spanish Riots of 1766', *Past and Present*, no. 59 (1973), which appear in a somewhat longer version in her book, cited above; Pierre Vilar's not always convincing analysis, 'El "Motín de Esquilache" y las "crisis del antiguo régimen" ', *Revista de Occidente*, 107 (1972); and the important study by Teófanes Egido, 'Madrid, 1766: "Motines de corte" y oposicion al gobierno', *Cuadernos de Investigación Histórica*, 3 (1979). On the church and clergy, William J. Callahan's *Church, Politics, and Society in Spain, 1750–1874* (Cambridge, Mass., 1984) in its early chapters provides a unique and perceptive synthesis. Also extremely valuable are the contributions on regalism and the Jesuits by Teófanes Egido and Antonio Mestre Sanchis in the latter's *La Iglesia en la España de los siglos XVII y XVIII*, vol. IV of Ricardo García Villoslada (ed.) *Historia de la Iglesia en España* (Madrid, 1978) and the study of Jansenism by Maria Giovanna Tomsich, *El jansenismo en España. Estudio sobre ideas religiosas en la segunda mitad del siglo XVIII* (Madrid, 1972). In *El obispo de Barcelona Josep Climent i Avinent (1706–1781). Contribución a la historia de la teologia pastoral tarraconense en le siglo XVIII* (Barcelona, 1978), Francesc Tort Mitjans examines the career of a leading Jansenist reformer and friend of Roda while C. C. Noel in 'Opposition to Enlightened Reform in Spain: Campomanes and the Clergy, 1765–1775', *Societas*, (winter, 1973), examines several confrontations between reformers and their clercial opponents.

University and *colegio mayor* reform have been the focus of considerable recent research. Richard L. Kagan in *Students and Society in Early Modern Spain* (Baltimore, 1974) discusses the decline of the universities and other educational institutions and the reforms of the Caroline period. George M. Addy, *The Enlightenment in the University of Salamanca* (Durham, N.C., 1966) and Antonio Alvarez de Morales, *La 'ilustración' y la reforma de la universidad en la España del siglo XVIII* (Madrid, 1971) deal in greater detail with and with good insight into the reform process, while José García Lasaosa, *Planes de reforma de estudios de la Universidad de Zaragoza en la segunda mitad del siglo XVIII* (Zaragoza, 1978) provides a detailed examination of a little-known series of curriculum reforms. Interest in the economic societies has produced an even larger number of recent works. Among the most helpful are: Robert Jones Shafer, *The Economic Societies in the Spanish World (1763–1821)* (Syracuse, 1958), perhaps the best general study in any language, and particularly valuable for its examination of the American societies and their background; Ramón Carande Tovar, 'El despotismo ilustrado de los Amigos del País', in his *Siete Estudios de Historia de Espāna* (Barcelona, 1969) and Gonzalo Anes Alvarez, 'Coyuntura económica e "ilustración". Las socièdades de amigos del país', in his *Economía e "ilustración" en la España del siglo XVIII* (Barcelona, 1969) – both general studies by eminent historians; Jorge and Paula de Demerson, 'La decadencia de las Reales Sociedades de Amigos del País,' *BOCES XVIII*, 4–5 (1977), an important study based on archival materials; and, as an example of an outstanding study of one of the most successful societies, José Francisco Forniés Casals, *La Real Sociedad Económica Aragonesa de Amigos del País en el período de la ilustración (1776–1808): Sus relaciones con el artesanado de la industria* (Madrid, 1978).

Studies of economic reform are less numerous than might be expected, but among the most revealing are the well-known study by Marcelin Defourneaux, *Pablo de Olavide ou l'Afrancesado (1725–1803)* (Paris, 1959) valuable for, among other things, the author's analysis of the Sierra Morena colonies. The colonisation movement is set in its longer term context by Ana Olivera Poll and Antonio Abellán García in 'Las nuevas poblaciones del siglo XVIII en España,' *Hispania*, 46 (1986), and Jean-Pierre Amalric, 'En el siglo XVIII: ¿Una agricultura agarrotada?' in B. Bennassar *et al. Orígines del atraso económico español* (Barcelona, 1985) makes clear the agricultural progress that was being made in some regions. William J. Callahan's important book, *Honor, Commerce and Industry in Eighteenth-Century Spain* (Boston, 1972) reveals the extent to which economic progress was linked to changing conceptions of honour, while James Clayburn Laforce, *The Development of the Spanish Textile Industry 1750–1800* (Berkeley, 1965) clarifies the crown's role in this important sector. The most outstanding discussion of economic reforms and changes however is contained in the collection edited by Roberto Fernández cited above.

American reforms have until recently, attracted relatively little attention. But especially helpful are Stanley J. Stein and Barbara H. Stein, *The Colonial Heritage of Latin America. Essays on Economic Dependence in Perspective* (New York, 1970), in establishing the complex background in several provocative chapters, and Miguel Artola, 'America en el pensamiento español del siglo XVIII', *Revista de Indias*, 29 (1969), in elucidating the evolution of American policy. Geoffrey Walker, *Spanish Politics and Imperial Trade 1700–1789* (London, 1979) discusses Caroline reform in less detail than might be expected, but Mark. A. Burkholder and D. S. Chandler in *From Impotence to Authority. The Spanish Crown and the American Audiencias 1687–1808* (Columbia, Mo. 1977), set a range of important reforms in their wider context in this suggestive book. See also the article by Stanley J. Stein, 'Bureaucracy and Business in the Spanish Empire, 1759–1804: Failure of a Bourbon Reform in Mexico and Peru', *Hispanic American Historical Review*, 61 (1981) and the ensuing debate between this author and Jacques A. Barbier and Mark A. Burkholder, *ibid.*, 62 (1982). Finally, several significant works have been published on less publicised reforms. Jacques Soubeyroux, *Pauperisme et rapports sociaux à Madrid au XVIIIeme siècle* (2 vols, Lille and Paris, 1978) and William J. Callahan, 'The problem of confinement: an aspect of poor relief in eighteenth-century Spain', *Hispanic American Historical Review*, 51 (1971) deal with poverty, charity and the reform of the latter; Pilar Cuesta Pascal, 'Los alcaldes de barrio en el Madrid de Carlos III y Carlos IV', *Anales del Instituto de Estudios Madrileños*, 19 (1982) examines this important institution in a suggestive essay; Miguel Artola in a long and perceptive book examines the mysterious royal finances and their reform, *La Hacienda del antiguo régimen* (Madrid, 1982); and José P. Merino Navarro in *La armada española en el siglo XVIII* (Madrid, 1981), analyses the numerous reforms of the period which strengthened the navy and related industries. In such a brief essay it is impossible to present a full roster of the many significant contributions made during the last several decades to the renaissance of eighteenth-century historical studies in Spain. The ambitious student is referred to Luis Suárez Fernández *et al.* (eds) *Historia General de España y América* (Madrid), vols. X-1, X-2 and XI-1 (1983–84) where up-to-date texts written by a number of outstanding scholars and especially helpful bibliographical essays will be found.

5. REFORM IN THE HABSBURG MONARCHY

The last generation has seen a proliferation of research and publications on the reform era in the Habsburg Monarchy. Happily, the quantity and quality of writing in English has also improved significantly and a substantial number of items can now be recommended. The list is headed by two of the most important books in any language on the reign of the Empress: Derek Beales, *Joseph II,* Vol I: *In the Shadow of Maria Theresa, 1741–1780* (Cambridge, 1987), the first volume of what will be the standard biography and now the most accessible introduction to the period up to 1780; and P. G. M. Dickson, *Finance and Government Under Maria Theresia, 1740–1780* (2 vols, Oxford, 1987), a remarkable and pathbreaking study of all aspects of internal developments and a book with far wider implications than the title might suggest. Derek Beales has also written an important article on the way in which forged documents have frequently misled historians: 'The false Joseph II', *Historical Journal* 18 (1975), 467–95. The best discussion of the period after 1780 in English is that in T. C. W. Blanning, *Joseph II and Enlightened Despotism* (London, 1970). P. P. Bernard, *Joseph II* (New York, 1968) is less convincing, particularly on the motivation of his reforms. The most authoritative study of the Emperor's hectic personal rule is the pioneering and widely-influential Paul von Mitrofanov,

Josef II: Seine politische und kulturelle Tätigkeit (2 vols, German transl. of 1907 Russian edn, Vienna, 1910), while the catalogue of the 1980 Melk exhibition, *Österreich zur Zeit Kaiser Josephs II*, ed. K. Gutkas (Vienna, 1980) contains a valuable and up-to-date series of articles. There is nothing as good on the Empress: W. J. McGill, *Maria Theresa* (New York, 1972) is a useful brief life, while Eugen Guglia, *Maria Theresia: Ihr Leben und Ihre Regierung* (2 vols, Munich, 1917) is a detailed though rather conventional biography. The imperial family have all but monopolised attention, at least in English, though a certain amount can be gleaned from Frank T. Brechka, *Gerard van Swieten and his World 1700–1772* (The Hague, 1970), while Joseph von Sonnenfels is discussed in pt IV of Robert A. Kann, *A Study in Austrian Intellectual History: From Late Baroque to Romanticism* (London, 1960), and P. P. Bernard has written a short study of 'The *philosophe* as public servant: Tobias Philip Gebler', *East European Review Quarterly*, 7 (1973) 41–51. There is an important monograph on Habsburg government in Galicia by Horst Glassl, *Das österreichische Einrichtungswerk in Galizien 1772–1790* (Wiesbaden, 1975) and a rather less satisfactory survey of Austrian rule in the Southern Netherlands by W. W. Davis, *Joseph II: an imperial reformer for the Austrian Netherlands* (The Hague, 1974) which deals with the 1780s; while chapters 11, 14 and 15 of Dino Carpanetto and Giuseppe Ricuperati, *Italy in the Age of Reason 1685–1789* (London, 1987) provide an up-to-date Italian view of Habsburg rule in the peninsula. Two recent collections which together give a reasonably comprehensive picture of the present state of research are the lavishly-illustrated *Maria Theresia und Ihre Zeit*, ed. Walter Koschatzky (Salzburg and Vienna, 1979) and *Österreich im Europa der Aufklärung: Kontinuität und Zäsur in Europa zur Zeit Maria Theresias und Josephs II*, ed. R. G. Plaschka and Grete Klingenstein (2 vols, Vienna, 1985), the second volume of which contains (pp. 969–1051) a very full and quite up-to-date bibliography of the decades from 1740 to 1790. Only a few of the more important German titles are included in the list that follows.

R. J. W. Evans, *The Making of the Habsburg Monarchy 1550–1700* (Oxford, 1979) is fundamental for any understanding of the polity overhauled after 1740. A rather different view of the reform era is provided by the sprightly and well-illustrated, if now rather dated, E. Wangermann, *The Austrian Achievement 1700–1800* (London, 1973) and the same author's more assertive 'Reform Catholicism and political radicalism in the Austrian Enlightenment' in *The Enlightenment in National Context*, eds. Porter and Teich. The intellectual environment can best be approached through Derek Beales, 'Christians and "*philosophes*": the case of the Austrian Enlightenment' in *History, Society and the Churches: essays in honour of Owen Chadwick*, eds D. Beales and Geoffrey Best (Cambridge, 1985), while Grete Klingenstein, *Staatsverwaltung und kirchliche Autorität im 18. Jahrhundert* (Munich, 1970) is a fundamental discussion not merely of its avowed subject – censorship – but of the whole environment of reform. Chapter 4 of Keith Tribe, *Governing Economy: the reformation of German economic discourse 1750–1840* (Cambridge, 1988) provides an important discussion of cameralism in Austria, while something of the intellectual flavour of the period is conveyed by P. P. Bernard, *Jesuits and Jacobins: Enlightenment and Enlightened Despotism in Austria* (Urbana, Ill. 1971).

Social and economic themes are only beginning to receive adequate attention. The opening chapters of David F. Good, *The Economic Rise of the Habsburg Empire 1750–1914* (Berekeley and Los Angeles, 1984) and the rather theoretical John Komlos, 'Institutional change under pressure: Enlightened government policy in the eighteenth-century Habsburg monarchy', *Journal of European Economic History*, 15 (1986) are more favourable in their assessment of the effects of government intervention than Hermann Freudenberger, 'An industrial momentum in the Habsburg monarchy', *ibid*, 12 (1983) and *idem*, 'State intervention as an obstacle to economic growth in the Habsburg monarchy', *Journal of Economic History*, 27 (1967). Freudenberger has also produced a significant case-study: *The Industrialization of a Central European City: Brno and the fine woollen industry in the 18th century* (Edington, Wilts., 1977). The collection of essays edited by Herbert Matis, *Von der Glückseligkeit des Staates: Staat, Wirtschaft und Gesellschaft in Österreich im Zeitalter des aufgeklärten Absolutismus* (Berlin, 1981) contains much important material, though at times the discussion is rather theoretical. Tariff policy is explored by Helen P. Liebel, 'Free Trade and protectionism under Maria Theresa and Joseph II', *Canadian Journal of History*, 14 (1979), while one of the Emperor's most celebrated initiatives is the subject of P. P. Bernard, 'The limits of absolutism: Joseph II and the *Allgemeines Krankenhaus*', *Eighteenth-Century Studies*, 9 (1975–76).

The essential starting-point for the administrative reforms is the discussion in Dickson, *Finance and Government*, vol. I, chs 9–11, which significantly modifies the picture to be derived from the older German literature and, above all, the writings of Friedrich Walter: *Die österreichische Zentralverwaltung in der Zeit Maria Theresias* (Vienna, 1938) and his shorter study, *Die theresianische Staatsreform von 1749* (Vienna, 1958); see also Franz A. J. Szabo, 'Haugwitz, Kaunitz and the structure of government in Austria under Maria Theresia', *Historical Papers* (Canada) (1979) for a brief if rather traditional survey. There is an important study of legal reform by Henry Strakosch, *State Absolutism and the Rule of Law: the struggle for the codification of civil law in Austria, 1753–1811* (Sydney, 1967) and a more anecdotal one by P. P. Bernard, *The Limits of Enlightenment: Joseph II and the law* (Urbana, Ill., 1979). One out-of-the-way aspect of the Emperor's activities is outlined by Hans Wagner, 'The pension payments and legal claims of Maria Theresa and their withdrawal by Joseph II' in *Intellectual and Social Developments in the Habsburg Empire from Maria Theresa to World War I: essays dedicated to Robert A. Kann*, eds S. B. Winters and Joseph Held (Boulder, Co., 1975).

The best introduction to the vexed and difficult subject of religious reform is now the discerning account in Chapter 14 of Beales, *Joseph II*, while a wider perspective is provided by Owen Chadwick, *The Popes and European revolution* (Oxford, 1981). To take the subject further involves reading the fundamental discussions in German. See, in particular, Klingenstein, *Staatsverwaltung und kirchliche Autorität*; Peter Hersche, *Der Spätjansenismus in Österreich* (Vienna, 1977), a splendid exploration of the wide-ranging influence of Jansenism; a collection of essays edited by Elisabeth Kovács, *Katholische Aufklärung und Josephinismus* (Munich, 1979); E. Zlabinger, *L. A. Muratori und Österreich* (Innsbruck, 1970); E. Winter, *Der Josefinismus* (Berlin, 1962); and F. Maass, *Der Frühjosephinismus* (Vienna, 1969). Good introductions to the subject of toleration are Charles H. O'Brien, *Ideas of Religious Toleration at the Time of Joseph II* (Philadelphia: *Transactions of the American Philosophical Society*, n.s. 59:7 [1969]) and P. P. Bernard, 'Joseph II and the Jews: the origins of the toleration patent of 1782', *Austrian History Yearbook*, IV–V (1968–9). There is a fine new book on the educational reforms by James van Horn Melton, *Absolutism and the Eighteenth-Century Origins of Compulsory Education in Prussia and Austria* (Cambridge, 1988), though some of its broader arguments are controversial, while P. J. Adler, 'Habsburg school reform among the Orthodox minorities, 1770–1780', *Slavic Review*, 33 (1974) and D. Kosáry, 'Les réformes scolaires de l'absolutisme éclairé en Hongrie entre 1765 et 1790', *Etudes Historiques Hongroises* (1980) are good studies of particular areas. The story is carried into the 1780s by Ernst Wangermann, *Aufklärung und Staatsbürgerliche Erziehung: Gottfried van Swieten als Reformator des Österreichischen Unterrichtwesens 1781–1791* (Vienna, 1978).

The social, economic and legal position of the peasantry is outlined by Dickson, *Finance and Government*, vol. I, ch. 6, while detailed information and a wider framework are provided by Jerome Blum, *The End of the Old Order in Rural Europe* (Princeton, 1978), though the overall perspective is not totally convincing. A stimulating discussion of the first phase is Helen Liebel-Weckowicz and Franz J. Szabo, 'Modernization forces in Maria Theresa's peasant policies, 1740–1780', *Histoire Sociale/Social History*, 15 (1982). There is a useful study by E. M. Link, *The Emancipation of the Austrian Peasant 1740–1798* (New York, 1949) and a rather better introduction to the *Raab* system by W. E. Wright, *Serf, Seigneur and Sovereign: Agrarian Reform in Eighteenth-Century Bohemia* (Minneapolis, 1966), but neither has the stamp of quality of the classic account in the first volume of Karl Grünberg, *Die Bauernbefreiung und die Auflösung des gutsherrlichbäuerlichen Verhältnisses in Böhmen, Mähren und Schlesien* (2 vols, Leipzig, 1893–94). Glimpses of the opposition within the government to Joseph II's reforms are contained in Helen P. Liebel-Weckowicz, 'Count Karl von Zinzendorf and the Liberal revolt against Joseph II's economic reforms, 1783–1790' in *Sozialgeschichte Heute:*

Festschrift für Hans Rosenberg zum 70. Geburtstag, ed. H.-U. Wehler (Göttingen, 1974).

6. MARIA THERESA AND HUNGARY

The main contours of the political history of Hungary under Maria Theresa remain those established by nineteenth-century writers. Besides the pedestrian narrative of J. Mailáth, *Geschichte des östreichischen Kaiserstaates*, vol. V (Hamburg, 1850), interesting because of the author's family connections with government, the two chief spokesmen on the Hungarian side were both liberals and sympathetic to much of what the Queen was trying to achieve: M. Horváth, *Magyarország történelme*, vol. VII (2nd edn, Budapest, 1873), also available in an earlier version as *Geschichte der Ungarn*, vol. II (Pest, 1855); H. Marczali, *Mária Terézia* (Budapest, 1891). See also the latter's general survey, *Hungary in the Eighteenth Century* (Cambridge, 1910), a translation of the first volume of Marczali's trilogy on Joseph II, mentioned below. The major twentieth-century synthesis, B. Hóman and Gy. Szekfű's *Magyar történet*, vol. IV (by Szekfű) 2nd edn, Budapest, 1935), paints much the same picture in more sombre colours. The relevant volume (IV, ed. E. H. Balázs *et al.*), covering the years 1686–1790, in the new *Magyarország története* is urgently awaited. Meanwhile the classic Austrian exposition by A. von Arneth, *Geschichte Maria Theresias*, 10 vols (Vienna, 1863–79), established the place of Hungary within the events of the Monarchy as a whole, without devoting much attention to her domestic evolution; and his colleagues largely ignored the problem: F. Krones, *Ungarn unter Maria Theresia und Joseph II* (Graz, 1871), despite its title, is slight and mainly ecclesiastical. Not until the bicentenary of Maria Theresa's death did sustained interest revive, now in the form of collective volumes: *Ungarn und Österreich unter Maria Theresia und Joseph II*, ed. A. M. Drabek *et al.* (Vienna, 1982), rather scrappy; G. Mraz (ed.) *Maria Theresia als Königin von Ungarn* (Eisenstadt, 1984), more extensive, with some valuable contributions; *Österreich im Europa der Aufklärung*, ed. R. G. Plaschka *et al.*, 2 vols (Vienna, 1985) larger still, unwielding and incoherent, but helpful and with a good bibliography. The latest work on the period, P. G. M. Dickson's *Finance and Government under Maria Theresia, 1740–1780*, 2 vols (Oxford, 1987), raises our understanding of the fiscal and administrative structure of central government onto a new plane, and carries large implications for the study of Hungary's place in that structure. The interpretation in the present chapter also builds on my own analysis of the preceding era: R. J. W. Evans, *The Making of the Habsburg Monarchy, 1550–1700. An Interpretation* (2nd edn, Oxford, 1984). For Transylvania, only marginally addressed here, see the good new account, with excellent bibliography, in *Erdély története*, vol. 2: *1606-tól 1830-ig*, eds L. Makkai and Z. Szász (Budapest, 1986). The main literature on the *Urbarium* is D. Szabó, *A magyarországi urbérrendezés története Mária Terézia korában* (Budapest, 1933), first volume only of a massive compilation, a second volume being completed but still unpublished; K. Rebro, *Urbárska regulácia Márie Terézie a poddanské úpravy Jozefa II na Slovensku* (Bratislava, 1959), which concentrates on the Slovakian (= north Hungarian) counties; and I. Felhő *et al.* (ed.) *Az úrbéres birtokviszonyok Magyarországon Mária Terézia korában*, vol. I: *Dunántúl* (Budapest, 1970), mainly a statistical analysis of western Hungary. For all aspects of cultural and intellectual life in mid-eighteenth-century Hungary consult the new standard work by D. Kosáry, *Művelődés a XVIII. századi Magyarországon* (Budapest, 1980), of which an abstract is now available as *Culture and Society in Eighteenth-Century Hungary* (Budapest, 1987). The *Ratio Educationis* has been printed and discussed by I. Mészáros, *Ratio Educationis: az 1777-i és az 1806-i kiadás magyar nyelvű fordítása* (Budapest, 1981). For the nationality issue see the Bibliography to the next chapter, below. The best and most detailed account of Joseph II's governance of Hungary is still H. Marczali's *Magyarország története II József korában*, 3 vols (Budapest, 1885–88). Another older book which retains its importance is the analysis and documentation of the Toleration Patent by E. Mályusz, *A türelmi rendelet*, vol. II: *Il József és a magyar protestantizmus*; vol. II: *Iratok a türelmi rendelet történetéhez* (Budapest, 1939). The most significant recent research is by L. Hajdu, *II József igazgatási reformjai Magyarországon* (Budapest, 1982). In English, B. K. Király, *Hungary in the Late Eighteenth Century* (New York, 1969), is informative, but not always accurate or judicious; while D. Beales, *Joseph II*, vol. I: *In the Shadow of Maria Theresa, 1741–1780* (Cambridge, 1987), sets the scene magisterially for his next volume in which Hungarian affairs will be much more prominent.

7. JOSEPH II AND NATIONALITY IN THE HABSBURG LANDS

The only full and general account to date of Joseph's policies, P. Mitrofanov, *Josef II. Seine politische und kulturelle Tätigkeit* (2 vols, Vienna, 1910), reflects the perceptions of contemporary government and treats his measures on 'ethnic' issues strictly in the context of the rest of his legislation. To a Russian, that may have come naturally; most historians from the central European nationalities concerned have – rightly *and* wrongly – forsworn such bureaucratic objectivity. The otherwise outstanding work of Marczali, for example (see Bibliography to Ch. 6) displays a marked Magyar bias. Little comparative work has therefore been done on national sentiment in the Habsburg lands before open conflict began to emerge from the 1790s. The best starting-point is the splendid but neglected survey of the linguistic ancien regime in Hungary by D. Rapant, *K počiatkom maďarizácie*, vol. I: *Vývoj rečovej otázky v Uhorsku v rokoch 1740–90* (Bratislava, 1927), far wider than its main title – 'The

Beginnings of Magyarisation' – suggests. See also E. Arató, *A feudális nemzetiségtől a polgári nemzetig* (Budapest, 1975); and the fascinating collection on the polyglot activity of the Buda University Press: *Typographia Universitatis Hungaricae Budae, 1777–1848*, ed. P. Király (Budapest, 1983). There is a corresponding wealth of investigations for each nationality once the 'threshold' of nationhood had been crossed, many of them making unwarranted assumptions about the character of earlier loyalties. The best way of avoiding *parti pris* is to write about other people's nationalism: again the scrupulous but slightly pedestrian Arató, *A nemzetiségi kérdés Magyarországon, 1790–1840* (2 vols, Budapest, 1960), has led the way. For an equally Marxist, but more ambitious approach, consult M. Hroch, *Die Vorkämpfer der nationalen Bewegung bei den kleinen Völkern Europas* (Prague, 1968), reworked as *Social Preconditions of National Revival in Europe. A comparative analysis of the social composition of patriotic groups among the smaller European nations* (Cambridge, 1985). Another study in English by an east-European historian is J. Chlebowczyk, *On Small and Young Nations in Europe: Nation-forming processes in ethnic borderlands in East-Central Europe* (Wrocław, 1980). For the Magyars see the useful collection of texts in M. C. Ives (ed.) *Enlightenment and National Revival: Patterns of interplay and paradox in late-eighteenth-century Hungary* (London, 1979). The classic credo of a hard-bitten Hungarian Josephinist who subsequently became a fervent and widely-influential nationalist is the autobiography of F. Kazinczy, in his *Művei*, ed. M. Szauder, 2 vols (Budapest, 1979), vol. I, pp. 209–418. Kazinczy retained enough cosmopolitan traits to attract criticism from the next generation of more chauvinist Magyars; he himself condemned as a lukewarm patriot his fellow liberal reformer Berzeviczy, the subject of an important monograph by E. H. Balázs, *Berzeviczy Gergely a reformpolitikus, 1763–95* (Budapest, 1967). In the spirit of Berzeviczy, M. Csáky argues, in *Ungarn und Österreich unter Maria Theresia und Joseph II*, ed. A. M. Drabek et al. (Vienna, 1982) pp. 71–89, and in *Wegenetz europäischen Geistes. . . .*, ed. R. G. Plaschka and K. Mack (Vienna, 1983) pp. 356–69, interestingly, though perhaps a little too sanguinely, for the active survival, until 1800 and beyond, of common ground in the patriotism of Magyars and the other nationalities of Hungary. Accessible insights into the proto-nationalism of the Serbs are provided by D. Obradović, *Life and Adventures*, transl. and ed. G R. Noyes (Berekeley and Los Angeles, 1953), and by the early chapters of D. Wilson, *The Life and Times of Vuk Stefanović Karadžić* (Oxford, 1970). For the Rumanians the best local guide is D. Prodan, *Supplex Libellus Valachorum* (Bucharest, 1971), and the most measured outsider's views come from K. Hitchins, most recently in his *L'Idée de nation chez les Roumains de Transylvanie, 1691–1849* (Bucharest, 1987). On the Czechs, A. S. Myl'nikov, *Vznik národně osvícenské ideologie v českých zemích 18. století* (Prague, 1974), overlapping with his *Épokha prosveschcheniya v Cheshskikh zemlyach* (Moscow, 1977), the first Russian since Mitrofanov to take a hard look at the eighteenth century in Habsburg Europe, displays some exaggeration and some crudeness, but he is also thorough and perceptive, and his arguments are firmer than those of W. Schamschula, *Die Anfänge der tschechischen Erneuerung und das deutsche Geistesleben* (Munich, 1973), who covers much of the same ground. The latest native treatment is by J. Haubelt, *České osvícenství* (Prague, 1986). F. Fancev's collection of documents, *Dokumenti za naše podrijetlo hrvatskoga preporoda* (Zagreb, 1933), is helpful evidence for the Croats, though his attempt to pursue the roots of their modern national awakening back into the eighteenth century has met much justified criticism. The best general introduction now is the massively erudite and circumspect, but intellectually unwieldy W. Kessler, *Politik, Kultur und Gesellschaft in Kroatien und Slawonien in der ersten Hälfte des 19. Jahrhunderts* (Munich, 1981). There appears to be no full account, outside the literary and linguistic context, of the Slovene circle around Zois. But a sketch may be found in J. Pogačnik, *Bartholomäus Kopitar, Leben und Werk* (Munich, 1978). Slovak self-perceptions are brought together in the major collection edited by J. Tibenský, *K počiatkom slovenského národného obrodenia* (Bratislava, 1964); compare the shrewd but sometimes mischievous approach of L. Gogolák, *Beiträge zur Geschichte des slowakischen Volkes*, vols I and II (Munich, 1963–69), and the balanced *compte-rendu* by P. Brock, *The Slovak National Awakening* (Toronto and Buffalo, 1976). The weakest nationality of all in the period, quaintly enough, was that of the Austrian Germans (or should I write 'German Austrians'?) – but any effort to examine that issue would break the bounds of this modest chapter and its bibliography.

8. THE SMALLER GERMAN STATES

Two books on Germany that discuss the problems of government and society in a general way are W. H. Bruford, *Germany in the Eighteenth Century: the social background of the literary revival* (Cambridge, 1935) and Klaus Epstein, *The Genesis of German Conservatism* (Princeton 1966). Three articles offering different interpretations of enlightened despotism in the German states are Helen P. Liebel, 'Enlightened Despotism and the crisis of society in Germany', *Enlightenment Essays*, 1 (1970) 151–68; Charles W. Ingrao, 'The problem of "Enlightened Absolutism" and the German States', *Journal of Modern History*, 58 (1986), Supplement; and Eberhard Weis, 'Enlightenment and Absolutism in the Holy Roman Empire: thoughts on Enlightened Absolutism in Germany,' ibid. For a closer study of the relationship between the *Aufklärung* and politics, see H. B. Nisbet, '*Was ist Aufklärung?*: The concept of Enlightenment in eighteenth-century Germany', *Journal of European Studies*, 12 (1982); Geraint Parry, 'Enlightened Government and its critics in eighteenth-century Germany', *Historical Journal*, 6 (1963); and the chapters on Catholic and Protestant Germany by T. C. W. Blanning and Joachim

Whaley in Roy Porter and Mikulás Teich (eds) *The Enlightenment in National Context* (Cambridge 1981). Among the plethora of titles on cameralism Albion W. Small, *The Cameralists: The Pioneers of German Social Polity* (Chicago 1909) is still a useful study of the leading cameralists. More recent treatments include Keith Tribe, 'Cameralism and the science of government,' *Journal of Modern History*, 56 (1984), 263–84 and *Governing Economy: the reformation of German economic discourse 1750–1840* (Cambridge, 1988); Marc Raeff, 'The well-ordered police state and the development of modernity in seventeenth- and eighteenth-century Europe: an attempt at a comparative approach,' *American Historical Review* 80 (1975), and *The Well-Ordered Police State: Social and Institutional Change through Law in the Germanies and Russia, 1600–1800* (New Haven, Conn. 1983); and Jutta Brückner, *Staatswissenschaften, Kameralismus und Naturrecht* (Munich 1977). Hans-Joachim Braun, 'Economic theory and policy in Germany 1750–1800', *Journal of European Economic History*, 4 (1975) is also useful.

A great number of books have appeared on individual German states. The first such study in English by Helen P. Liebel, *Enlightened Bureaucracy versus Enlightened Despotism in Baden, 1750–1792* (*American Philosophical Society, Transactions*, Philadelphia, 1965) stresses the role of Baden's ministers while Charles W. Ingrao, *The Hessian Mercenary State: Ideas, Institutions and Reform under Frederick II, 1760–1785* (New York and Cambridge 1987) and ' "Barbarous strangers": Hessian state and society during the American Revolution,' *American Historical Review* 87 (1982), portrays reform in Hesse-Cassel as a co-operative effort between crown, officials and estates. Jonathan B. Knudsen, *Justus Möser and the German Enlightenment* (New York and Cambridge 1986) links a leading conservative figure with enlightened reform in the secularised bishopric of Osnabrück. James A. Vann, *The Making of a State: Württemberg 1593–1793* (Ithaca, NY, 1984) presents a necessarily unflattering view of Charles Eugene, as does the somewhat breezy, but useful Adrien Fauchier-Magnan, *The Small German Courts in the Eighteenth Century* (London, 1958), half of which focuses on Württemberg. For Bavaria, there is Waltraud Müller, *'Zur Wohlfahrt des gemeinen Wesens': Ein Beitrag zur Bevölkerungs- und Sozialpolitik Max III. Joseph (1745–1777)* (Munich 1984). Max Spindler, (ed.) *Handbuch der bayerischen Geschichte*, 2–5 (Munich, 1969–71) treats Bavaria and the smaller principalities incorporated in today's Bundesstaat. Sigisbert Conrady, 'Die Wirksamkeit König Georgs III. für die hannoverschen Kurlande', *Niedersächsiches Jahrbuch*, 39 (1967), offers a fascinating revision of our view of George III in his capacity as Hanoverian elector.

For the smaller states there are Wolfram Fischer, *Das Fürstentum Hohenlohe im Zeitalter der Aufklärung* (Tübingen, 1958); Volker Wehrmann, *Die Aufklärung in Lippe: Ihre Bedeutung für Politik, Schule, und Geistesleben* (Detmold, 1972); and the older Fritz Hartung, *Das Großherzogtum Sachsen unter der Regierung Carl Augusts, 1775–1828* (Weimar, 1923), by one of the major early proponents of the concept of enlightened absolutism. For agrarian reform in the Danish dependency of Schleswig-Holstein, there is Wolfgang Prange, *Die Anfänge der großen Agrarreformen in Schleswig-Holstein bis 1771* (Neumünster, 1971).

The Catholic prince-bishoprics are well-covered by recent scholarship. T. C. W. Blanning, *Reform and Revolution in Mainz*, 1743–1803 (Cambridge, 1974) stresses the struggle between enlightened regimes and their conservative subjects. For Cologne and its sister bishopric of Münster there is Max Braubach, *Maria Theresias jüngster Sohn Max Franz: Letzter Kurfürst von Köln und Fürstbischof von Münster* (Vienna and Munich, 1961) and Alwin Hanschmidt, *Franz von Fürstenberg als Staatsmann: Die Politik des münsterischen Ministers, 1762–1780* (Münster Aschendorff, 1969). Hildegunde Flurschütz, *Die Verwaltung des Hochstifts Würzburg unter Franz Ludwig von Erthal (1779–1795)* (Würzburg, 1965) and Alfred Heggen, *Staat und Wirtschaft im Fürstentum Paderborn im 18. Jahrhundert* (Paderborn, 1978) provide forthright studies of government policies. Dorothea Wachter, *Degen & Krummstab: Clemens Wenzeslaus, Prinz von Sachsen, Königl. Prinz von Polen & Litauen, Kurfürst und Erzbischof von Trier, Fürstbischof von Augsburg, 1739–1812* (Kempten, 1978) is a fragmented and pedestrian examination of Trier's important archbishop-elector. Finally, Max Braubach, 'Die kirchliche Aufklärung im katholischen Deutschland im Spiegel des "Journal von und für Deutschland" (1784–1792)', *Historisches Jahrbuch*, 54 (1934) presents contemporary commentary on the spread of Enlightenment ideas and reform in all of Germany's Catholic principalities.

9. THE DANISH REFORMERS

There is only a limited amount of useful secondary material available in English on the period of Enlightened Absolutism in Denmark. An excellent survey in a wider historical context is found in S. P. Oakley, *The Story of Denmark* (London, 1972), whilst a full Scandinavian perspective over a narrower period is provided by H. Arnold Barton, *Scandinavia in the Revolutionary Era, 1760–1815* (Minneapolis, 1986), which also gives detailed references and a good bibliography of material in several languages. A different angle of approach is that adopted by Lawrence J. Baack, 'State service in the eighteenth century: the Bernstorffs in Hanover and Denmark', *International History Review*, I (1979). Baack has also published a brief conventional survey of *Agrarian Reform in Eighteenth-Century Denmark* (Lincoln, Nebraska, 1977), while H. Arnold Barton has written a historiographical survey: 'The Danish agrarian reforms, 1784–1814, and the historian', *Scandinavian Economic History Review*, 36 (1988). There is a short essay on 'Struensee and the Enlightenment' by H. S. Commager in his volume of essays entitled *The search for a usable past* (New York, 1967). Of contemporary material in a western language, the most important is Elie Reverdil, *Struensee et la cour de Copenhague 1760–1772* (Paris, 1858).

Amongst recent surveys of the period in Danish the best is undoubtedly Ole Feldbæk, *Danmarks historie*, vol. IV: *Tiden 1730–1814* (Copenhagen, 1982), which combines clarity in thematic treatment with full bibliographies and comments on primary sources. On the wider social history context see Hans Chr. Johansen, *En samfundsorganistation i opbrud 1700–1870: Dansk socialhistore*, vol. IV (Copenhagen, 1979), which has a full bibliography; see also E. Helmer Pedersen, 'Dansk landbrugsudvikling i det 18. århundrede', *Bol og By*, 2/V (1983).

The founder of modern Danish agrarian history is Fridlev Skrubbeltrang, whose *Husmand og inderste: studier over sjællandske landboforhold i perioden 1660–1800* (Copenhagen, 1940; repr. 1974) is of fundamental importance in bringing attention to a range of issues, not just the predicament of the smallholders alone. In *Det danske Landbosamfund 1500–1800* (Copenhagen, 1978) Skrubbeltrang summed up both his own lifetime's work and that of many who have followed him. Not much of his work is available in English, but see the first part of his survey of *Agricultural Development and Rural Reform in Denmark* (Rome: FAO, 1953), and his brief article on 'Developments in tenancy in 18th-century Denmark as a move towards peasant proprietorship', in *Scandinavian Economic History Review*, 9 (1961). Claus Bjørn has pioneered the study of peasant *mentalités*, notably in his *Bonde, herremand, konge: bonden i 1700-tallets Danmark* (Copenhagen: Gyldendal, 1981) which also dispels the myth of an apathetic and unshakeably tradition-bound peasantry. His article on 'The peasantry and agrarian reform in Denmark', in *Scandinavian Economic History Review*, 25 (1977), is of fundamental importance. Gunnar Olsen, in *Hovedgård og bondegård: studier over stordriftens udvikling i Danmark i tiden 1525–1774* (Copenhagen, 1957; repr. 1975) examined the spread of demesne land. A large number of regional studies have thrown light on the process of enclosures, on peasant–seigneur relations and on many other aspects: amongst them one may mention Jens Holmgaard, 'De nordsjællandske landboreformer og statsfinanserne', *Erhvervshistorisk Årbog*, 6 (1954); Jørgen Dieckmann Rasmussen, *Bønderne og udskiftningen* (Copenhagen, 1977); Margit Mogensen, *Fæstebønderne i Odsherred* (Copenhagen: Lokalhist. Afd., 1974); Birte Stig Jørgensen, 'Bønderne og udskiftningen på Hørsholm Amt' in *Landbohistoriske studier tilegnede Fridlev Skrubbeltrang*, ed. S. Gissel (Copenhagen, 1970); and many others. Labour services were critical in this period, and much work has been done on it recently: see notably F. Skrubbeltrang, 'M. H. Løvenskiolds hoveridagbog 1795–97', in *Bol og By*, 7 (1973); and Birgit Løgstrup, 'Markdrift og hoveri på Løvenborg 1771–72', *Bol og By*, 8 (1974). On the historiographical trend see especially T. Kjærgaard, 'The farmer interpretation of Danish history', *Scandinavian Journal of History*, 10 (1985).

Some progress has also been made towards a reinterpretation of landowner attitudes. C. Bjørn, 'Den jyske proprietærfejde 1790–91', *Historie – Jyske Samlinger* 13 (1979), includes a considerable number of citations from contemporary sources. More recently M. Mogensen and P. E. Olsen have edited *Godsejerrøster: landøkonomiske indberetninger fra Roskilde Amt 1735–1770* (Copenhagen, 1984).

Enlightened absolutism as such, and in particular the relationship between local problems and incentives, central government objectives, and the means of practical enforcement, has not been studied in as much depth recently. Amongst older standard works on the period must be cited Edvard Holm's monumental *Danmark-Norges historie fra den store nordiske krigs afslutning til rigernes adskillelse*, vols I–VII (Copenhagen, 1891–1912); and his *Kampen om landboreformerne i Danmark* (Copenhagen, 1888; repr. 1974). Also of value on the press and public opinion is his study of *Den offentlige mening og statsmagten i den dansk-norske stat i slutningen af det 18. århundrede* (Copenhagen, 1888; repr. 1975). Interesting material on the atmosphere in Copenhagen, including reactions to the French Revolution, is provided in Bent Blüdnikow, 'Folkelig uro i København', *Fortid og Nutid*, 33 (1986). A general survey of a dramatic phase is S. Cedergreen Bech, *Struensee og hans tid* (Copenhagen, 1972). Hans Jensen, *Dansk jorpolitik 1757–1919*, vol. I (Copenhagen, 1936; repr. 1975) has rightly been criticised for tending to treat central government in a rather idealised vacuum, but has much of value within those limitations.

Some of the source material is available in older or more recent printed editions. Of particular interest is the record of the Great Agrarian Commission itself, *Den for Landbovæsenets nedsatte Kommissions Forhandlinger*, vols I–II (Copenhagen, 1788–89). Thorkild Kjærgaard, *Konjunkturer og afgifter: C. D. Reventlows betænkning af 11.februar 1788 om hoveriet* (Copenhagen, 1980) gives a complete text of Reventlow's early ideas of how to tackle the question of labour services, with full editorial comments.

10. FREDERICK THE GREAT AND ENLIGHTENED ABSOLUTISM

Two useful and concise histories of Prussia are E. J. Feuchtwanger, *Prussia: myth and reality. The role of Prussia in German history* (London, 1970) and H. W. Koch, *A History of Prussia* (London, 1978). The most substantial history of Germany in the period is Hajo Holborn's *History of Modern Germany*, vol. II (London, 1965), which is stodgy but thoughtful. On the international context, there is Derek McKay and H. M. Scott's excellent *The Rise of the Great Powers* (London, 1983). Until Theodor Schieder's recent biography can be translated, the best study in English is Gerhard Ritter's *Frederick the Great: an historical profile* (London, 1968). Dull but very informative is Walther Hubatsch's *Frederick the Great. Absolutism and administration* (London, 1975). There are several helpful collections of documentary sources available, notably:

C. A. Macartney (ed.) *The Habsburg and Hohenzollern Dynasties in the Seventeenth and Eighteenth Centuries* (London, 1970) and A. Lentin (ed.) *Enlightened Absolutism (1760–1790)* (Newcastle-upon-Tyne, 1985). The prime source is Frederick's own *Posthumous Works* (London, 1789). Another important text available in translation is Frederick's *The Refutation of Machiavelli's Prince or Anti-Machiavel*, ed. Paul Sonnino (Athens, Ohio, 1981). Most helpful on the political and administrative structure of Prussia are W. L. Dorn, 'The Prussian bureaucracy in the eighteenth century', *Political Science Quarterly*, 46 (1931) and 47 (1932), part of which was also reprinted in Peter Paret (ed.) *Frederick the Great* (New York, 1970); Hubert C. Johnson, *Frederick the Great and his Officials* (New Haven, 1975) and Hans-Eberhard Mueller, *Bureaucracy, Education and Monopoly: civil service reforms in Prussia and England* (Berkeley and Los Angeles, 1984). Stimulating, influential but tendentious is Hans Rosenberg's *Bureaucracy, Aristocracy and Autocracy: the Prussian experience, 1660–1815* (Cambridge, Mass., 1958); a useful corrective is C. B. A. Behrens, *Society, Government and the Enlightenment: the experience of eighteenth-century France and Prussia* (London, 1985). On the military aspects in general and militarism in particular, there is a host of titles; particularly helpful are: Michael Howard, *War in European History* (Oxford, 1976); Hew Strachan, *European Armies and the Conduct of War* (London, 1983); Martin Kitchen, *A Military History of Germany from the Eighteenth Century to the Present Day* (London, 1975); Gerhard Ritter, *The Sword and the Sceptre: the problem of militarism in Germay*, vol. I (London, 1972); Gordon Craig, *The Politics of the Prussian Army* (Oxford, 1964). Two thorough studies of Frederick as a military man are Christopher Duffy's *The Army of Frederick the Great* (Newton Abbot, 1974) and *Frederick the Great: a military life* (London, 1985). On the Enlightenment in Prussia, the crucial text is Kant's 'What is Enlightenment?', which can be found in a good English translation in Hans Reiss (ed.) *Kant's Political Writings* (Cambridge, 1970). Helpful secondary works on the Prussian Enlightenment are Klaus Epstein, *The Genesis of German Conservatism* (Princeton, 1966), H. B. Nisbet, '*Was ist Aufklärung?* The concept of Enlightenment in eighteenth-century Germany', *Journal of European Studies*, 12 (1982) and James van Horn Melton, 'From Enlightenment to revolution: Hertzberg, Schlözer, and the problem of despotism in the late *Aufklärung*', *Central European History*, 12 (1979). On Frederick's Enlightened Absolutism, Perry Anderson's Marxist interpretation, in his *Lineages of the Absolutist State* (London, 1975), can be counter-balanced by Fritz Hartung's *Enlightened Despotism* (Historical Association Pamphlet, G. 36). A recent study of importance is James Van Horn Melton, *Absolutism and the Eighteenth-Century Origins of Compulsory Schooling in Prussia and Austria* (Cambridge, 1988).

In German, the best biography is Theodor Schieder's *Friedrich der Grosse. Ein Königtum der Widersprüche* (Frankfurt-am-Main, Berlin and Vienna, 1983), but still important is Reinhold Koser's *Geschichte Friedrichs des Grossen* (Stuttgart and Berlin, 1914). Another older work which still must be read is Otto Hintze, *Die Hohenzollern und ihr Werk* (8th edn, Berlin, 1916). More hostile to Frederick is the study by the East German Ingrid Mittenzwei: *Friedrich II von Preussen. Eine Biographie* (East Berlin, 1979). Also Prussophobe in tone is the general history of Prussia by Günter Vogler and Klaus Vetter, *Preussen von den Anfängen bis zur Reichsgründung* (Cologne, 1981). A particularly valuable collection of source material is Otto Bardong's *Friedrich der Grosse* (Darmstadt, 1982). Also invaluable is Horst Steinmetz (ed.) *Friedrich II, König von Preussen und die deutsche Literatur des 18. Jahrhunderts* (Stuttgart, 1985). Two patchy but indispensable collections of articles are *Preussen – Versuch einer Bilanz* (5 vols, Hamburg, 1981) and Otto Büsch and Wolfgang Neugebauer (eds) *Moderne preussische Geschichte* (2 vols, Berlin and New York, 1981). On militarism, the most important study is Otto Büsch, *Militärsystem und Sozialleben im alten Preussen 1713–1807* (new edn, Frankfurt-am-Main, 1981). Two important studies of the Prussian bureaucracy and its intellectual milieu, with forbidding titles but important contents, are Wilhelm Bleek, *Von der Kameralausbildung zum Juristenprivileg. Studium, Prüfung und Ausbildung der höheren Beamten des allgemeinen Verwaltungsdienstes in Deutschland im 18. und 19. Jahrhundert* (Berlin, 1972) and Eckhart Hellmuth, *Naturrechtsphilosophie und bürokratischer Werthorizont. Studien zur preussischen Geistes- und Sozialgeschichte des 18. Jahrhunderts* (Göttingen, 1985). Of the numerous articles on the problem of enlightened absolutism, particularly important are: Hans von Voltelini, 'Die naturrechtlichen Lehren und die Reformen des 18. Jahrhunderts', *Historische Zeitschrift*, 105 (1910); E. Walder, 'Zwei Studien über den aufgeklärten Absolutismus', *Schweizer Beiträge zur allgemeinen Geschichte*, 15 (1957); Ingrid Mittenzwei, 'Über das Problem des aufgeklärten Absolutismus', *Zeitschrift für Geschichtswissenschaft*, 18 (1970); Volker Sellin, 'Friedrich der Grosse und der aufgeklärte Absolutismus', U. Engelhardt (ed.), *Soziale Bewegung und politische Verfassung. Beiträge zur Geschichte der modernen Welt* (Stuttgart, 1976); Gottfried Niedhart, 'Aufgeklärter Absolutismus oder Rationalisierung der Herrschaft', *Zeitschrift für historische Forschung*, 6:ii (1979); Horst Möller, 'Wie aufgeklärt war Preussen?', Hans-Jürgen Puhle and Hans-Ulrich Wehler (eds) *Preussen im Rückblick* (Göttingen, 1980); Klaus Deppermann, 'Der preussische Absolutismus und der Adel. Eine Auseinandersetzung mit der marxistischen Absolutismustheorie', *Geschichte und Gesellschaft*, 8 (1982).

11. CATHERINE THE GREAT

Fortunately for English-speaking readers, the last twenty-five years have seen the publication in Great Britain and the USA of a large number of works of high scholarly quality on the reign of Catherine II. More attention indeed has been devoted to this period than in the Soviet

Union. A useful introduction is P. Clendenning and R. Bartlett, *Eighteenth Century Russia: A select bibliography of works published since 1955* (Newtonville, Mass., 1981). In Russia the starting point to a new approach has been the article by N. M. Druzhinin, 'Prosveshchennyy absolyutizm v Rossii' in the very useful symposium *Absolyutizm v Rossii*, ed. B. B. Kafengauz (Moscow, 1964). Subsequently there has been considerable discussion among Soviet historians on the problem of agreeing on a Marxist concept of Enlightened Absolutism applicable to Russia as well as to Western Europe. A general account of the discussion can be found in H.-J. Torke, 'Die neuere Sowjethistoriographie zum Problem des russischen Absolutismus' in *Forschungen zur Osteuropäischen Geschichte*, 20 (1973). A brief account in English is given by M. S. Anderson, in *Historians and Eighteenth Century Europe, 1715–1789* (Oxford, 1979) pp. 169 *et seq.*

For a general survey of Catherine's reign, the most up-to-date work is Isabel de Madariaga, *Russia in the Age of Catherine the Great*, (London, 1981) which discusses previous scholarship and provides new interpretations of many aspects of Catherine's reign. For the legislative commission see P. Dukes, *Catherine the Great and the Russian Nobility* (Cambridge, 1967). G. Sacke's *Die Gesetzgebende Kommission Katharinas II: Ein Beitrag zur Geschichte des Absolutismus in Russland* (Breslau, 1940) is still worth consulting though very out-of-date in its interpretation. The official contemporary English translation of Catherine's instruction was reprinted in W. F. Reddaway (ed.) *Documents of Catherine the Great. The Correspondence with Voltaire and the Instruction of 1767 in the English text of 1768* (Cambridge, 1931). It is preferable to later translations which claim to be closer to the Russian, since it should be remembered that the original text was written in French, not in Russian. The demands of the towns have been studied by F.-X. Coquin, *La Grande Commission Legislative (1767–1768). Les Cahiers de doléance urbains* (Louvain and Paris, 1972).

The Origins of the Russian Intelligentsia: The eighteenth-century nobility by Marc Raeff (New York, 1966) is an essential introduction to the subject. The structure and political claims of the nobility are discussed in R. E. Jones, *The Emancipation of the Russian Nobility, 1762–1785* (Princeton, 1973), while J. M. Hittle in *The Service City: state and townsmen in Russia, 1600–1800* (Harvard, 1979), analyses the problems of the towns. There is an excellent portrait of the city of Moscow in John T. Alexander, *Bubonic Plague in Early Modern Russia: Public health and urban disaster*, (Baltimore, 1980). There is no single work devoted to the peasantry, but it is treated by Jerome Blum, *Lord and Peasant in Russia from the Ninth to the Nineteenth Century*, (Princeton, 1961), which is rather out of date and should be supplemented by the relevant chapters in Isabel de Madariaga, *Russia in the Age of Catherine the Great* and by *idem*, 'Catherine II and the Serfs: A reconsideration of some problems' in *Slavonic and East European Review*, 52 (1974). There is no modern English study of the peasant commune, but in Russian see V. A. Aleksandrov, *Sel'skaya Obshchina v Rossii (XVII-nachalo XIX vv)* (Moscow, 1976). For relationships on the land see M. Confino, *Domaines et seigneurs en Russie vers la fin du XVIIIe-siècle. Etude de structures agraires et de mentalités économiques* (Paris, 1963).

The standard Russian work on the Pugachev revolt is V. V. Mavrodin, *Krest'anskaya voyna v Rossii* (3 vols, Leningrad, 1961–70); more accessible is J. T. Alexander, *Emperor of the Cossacks* (Lawrence, Kansas, 1973) and *idem*, *Autocratic Politics in a National Crisis: The Imperial Russian government and Pugachev's revolt*, (Bloomington, Ind., 1969).

The administrative reforms of Catherine have been treated in Isabel de Madariaga, *Russia in the Age of Catherine the Great*, and also in J. P. LeDonne, in a somewhat ideologically coloured way, in *Ruling Russia: Politics and administration in the Age of Absolutism, 1762–1796*, (Princeton, 1984. See also R. E. Jones, *Provincial Development in Russia, Catherine II and Jacob Sievers* (Rutgers, NJ, 1984); and J. M. Hartley, 'Town Government in Saint Petersburg Guberniya after the Charter to the Towns of 1785', *Slavonic and East European Review*, 62 (1984). An interesting comparative approach will be found in M. Raeff, *The Well-Ordered Police State. Social and institutional change through law in the Germanies and Russia. 1600–1800* (New Haven, 1983). By the same author there is also an important article on Catherine's borrowings from Blackstone, 'The Empress and the Vinerian Professor', *Oxford Slavonic Papers*, 7 (1974). The evolution of the Russian bureaucracy is dealt with by R. Givens, 'Eighteenth-century Nobiliary Career Patterns and Provincial Government' and W. M. Pintner, 'The Evolution of Civil Officialdom, 1755–1855', both published in *Russian Officialdom*, eds W. M. Pintner and D. K. Rowney (London, 1980). There is a brief description of the judicial reform of 1775 in F. B. Kaiser, *Die Justizreform von 1864: Zur Geschichte der russischen Justiz von Katharina II bis 1917* (Leiden, 1972) which should be supplemented by J. M. Hartley, 'Catherine's Conscience Court – an English Equity Court?' in *Russia and the West in the Eighteenth Century*, ed. A. G. Cross (Newtonville, Mass., 1983) and Isabel de Madariaga 'Penal policy in the age of Catherine II', in *La 'Leopoldina' – Criminalità e Giustizia Criminale nelle Reforme del Settecento Europeo* (Università degli Studi di Siena, forthcoming).

J. L. Black in *Citizens for the Fatherland. Education, educators and pedagogical ideals in eighteenth century Russia*, (New York, 1979) reprints the *Book on the Duties of Man and Citizen* issued for schools in St Petersburg in 1783, and can be supplemented by M. Okenfuss, 'Education and Empire: School reform in Enlightened Russia', *Jahrbücher für Geschichte Osteuropas*, 27 (1979) and Isabel de Madariaga, 'The Foundation of the Russian Educational System by Catherine II', *Slavonic and East European Review*, 57 (1979).

Freedom of Expression in Eighteenth Century Russia by K. Papmehl (The Hague, 1971) deals with government policy while G. Marker, *Publishing,*

Printing and the Origins of Intellectual Life in Russia, 1700–1800 (Princeton, 1985) provides the background to intellectual life. Catherine's relations with Diderot are dealt with in Isabel de Madariaga, 'Catherine and the *philosophes*', in *Russia and the West in the Eighteenth Century*, ed. Cross; Diderot's *Mémoires pour Catherine II* have been edited by P. Vernière, (Paris, 1966) and a selection of her correspondence with Voltaire appears in A. Lentin (ed.) *Voltaire and Catherine the Great. Selected correspondence* (Cambridge, 1974). The same author has translated and edited M. M. Shcherbatov's *On the Corruption of Morals in Russia*, (Cambridge, 1969) (marred only by a misleading appendix on the Russian nobility). On Shcherbatov see also Isabel de Madariaga, 'Catherine II and Montesquieu between Prince M. M. Shcherbatov and Denis Diderot' in the second volume of *L'Età dei Lumi, Studi storici sul settecento europeo in onore di Franco Venturi* (Naples, 1985). For Radishchev, the best approach is still A. MacConnell, *A Russian Philosophe, Alexander Radishchev, 1749–1802* (The Hague, 1964) while W. G. Jones brings the portrayal of Novikov up-to-date in two important articles, 'The Morning Light Charity Schools 1777–80', *Slavonic and East European Review*, 56 (1978), and 'The Polemics of the 1769 Journals: A reappraisal', *Canadian American Slavic Studies*, 16, (1982). This issue of *Canadian American Slavic Studies* together with J. G. Garrard (ed.) *The Eighteenth Century in Russia*, (Oxford, 1973) contain a number of excellent articles on intellectual and artistic trends in Russia.

Religious life has been somewhat neglected, but see K. Papmehl, *Metropolitan Platon of Moscow* (Newtonville, Mass., 1983); G. Freeze, *The Russian Levites: Parish clergy in the eighteenth century*, (Cambridge, Mass., 1977); and Pia Pera, 'Dispotismo Illuminato e dissenso religioso: i vecchi credenti nell'età di Caterina II', *Rivista Storica Italiana*, 97 (1985).

Notes and References

INTRODUCTION: THE PROBLEM OF ENLIGHTENED
ABSOLUTISM   *H. M. Scott*

1. See, in particular, the surveys by Geoffrey Bruun, *The Enlightened Despots* (New York, 1929; 2nd edn, 1967) and Leo Gershoy, *From Despotism to Revolution 1763–1789* (New York, 1944). This older view was reflected in certain textbooks published in the 1960s: R. W. Harris, *Absolutism and Enlightenment* (London, 1964); James L. White, *The Origins of Modern Europe* (London, 1964); Stuart Andrews, *Eighteenth-Century Europe* (London, 1965); and R. J. White, *Europe in the Eighteenth Century* (London, 1965) all commit themselves to the view that enlightened despotism did exist.

2. *Europe in the Eighteenth Century 1713–1783* (London, 1961) pp. 121–9.

3. *The American and French Revolutions 1763–93*, ed. A. Goodwin (Cambridge, 1965) pp. 16, 19.

4. Ibid., pp. 296, 297, 331, 361.

5. *Le despotisme éclairé* (Paris, 1968), especially p. 354.

6. See, e.g., G. Rudé, *Europe in the Eighteenth Century* (London, 1972) pp. 97–101.

7. 'Enlightened despotism', *Historical Journal*, 18 (1975), 402, 408. Her remarks were more notable in that she was ostensibly reviewing two works which make a powerful case for the *existence* of enlightened absolutism, at least in a German context: these were T. C. W. Blanning, *Reform and Revolution in Mainz 1743–1803* (Cambridge, 1974) and K. O. Freiherr von Aretin (ed.) *Der aufgeklärte Absolutismus* (Cologne, 1974). Subsequently, Miss Behrens seems to have changed her mind: see her more recent *Society, Government and the Enlightenment: the experiences of eighteenth-century France and Prussia* (London, 1985) p. 178; cf above, p. 265.

8. 'The false Joseph II', *Historical Journal*, 18 (1975) 467–95.

9. This seems to be the thrust of his 'Sur le système du despotisme éclairé', *Les Lumières en Hongrie, en Europe Centrale et en Europe Orientale*, eds B. Köpeczi, E. Bene and I. Kovács (Budapest, 1977) pp. 19–29.

10. 'Enlightened despotism and the crisis of society in Germany', *Enlightenment Essays*, 1 (1970) 151–68.

11. *Europe in the Eighteenth Century* (2nd edn, London, 1976) pp. 171–3. This view is endorsed in the third edition (London, 1987): see especially pp. 206–8.

12. Herbert H. Rowen, 'Louis XIV and Absolutism' in John C. Rule (ed.) *Louis XIV and the Craft of Kingship* (Columbus, Ohio, 1969) p. 312.

13. See the valuable article by François Bluche, 'Sémantique du despotisme éclairé', *Revue historique du droit français et étranger*, 56 (1978) 79–87.

14. See R. Koebner, 'Despot and despotism: vicissitudes of a political term', *Journal of the Warburg and Courtauld Institutes*, 14 (1951) 275–302.

15. In a much-quoted letter to the elder Mirabeau, 26 July 1767: this is printed in C. E. Vaughan (ed.) *The Political Writings of Jean-Jacques Rousseau* (2 vols, Cambridge, 1915) vol. II, pp. 159–62.

16. See the influential and thoroughly sceptical article of R. Derathé, 'Les philosophes et le despotisme' in P. Francastel, *Utopie et institutions au XVIIIᵉ siècle: le pragmatisme des lumières* (Paris and The Hague, 1963) pp. 57–75.

17. Alfred Cobban, *In Search of Humanity: the role of the Enlightenment in modern history* (London, 1960) pp. 161–7, for the way many eighteenth-century thinkers, particularly in France, rejected 'enlightened despotism'. For a more extreme statement of this view, see Peter Gay, *The Party of Humanity: studies in the French Enlightenment* (London, 1964 edn) pp. 274–5.

18. Charles G. Stricklen, Jr, 'The *philosophe's* political mission: the creation of an idea, 1750–1789', *Studies on Voltaire and the Eighteenth Century*, 86 (1971) 137–228, esp. 139, 143–4 and 158.

19. His ideas were first put forward in 'Umrisse zur Naturlehre der drei Staatsformen', *Allgemeine Zeitschrift für Geschichte*, 7 (1847) 79–88, 322–65 and 436–73 (see especially 451), and subsequently developed in his *Geschichte der National-Ökonomik in Deutschland* (Munich, 1874) pp. 380–429 (especially pp. 380–4). According to Leonard Krieger, *An Essay on the Theory of Enlightened Despotism* (Chicago, 1975) p. 94, n. 2, the 'most elaborate presentation' of Roscher's ideas was in his later *Politik: Geschichtliche Naturlehre der Monarchie, Aristokratie, und Demokratie* (2nd edn, Stuttgart, 1893) pp. 250–1, 281–99, but I have been unable to trace this book.

20. Krieger, *Essay on the Theory of Enlightened Despotism*, p. 18, quoting Treitschke's *Deutsche Geschichte in neunzehnten Jahrhundert* (4th edn, Leipzig, 1886), vol. I, p. 70.

21. 'Die Epochen der Absoluten Monarchie in der Neueren Geschichte', *Historische Zeitschrift*, 61 (1889) 246–87. Forty years later, this view was explicitly upheld by Fritz Hartung, in an article with the identical title which was also published in the *Historische Zeitschrift*, 45 (1932) 46–52.

22. F. Hartung, *Enlightened Despotism* (Historical Association Pamphlet, London, 1957). This first appeared in German in the

Historische Zeitschrift, 180 (1955); Aretin (ed.) *Der aufgeklärte Absolutismus.*

23. The widely-used and frequently-reprinted textbook of A. H. Johnson, *The Age of the Enlightened Despot 1660–1789* (London, 1909) used 'enlightened despot' as a synonym for 'monarch' and characterised the period as one of continuously expanding monarchical power at home and abroad. It shows little acquaintance with German scholarship. One faint echo was in two sketchy articles written by the American scholar George Matthew Dutcher, but these seem to have had little impact: 'The enlightened despotism' in *Annual Report of the American Historical Association for the year 1920* (Washington, DC, 1925) pp. 189–98; and 'Further considerations on the origins and nature of the enlightened despotism' in *Persecution and Liberty: essays in honor of George Lincoln Burr* (New York, 1931) pp. 375–403.

24. Cf. *Bulletin of the International Committee of Historical Sciences* (hereafter *BICHS*), 9 (1937) 189.

25. It can be followed in *BICHS*, 1 (1926–29) 601–12; 2 (1930) 71–2, 533–52; 3 (1931) 246–9; 5 (1933) 701–804; 8 (1936) 541–3; 9 (1937) 3–225 and 519–36.

26. *BICHS*, 1 (1926–29) iii–iv, vi–vii.

27. *BICHS*, 1 (1926–29) 601.

28. 'Le despotisme éclairé, de Frédéric II à la Révolution', *BICHS*, 9 (1937) 181–225.

29. 'Les intendants du XVIIIᵉ siècle et le despotisme éclairé', *Revue d'histoire du droit* (1929).

30. *BICHS*, 1 (1926–29) 604.

31. See the influential article by Charles Morazé, 'Finance et despotisme: Essai sur les despotes éclairés', *Annales: économies, sociétés, civilisations*, 3 (1948) 279–96.

32. *Die Philosophie der Aufklärung* (Tübingen, 1932).

33. Helen P. Liebel, *Enlightened Bureaucracy versus Enlightened Despotism in Baden, 1750–1792* (*Transactions of the American Philosophical Society*, 55:5; Philadelphia, 1965) pp. 47–50.

34. One of the most extreme rejections is that in Roland Mousnier and Ernest Labrousse, *Le XVIIIᵉ siècle* ('Histoire générale des civilisations', Paris, 1953) p. 173.

35. The principal such books and articles are listed in the national bibliographies at the end of this volume, above, pp. 317–41.

36. See, in particular, the generally valuable collection edited by Roy Porter and Mikuláš Teich, *The Enlightenment in National Context* (Cambridge, 1981).

37. There is a brief introduction to his work by Stuart Woolf in Franco Venturi, *Italy and the Enlightenment: studies in a cosmopolitan century* (London, 1972) pp. vii–xiv.

38. For this approach see his 'The European Enlightenment' in Venturi, *Italy and the Enlightenment*, pp. 1–32, and his *Utopia and Reform in the Enlightenment* (Cambridge, 1971). One of the most remarkable

attempts to integrate the history of ideas with the study of political developments is Furio Diaz, *Filosofia e politica nel settecento francese* (Turin, 1962), which is devoted to the period 1749–76 in France.

39. Exemplified by the title of his great work, *Settecento riformatore* (Turin, 1969–), which has now reached vol. V, pt i.

40. See, in particular, three articles by the East German historian Ingrid Mittenzwei: 'Über das Problem des aufgeklärten Absolutismus', *Zeitschrift für Geschichtswissenschaft*, 18 (1970) 1162–72; 'Der "aufgeklärte" Absolutismus in den deutschen Territorialstaaten', *Geschichtsunterricht und Staatsburgerkunde*, 14 (1972) 1112–21; and 'Theorie und Praxis des aufgeklärten Absolutismus in Brandenburg-Preussen', *Jahrbuch für Geschichte*, 6 (1972) 53–106; and the two volumes which have their origins in the 'Colloque de Mátrafüred': *Les lumières en Hongrie, en Europe Centrale et en Europe Orientale*, eds B. Köpeczi, E. Bene and I. Kovács (Budapest, 1977) pp. 17–86; and *L'absolutisme éclairé*, eds B. Köpeczi, A. Soboul, E. H. Balázs and D. Kosáry (Budapest and Paris, 1985).

41. See, e.g., the informative article of Horst Möller, 'Die Interpretation der Aufklärung in der Marxistische-Leninistischen Geschichtsschreibung', *Zeitschrift für Historische Forschung*, 4 (1977) 438–72.

42. See his *The Well-Ordered Police State: social and institutional change through law in the Germanies and Russia, 1600–1800* (New Haven, Conn., 1983) and the earlier brief statement of his views, 'The well-ordered police state and the development of modernity in seventeenth and eighteenth-century Europe', *American Historical Review*, 80 (1975) 1221–43.

43. Cf. Krieger, *An Essay on the Theory of Enlightened Despotism*, pp. viii–ix.

44. This emerges from the pioneering P. G. M. Dickson, *Finance and Government under Maria Theresia 1740–1780* (2 vols, Oxford, 1987).

45. See the illuminating study by James C. Riley, *The Seven Years War and the Old Regime in France: the economic and financial toll* (Princeton, 1986).

46. Their importance was first made clear by H. von Voltelini, 'Die naturrechtlichen Lehren und die Reformen des 18. Jahrhunderts', *Historische Zeitschrift*, 105 (1910) 65–104. See more recently Hermann Conrad, *Staatsgedanke und Staatspraxis des aufgeklärten Absolutismus* (Opladen, 1971).

47. Quoted in translation by Eberhard Weis, 'Enlightenment and Absolutism in the Holy Roman Empire: thoughts on Enlightened Absolutism in Germany', *Journal of Modern History*, 58 (1986), Supplement, 197.

48. Quoted in translation in R. Burr Litchfield, *Emergence of a Bureaucracy: the Florentine patricians 1530–1790* (Princeton, 1986) p. 331.

49. The Italian inspiration of religious reform in Portugal is given rather more emphasis by Samuel J. Miller, *Portugal and Rome c. 1748–1830: an aspect of the Catholic Enlightenment* (Rome, 1978) than by Professor Maxwell, below, ch. 3.

50. I hope to provide a much fuller account of the links between the Enlightenment and diplomacy on a future occasion. The short section which follows amplifies and develops, and in some measure contradicts, what I wrote in D. McKay and H. M. Scott, *The Rise of the Great Powers 1648–1815* (London, 1983), especially pp. 213–14.

51. A brief guide is provided by Felix Gilbert, 'The "New Diplomacy" of the eighteenth century' in his collected essays, *History: choice and commitment* (Cambridge, Mass., 1977) pp. 323–50.

52. A. Wandruszka, *Leopold II* (2 vols, Vienna, 1963–5).

53. Lawrence J. Baack, 'State service in the eighteenth century: the Bernstorffs in Hanover and Denmark', *International History Review*, 1 (1979) 323–48, especially 330–48.

54. There is a valuable article by Orville T. Murphy, 'Charles Gravier de Vergennes: portrait of an old régime diplomat', *Political Science Quarterly*, 83 (1968) 400–18.

55. There is an authoritative and penetrating discussion in Grete Klingenstein, *Der Aufstieg des Hauses Kaunitz* (Göttingen, 1975), pt II, especially pp. 170–1.

56. This is apparent from the exemplary study by Horst Glassl, *Das österreichische Einrichtungswerk in Galizien (1772–1790)* (Wiesbaden, 1975).

57. See the excellent article by Thadd E. Hall, 'Thought and practice of enlightened government in French Corsica', *American Historical Review*, 74 (1969) 880–905.

1. SOCIAL FORCES AND ENLIGHTENED POLICIES   *Derek Beales*

1. See A. Soboul and P. Goujard, *L'Encyclopédie ou Dictionnaire raisonné des Sciences, des Arts et des Métiers: Textes choisis* (Paris, 1984), incorporating the essence of Soboul's edition of 1952. Goujard's (later) contributions often take a different line. Cf. Soboul's own essay 'Les philosophes et la révolution' in J. Le Goff and B. Köpeczi, *Intellectuels français, intellectuels hongrois, XIII–XXe siècles* (Budapest, 1985) pp. 113–32.

This essay is a substantially revised and enlarged version of a paper which has already appeared under the same title in *Seventh International Congress on the Enlightenment: introductory papers. Budapest, 26 July–2 August 1987* (Voltaire Foundation, Oxford, 1987) pp. 33–43. I am most grateful to Dr Andrew Brown and to the editor of the conference proceedings, Professor Haydn Mason, for permission to publish this article here.

I acknowledge generous help from Dr J. Black and Dr T. C. W. Blanning.

2. See especially his contributions to B. Köpeczi, E. Bene and I. Kovacs, *Les Lumières en Hongrie, en Europe centrale et en Europe orientale. Actes du troisième Colloque de Mátrafüred 28 septembre–2 octobre 1975* (Budapest, 1977) and to B. Köpeczi, A. Soboul, E. H. Balázs and D. Kosáry (eds) *L'absolutisme éclairé* (Budapest, 1985).

3. E.g. J. V. H. Melton, 'Arbeitsprobleme des aufgeklärten Absolutismus in Preußen und Österreich', *Mitteilungen des Instituts für Österreichische Geschichte [MIÖG]*, 90 (1982) 49–75.

4. For the content and importance of the *Discours préliminaire* see e.g. E. Cassirer, *The Philosophy of the Enlightenment* (Boston, 1955) pp. 8, 223–5; A. M. Wilson, *Diderot* (Oxford, 1957) pp. 131–4; R. Grimsley, *Jean d'Alembert (1717–1783)* (Oxford, 1963) pp. 18–22; B. Cohen, 'The eighteenth-century origins of the concept of scientific revolution', *Journal of the History of Ideas [JHI]*, 27 (1976) 270–3.

5. The bibliography of Voltaire's *Essai sur les moeurs* is extremely complicated, but it was in 1756 that his account of general history first appeared in 'its full development' with *Essai sur les Moeurs* as part of its title. The clearest statement of the author's aims and view of historical explanation is to be found in *Remarques pour servir de supplément à l'Essai sur les Moeurs* (1763). See Voltaire, *Essai sur les moeurs* (ed. R. Pomeau, 2 vols, Paris, 1963), esp. vol. I, p. xvi; vol. II, pp. 904–6, 910–18.

6. C. Beccaria, *Dei delitti e delle pene* (1764; facsimile edn, Turin, 1964) p. 4.

7. See Cohen, *JHI* 27 (1976) 257–88. Among other writers of the Enlightenment who take this line are Chastellux in *De la félicité publique* (1772) and Condorcet in his *Esquisse d'un tableau historique des progrès de l'esprit humain* (1793).

8. E.g. Le Mercier de la Rivière, *L'ordre naturel et essentiel des sociétés politiques* (1767) (ed. E. Depitre, Paris, 1910), esp. ch. IX.

9. E.g. Holbach (see n. 64 below) in *Essai sur les préjugés* (1769).

10. Montesquieu, *The Spirit of the Laws*, transl. T. Nugent (London, 1949) pp. 81, 91.

11. A. Ferguson, *An Essay on the History of Civil Society* (ed. D. Forbes, Edinburgh, 1966). On the general issue see Cassirer, *Philosophy of the Enlightenment*, ch. V.

12. It is difficult to footnote a negative, but I believe that what I say in the text is borne out by Rousseau's works, esp. *A Discourse on Inequality* (transl. M. Cranston, Harmondsworth, 1984).

13. S. N. H. Linguet, *Théorie des lois civiles* (1767) (Paris, 1984), esp. Livre V. H.-U. Thamer, *Revolution und Reaktion in der französischen Sozialkritik des 18. Jahrhunderts. Linguet, Mably, Babeuf* (Frankfurt-am-Main, 1973), esp. p. 275.

14. F. M. Barnard (ed.) *J. G. Herder on Social and Political Culture* (Cambridge, 1969), e.g. pp. 4–5, 34–45, 201–5.

15. A vast bibliography could be compiled on this subject. I will refer only to Voltaire's frequent discussions, esp. in the *Essai sur les moeurs*, and to O. Chadwick, *The Popes and European Revolution* (Oxford, 1981) esp. chs. 3 and 5.

16. Another vast theme. See P. Gay, *Voltaire's Politics* (Princeton, 1959); D. Dakin, *Turgot and the Ancien Régime in France* (London, 1939); H. E. Strakosch, *State Absolutism and the Rule of Law* (Sydney, 1967);

F. Diaz, *Filosofia e politica nel Settecento francese* (2nd edn, Turin, 1962).

17. Frédéric II, *Essai sur les formes de gouvernement* (1777) in J. D. E. Preuss (ed.) *Oeuvres de Frédéric le Grand* (31 vols, Berlin, 1846–57) vol. IX, pp. 205–6; D. Beales, *Joseph II: in the shadow of Maria Theresa, 1741–1780* (Cambridge, 1987) ch. XI; J. J. Rousseau, *Considérations sur le gouvernement de Pologne* (1771) in C. E. Vaughan, *The Political Writings of Jean Jacques Rousseau* (2 vols, Cambridge, 1915) vol. II, pp. 497–9.

18. Cf. P. Gay, *The Enlightenment: an interpretation* (2 vols, London, 1967, 1970) vol. II: *The Science of Freedom*, esp. ch. X; H. Chisick, *The Limits of Reform in the Enlightenment* (Princeton, 1981).

19. See Derek Beales, *Joseph II*, chs 8, 11 and 12.

20. Some of the most interesting recent discussion of these rebellions is in F. Venturi, *Settecento riformatore* (5 vols to date, 1969–) vol. III, chs V and VI, vol. IV, tome 2, ch. VIII.

21. See Beales, *Joseph II*, pp. 237, 267–8, 350, 466.

22. T. W. Perry, *Public Opinion, Propaganda, and Politics in Eighteenth-Century England: a study of the Jew Bill of 1753* (Cambridge, Mass., 1962); J. Stevenson, *Popular Disturbances in England, 1700–1870* (London, 1979) esp. pp. 76–90.

23. U. Benassi, 'Guglielmo du Tillot', ch. VIII, 'Il commercio' in *Archivio storico per le province parmensi*, new series, vol. XXIII (1923) p. 112.

24. K. Tønnesson, 'Les exemples scandinaves: le Danemark' in Köpeczi etc., *L'absolutisme éclairé*, pp. 301–2.

25. P. A. F. Gérard, *Ferdinand Rapedius de Berg* (2 vols. Brussels, 1842–3) vol. I, pp. 228–9; C. A. Macartney, *The Habsburg Empire, 1790–1918* (London, 1968) p. 137.

26. G. V. Taylor cited in W. Doyle, *Origins of the French Revolution* (Oxford, 1980) p. 20. But see G. Chaussinand-Nogaret, *The French Nobility in the Eighteenth Century* (Cambridge, 1985) chs 7 and 8, where the emphasis is on the relative radicalism of the nobles or their willingness to join the Third Estate in radical demands.

27. See the discussion in Vienna on how to deal with the Bohemian revolt of 1775 reported in Beales, *Joseph II*, pp. 350–8.

28. Beales, *Joseph II*, pp. 197–8, 267.

29. Ibid. esp. ch. 8; Gérard, *F. Rapedius de Berg*, vol. II, p. 210.

30. See e.g. M. D. George, *England in Transition* (London, 1953) p. 77; J. McManners, *Death and the Enlightenment* (Oxford, 1981) pp. 105–11; P. G. M. Dickson, *Finance and Government under Maria Theresia, 1740–1780* (2 vols, Oxford, 1987).

31. Cf. E. A. Wrigley, 'The classical economists and the Industrial Revolution' in his *People, Cities and Wealth* (London, 1987) and D. N. Cannadine, 'The present and the past in the English Industrial Revolution, 1880–1980', *Past and Present*, 103 (1984) 131–72.

32. R. Pomeau, *La religion de Voltaire* (Paris, 1956); Kant, 'What is Enlightenment?' (1784) in H. Reiss (ed.) *Kant's Political Writings*

(Cambridge, 1971) p. 59. Cf. Gay, *The Enlightenment: an interpretation*.

33. This section is based on D. D. Bien, *The Calas Affair* (Princeton, 1960) and on Voltaire, *Traité sur la Tolérance* (in J. van den Heuvel (ed.) *L'affaire Calas* (Paris, 1975), esp. pp. 90–8).

34. See e.g. D. V. Kley, *The Jansenists and the Expulsion of the Jesuits from France, 1757–1765* (London, 1975).

35. The general point is powerfully supported in J. McManners, *Death and the Enlightenment*.

36. As Besterman points out (*Voltaire*, p. 110). *L'affaire Calas*, p. 106.

37. McManners, *Death and the Enlightenment*; the chapters on France in R. Crahay (ed.) *La tolérance civile* (Brussels, 1982).

38. G. Lewis, *The Second Vendée* (Oxford, 1978) esp. ch. II.

39. This paragraph is based on J. Whaley, *Religious Toleration and Social Change in Hamburg, 1529–1819* (Cambridge, 1985) and F. Kopitzsch, *Grundzüge einer Sozialgeschichte der Aufklärung in Hamburg und Altona* (2 pts, Hamburg, 1982).

40. Cf. Beales, *Joseph II*, p. 131.

41. Ibid., p. 470 and n. 106.

42. C. H. O'Brien, 'Ideas of religious toleration at the time of Joseph II', *Transactions of the American Philosophical Society*, new series, vol. LIX, pt 7 (1969); Crahay, *La tolérance civile*, esp. the article by H. Hasquin on marriage.

43. Beales, *Joseph II*, pp. 58, 61, 465–73.

44. Ibid., p. 472.

45. R. J. W. Evans, *The Making of the Habsburg Monarchy, 1550–1700* (Oxford, 1979) p. 405.

46. Articles on Belgium in Crahay, *La tolérance civile*.

47. As well as O'Brien's study, see A. W. Small, *The Cameralists* (Chicago, 1909); R. A. Kann, *A Study in Austrian Intellectual History* (New York, 1973); K.-H. Osterloh, *Joseph von Sonnenfels und die österreichische Reformbewegung im Zeitalter des aufgeklärten Absolutismus* (Lübeck, 1970).

48. G. Bodi, *Tauwetter in Wien* (Frankfurt-am-Main, 1977).

49. L. A. Hoffmann(?), *Zehn Briefe aus Oesterreich an den Verfasser der Briefe aus Berlin* (3rd edn, 1784) p. 160: 'Ich will nun sagen, daß diese ganze Reformation, die Einführung der Toleranz . . . u.s.w. nie so leicht würde seyn zu Stande gebracht worden, wenn die Broschürenmacher nicht die Köpfe des Volks mit so vielen kleinen Büchelchen belagert, und auf einigen Boden gerade das gesagt hätten, was das Volk zu wissen brauchte.'

50. Beales, *Joseph II*, pp. 470–1.

51. Cf. the thesis of R. Mandrou, *La raison du prince: l'Europe Absolutiste, 1649–1775* (Verviers, 1980). Professor D. Kosáry, in his introductory address to the Budapest conference for which this paper was originally written, made a similar distinction between the 'centre', i.e. England, France and Holland, and the 'periphery'.

52. E.g. H. S. Commager, *Jefferson, Nationalism, and the Enlightenment*

(New York, 1975); more critical, J. R. Pole, 'Enlightenment and the politics of American nature' in that generally useful collection, R. Porter and M. Teich (ed.) *The Enlightenment in National Context* (Cambridge, 1981) pp. 192–214; S. E. Ahlstrom, *A Religious History of the American People* (London, 1972) esp. ch. 23; J. G. A. Pocock, 'Enlightenment and revolution: the case of English-speaking North America' in *Seventh International Congress on the Enlightenment: introductory papers*, pp. 45–57. Dr M. D. Kaplanoff gave me some guidance here.

53. McManners, *Death and the Enlightenment*, chs XI and XII; M. Mac-Donald, 'The secularization of suicide in England, 1660–1800', *Past and Present* 111 (1986) 50–97, and 119 (1988) 158–70 (with D. T. Andrew); J. Innes and J. Styles, 'The crime wave: recent writing on crime and criminal justice in eighteenth-century England', *Journal of British Studies*, 25 (1986) 380–435.

54. The great authority of course, as for so much else, is F. Venturi, *Settecento riformatore*. Cf. C. Capra, 'Il "Mosé della Lombardia": la missione di Carlo Antonio Martini a Milano, 1785–1786' in C. Mozzarelli and G. Olmi, *Il Trentino nel Settecento fra Sacro Romano Impero e antichi stati italiani* (Bologna, 1985).

55. Among recent discussions of Russian Enlightenment see W. J. Gleason, *Moral Idealists, Bureaucracy, and Catherine the Great* (New Brunswick, NJ, 1981) and J. P. LeDonne, *Ruling Russia: politics and administration in the age of absolutism, 1762–1796* (Princeton, 1984).

56. This point of view has recently been argued strongly in the case of England by J. C. D. Clark, *English Society, 1688–1832* (Cambridge, 1985).

57. See the works by O'Brien and Whaley cited above and L. R. Lewitter, 'The partitions of Poland' in *New Cambridge Modern History*, vol. VIII (Cambridge, 1965), esp. pp. 338–9, 344.

58. This factor is seldom stressed in modern accounts, but seems overwhelming in the case of the Austrian Monarchy. See A. Ritter von Arneth, *Geschichte Maria Theresias* (10 vols, Vienna, 1863–79) vol. IX, esp. ch. IV.

59. W. H. McNeill, *Plagues and Peoples* (Harmondsworth, 1979) pp. 229–38; P. Razzell, *The Conquest of Smallpox* (Firle, 1977); D. Baxby, *Jenner's Smallpox Vaccine* (London, 1981); McManners, *Death and the Enlightenment*, pp. 46–7; Beales, *Joseph II*, p. 158.

60. For a recent discussion, C. B. A. Behrens, *Society, Government and the Enlightenment: the experiences of eighteenth-century France and Prussia* (London, 1985) esp. pp. 128–40, 152–75. See also the classic study by S. L. Kaplan, *Bread, Politics and Political Economy in the Reign of Louis XV* (2 vols, The Hague, 1976).

61. Melton, *MIÖG* (1982).

62. Cf. Beales, *Joseph II*, pp. 455–60.

63. Chisick, *The Limits of Reform in the Enlightenment*.

64. Frédéric II, *Examen de l'essai sur les préjugés* (1770) in *Œuvres*

philosophiques (Paris, 1985) pp. 361–85. Diderot, *Pages contre un tyran* (1771) in *Œuvres politiques*, ed. P. Vernière, pp. 135–48. J. Vercruysse, *Bibliographie descriptive des écrits du Baron d'Holbach* (Paris, 1971) accepts the *Essai* as Holbach's, but it is impossible to be sure.

65. See esp. M. Raeff, *The Well-Ordered Police State* (London, 1983).

66. D. Beales (ed.), 'Joseph II's "Rêveries"', *MIÖG*, 33 (1980) 156.

67. See W. Bleek, *Von der Kameralausbildung zum Juristenprivileg: Studium, Prüfung und Ausbildung der höheren Beamten des allgemeinen Verwaltungdienstes im 18. und 19. Jahrhundert* (Berlin, 1972) esp. p. 76.

68. See A. Ryan, 'The Marx problem book', *Times Literary Supplement*, 25 April 1986.

69. Cf. K. V. Thomas, *Religion and the Decline of Magic* (London, 1971).

70. Beales, *Joseph II*, pp. 456–7.

71. Chadwick, *The Popes and European Revolution*, pp. 374–90.

2. THE ITALIAN REFORMERS   *M. S. Anderson*

1. R. Burr Litchfield, *Emergence of a Bureaucracy: The Florentine Patricians, 1530–1790* (Princeton, 1986) p. 271.

2. O. Chadwick, *The Popes and European Revolution* (Oxford, 1981) p. 399.

3. Jansenism originated in mid-seventeenth-century France essentially as a body of theological doctrines on the question of predestination versus free-will. By the eighteenth century it had broadened to a set of attitudes stressing the independence of national or state churches against papal power, a general hostility to elaborate church ceremonial, and in some cases the rights of parish priests vis-à-vis the church hierarchy.

4. M. Schipa, *Il regno di Napoli al tempo di Carlo di Borbone* (Naples, 1923) vol. II, p. 121.

5. A. Wandruszka, *Leopold II* (Vienna, 1964–65) vol. I, p. 267.

6. P. Chorley, *Oil, Silk and Enlightenment* (Naples, 1965) p. 165.

7. D. Carpanetto and G. Ricuperati, *Italy in the Age of Reason, 1685–1789* (London and New York, 1987) p. 241.

8. F. Venturi, *Italy and the Enlightenment* (London, 1972) p. 251.

9. F. Venturi, *Settecento riformatore*: vol. i, *Da Muratori a Beccaria* (Turin, 1969) p. 559.

3. POMBAL   *Kenneth Maxwell*

1. Cited by Susan Schneider, *O marquês de Pombal e o vinho do Porto: dependência e subdesenvolvimento em Portugal no século XVIII* (Lisbon, 1980) p. 8.

2. Correspondence of Maria Theresa and the Countess of Oeiras,

appendix to John Athelstone Smith, *The Marquis of Pombal* (2 vols, London, 1843) vol. II, pp. 376–7.

3. Queen D. Maria I was declared mentally incompetent in 1799 when her second son D. João became Prince Regent. The Queen died in Rio de Janeiro in 1816.

4. Samuel J. Miller, *Portugal and Rome c. 1748–1830. An aspect of the Catholic Enlightenment* (Rome, 1978) p. 186.

5. *Brotéria, no bicentenario do Marquês de Pombal*, vol. 115, 2 vols, II, p. 127.

6. T. D. Kendrick, *The Lisbon Earthquake of 1755* (London, 1956).

7. Cited in C. R. Boxer, *Some Contemporary Reactions to the Lisbon Earthquake of 1755* (Lisbon, 1956).

8. Cited in A. R. Walford, *The British Factory* (Lisbon, 1940) p. 20.

9. Arthur William Costigan, *Sketches of Society and Manners in Portugal* (2 vols, London, 1788) vol. II, p. 29.

10. Instituto Histórico e Geográfico Brasileiro, arquive, Rio de Janeiro, 1-1-8 f.43.

11. Details of profitability from H. E. S. Fisher, *The Portugal Trade* (London, 1971).

12. There is a vast literature on Verney. For a brief introduction see A. A. Banha de Andrade, *Verney e a projeccão da sua obra*, which contains in an appendix extracts from Vernay's correspondence with Muratori (Lisbon, 1980).

13. *Instrucções inéditas de D. Luís da Cunha e Marco António de Azeredo Coutinho*, ed. Pedro de Avevedo e Antonio Baiao (Coimbra, 1930) p. 139.

14. Correspondencia entre o duque Manuel Teles da Silva e Sebastião José de Carvalho e Melo, in *Anais da Academia Portuguesa da Historia* (AAP) 2nd series vol. VI (Lisbon, 1955) pp. 313–15.

15. Regimento . . . casas de inspeção, 1 April 1751, Instituto histórico e Geográfico Brasileiro, arquivo, lata 71, doc. 17.

16. Carta secretissim para Gomes Freire de Andrada. Lisbon, 21 Sept. 1751. Marcos Carneiro de Mendonça, *O Marquês de Pombal e o Brasil* (São Paulo, 1960) p. 188.

17. *AAP*, pp. 419–20.

18. John Croft, *A Treatise on the Wines of Portugal*, Oporto Instituto do Vinho do Porto, 1940 (facsimile of 1788 edition).

19. Cited in Schneider, p. 169.

20. Estatutos da Junta do Commercio, 30 Sept. 1755 and Alvará porque . . . he por bem confirmar os estatutos da junta do commércio, 16 Dec. 1756.

21. Cited in Schneider, p. 186.

22. Cited by Jorge Borges de Macedo, *problemas de história da indústria Portuguesa no século XVIII* (Lisbon, 1963) p. 147.

23. Memorias de consul e fatoria Britannica, Biblioteca Nacional, Lisbon, Pombaline collection, codice 94, f.25v.

24. These quotations are from Schneider, p. 200.

25. For a comprehensive collection of documents relating to the *aula do comercio* see Marcos Carneiro de Mendonça, *Aula do Comercio* (Rio de Janeiro, 1982).

26. See for example the preamble to the royal decree setting the conditions for the raising of the *décima* (1762).

27. Cited by Miller, *Portugal and Rome*, p. 109.

28. *Relaçao Geral do estado da universidade, 1777* (Coimbra, 1983) p. 232.

29. Cited by António José Saraiva, *Inquisição a Cristas-Novos*, 4th edn (Porto, 1909) p. 317.

30. *Gazzetta Universale*, 3 April 1777.

31. Vasta exposição de motívos a Rainba a favor da extinção das companhias de comercio exclusivas ... por José Vasque da Cunha, Instituto Histórico a Geográfico, Brasilero, arquive, 1-1-8, p. 133.

32. Luís F. De Carvalho Dias, *A relaçao das fábricas de 1788* (Coimbra, 1955) p. 95.

33. Jacques Ratton, *Recordações*, p. 152.

34. Robert Southey, *Journal of a Residence in Portugal 1800–1801* (ed. A. Cabral, Oxford, 1900) pp. 137–9; and William Beckford, *The Journal of William Beckford in Portugal and Spain 1787–1788* (ed. Boyde Alexander, London, 1954) pp. 257–8.

35. Cited by Macedo in *Problemas da Indústria*, p. 216.

4. CHARLES III *Charles C. Noel*

1. Stanley J. Stein and Barbara H. Stein, *The Colonial Heritage of Latin America. Essays on economic dependence in perspective* (New York, 1970) pp. 10–21, 49–51.

2. Quoted in Miguel Artola, 'America en el pensamiento español del siglo XVIII', *Revista de Indias*, 19 (1969) 63.

3. Roberto Fernández (ed.) *España en el siglo XVIII. Homenaje a Pierre Vilar* (Barcelona, 1985) pp. 51–2. See also the essays, ibid., by Pedro Ruíz Torres and Carlos Martínez Shaw; and Jean-Pierre Amalric, '¿En el siglo XVIII: Una agricultura agarrotada?', in B. Bennassar et al. *Orígenes del atraso económico español* (Barcelona, 1985) pp. 15–79.

4. François López, 'La historia de las ideas en el siglo XVIII: concepciones antiguas y revisiones necesarias', *Boletín del Centro de Estudios del Siglo XVIII (BOCES XVIII)*, 3 (1975) 3ff.; José Mariá López Piñero, 'La introducción de la ciencia moderna en España', *Revista de Occidente*, 2nd series, (1966) 133–56; Vicente Peset Llorca, *La Universidad de Valencia y la renovación científica española, 1687–1727* (Castellón de la Plana, 1966); and Antonio Mestre, *El mundo intelectual de Mayáns* (Valencia, 1978).

5. Antonio Rodriguez Villa, *Don Cenón de Somodevilla, Marqués de la Ensenada* (Madrid, 1878); Felipe Abad León, *El marqués de la Ensenada. Su vida y su obra* (Madrid, 1985); Angela García Rives, *Fernando VI y*

doña Bárbara de Braganza (1748–1759). Apuntes sobre su reinado (Madrid, 1917); and the correspondence of Ricardo Wall and others: Archivo Histórico Nacional, Madrid, *Estado, legs.* 2483, 2532, 2625 and 2743.

6. Carvajal quoted in Didier Ozanam (ed.) *La diplomacia de Fernando VI. Correspondencia reservada entre D. José de Carvajal y el Duque de Huéscar, 1746–1749* (Madrid, 1975) p. 8; Rafael Olaechea, 'Política eclesiástica del gobierno de Fernando VI', in *La época de Fernando VI* (Oviedo, 1981) p. 205; and Rafael Olaechea and José A. Ferrer Benimeli, *El Conde de Aranda. Mito y realidad de un político aragonés* (Zaragoza, Libraria General, 1978) vol. II, p. 53.

7. *Los españoles de la ilustración* (Madrid, 1964) p. 27.

8. Antonio Domínguez Ortiz presented a dissenting view, *Sociedad española en el siglo XVIII* (Madrid, 1955) pp. 26–33; the contemporary descriptions are by British and Austrian ambassadors – Robert Liston (National Library of Scotland, Edinburgh, MS 5554 fol. 131), James Gray (Public Record Office, London, SP 93/13) and Ruvigny de Cosne (ibid., SP 94/160) and María del Carmen Velázquez, *La España de Carlos III de 1764 a 1776. Según los embajadores austriacos* (México, 1963), despatches of March, 1764. See also Vicente Rodríguez Casado, *La política y los politicos en el reinado de Carlos III* (Madrid, 1962) pp. 84, 153.

9. On aristocratic attempts to return to power, see Teófanes Egido López, *Opinión pública y oposición al poder en la España des siglo XVIII (1713–1759)* (Valladolid, 1971).

10. The works of Vicente Rodríguez Casado and Vicente Palacio Atard present slightly differing versions of the middle-class thesis.

11. Quoting Paul-J. Guinard, *La presse espagnole de 1737 à 1791. Formation et signification d'un genre* (Paris, 1973) p. 34.

12. By, for example, Vicente Rodríguez Casado, 'El intento español de "ilustración cristiana"', *Estudios Americanos*, 9 (1955) pp. 141–69; and Joël Saugnieux, 'Foi et lumières au XVIIIe siècle', in his *Foi et lumières dans l'Espagne du XVIIIe siècle* (Lyon, 1985) pp. 9–25.

13. On the radical thought of the 1780s see José Antonio Maravall, 'Las tendencias de reforma política en el siglo XVIII español', *Revista de Occidente*, 2nd series, 5 (1967) 53–82, and Antonio Elorza, *La ideología liberal en la ilustración española* (Madrid, 1970).

14. Roberto Fernández (ed.) *España en el siglo XVIII*, p. 33.

15. Miguel Artola, *Hacienda del antiguo régimen* (Madrid, 1982) pp. 261–99; Juan Hernández Andreu, 'Evolución histórica de la contribución directa en España desde 1700 a 1814', *Revista de Economía Política*, num. 61 (1972) 31–90; and Antonio Domínguez Ortiz (ed.) *Historia de Andalucía* (Madrid, 1980–), vol. VI, p. 176.

16. Teófanes Egido characterises Compomanes' regalism in *La iglesia en la España de los siglos XVII y XVIII*, vol. IV of Ricardo García Villoslada (ed.) *Historia de la iglesia en España* (Madrid, 1979) pp. 153–6.

17. Rafael Olaechea, 'El anticolegialismo del gobierno de Carlos III',

Cuadernos de Investigación. Geografía e Historia, 2, fasc. 2 (1976) pp. 53–90 and Janine Fayard, 'Los miembros del Consejo Real de Castilla (1746–1788)', *Cuadernos de Investigación Histórica*, 6 (1982) 109–36.

18. See the works listed in the Bibliography.

19. N. T. Phillipson, 'Culture and society in the 18th century province. The case of Edinburgh and the Scottish Enlightenment', in L. Stone (ed.) *The University in Society* (Princeton, 1974), II, pp. 407–10, presents a brief but illuminating analysis of provincial society.

20. Robert Jones Shafer, *The Economic Societies in the Spanish World (1763–1821)* (Syracuse, 1958) p. 73.

21. Olaechea and Benimeli, *El Conde de Aranda*, vol. II, p. 53.

5. REFORM IN THE HABSBURG MONARCHY   *H. M. Scott*

1. *Geschichte Maria Theresias* (10 vols, Vienna, 1863–79).

2. *Josef II: Seine politische und kulturelle Tätigkeit* (2 vols, German translation of the 1907 Russian edition, Vienna, 1910).

3. P. G. M. Dickson, *Finance and Government under Maria Theresia 1740–1780* (2 vols, Oxford, 1987) vol. II, pp. 8–9.

4. The classic study of these developments is R. J. W. Evans, *The making of the Habsburg Monarchy 1550–1700* (Oxford, 1979).

5. Dickson, *Finance and Government*, vol. I, p. 222.

6. I owe this point to an important and pioneering paper by Professor Grete Klingenstein (University of Graz) to a conference of British and Austrian historians of the eighteenth century in 1983, though unhappily it has never been published. The continuing role of the Estates in government is confirmed by Dr Dickson: *Finance and Government*, vols I & II *passim*.

7. Dickson, *Finance and Government*, vol. I, pp. 305–10.

8. Ibid., vol. I, pp. 318–19. These figures exclude Hungary.

9. Quoted by Derek Beales, *Joseph II*. vol. I: *In the Shadow of Maria Theresa, 1741–1780* (Cambridge, 1987) vol. I, p. 469.

10. Reliable figures for the scale of the dissolution have now been provided by Dickson, *Finance and Government*, vol. I, pp. 72–7.

11. The discussion which follows depends heavily on the important new study of James Van Horn Melton, *Absolutism and the Eighteenth-Century Origins of Compulsory Schooling in Prussia and Austria* (Cambridge, 1988), though not all of its larger arguments can be accepted without qualification.

12. Quoted by Beales, *Joseph II*, vol. I, p. 347.

6. MARIA THERESA AND HUNGARY   *R. J. W. Evans*

1. Arneth (Bibliog.), i, 256ff., drawing on G. Kolinovics, *Nova Ungariae Periodus . . . sive Comitiorum Generalium . . . in anno 1741 . . . narratio*, ed. M. G. Kovachich (Buda, 1790). Strictly speaking, for bizarre constitutional reasons, Maria Theresa was *King* of Hungary: cf. E. Holzmair in *Mitteilungen des Instituts für Österreichische Geschichtsforschung*, 72 (1964) 122–34.

2. The latest studies of overall policy are in *Österreich im Europa der Aufklärung* (Bibliog.) and Dickson (Bibliog.). For the economy, G. Otruba, *Die Wirtschaftspolitik Maria Theresias* (Vienna, 1963), is still useful. The classic Hungarian diatribe against 'colonialisation' by stealth is F. Eckhart, *A bécsi udvar gazdaságpolitikája Mária Terézia korában* (Budapest, 1922). For the impopulation see K. Schünemann, *Österreichs Bevölkerungspolitik unter Maria Theresia* (Berlin [1935]).

3. There is no proper history of either institution, but for the Lieutenancy Council see Gy. Ember, *A magyar királyi helytartótanács ügyintézésének története, 1724–1848* (Budapest, 1940), and I. Felhő and A. Vörös, *A helytartótanácsi levéltár* (Budapest, 1961). On officialdom cf. Z. Fallenbüchl in *Die Verwaltung*, 14 (1981) 329–50. Personal details have to be pieced together individually from such works as I. Nagy, *Magyarország családai*, 8 vols (Pest, 1857–68); C. von Wurzbach, *Biographisches Lexikon des Kaiserthums Oesterreich*, 60 vols (Vienna, 1856–91); and J. Szinnyei, *Magyar írók élete és munkái*, 14 vols (Budapest, 1891–1913).

4. I. Nagy, *A magyar kamara, 1686–1848* (Budapest, 1971); Z. Fallenbüchl in *Levéltári Közlemények*, 41 (1970) 259–336. For Balassa see I. Kállay in *Levéltári Közlemények*, 44–5 (1974) 247–69.

5. On Kempelen: I. Kőszegi and J. Pap, *Kempelen Farkas* (Budapest, 1955), and long entry in Wurzbach, *s.v.*

6. For the Austrian origins of the *Urbarium*, see various articles by Gy. Ember: originally in *Századok*, 69 (1935) 554–664, and in *A Bécsi Magyar Történeti Intézet Évkönyve*, 5 (1935) 103–49; much of this then rendered into German in *Acta Historica* (Budapest), 6 (1959) 105–53, 331–71; 7 (1960), 149–87; and again in summary form in *Ungarn und Österreich* (Bibliog.), 43–54.

7. For the diet: Arneth, *op. cit.*, VII, pp. 105ff. For the making of the *Urbarium*: Szabó (Bibliog.), K. Vörös in *Tanulmányok a parasztság történetéhez Magyarországon, 1711–1790*, ed. Gy. Spira (Budapest, 1952) pp. 299–383; summary by F. Eckhart in *Századok*, 90 (1956) 69–125, at 69ff. For resistance to it: D. Szabó, *A megyék ellenállása Mária Terézia úrbéri rendeleteivel szemben* (Budapest, 1934).

8. Quoted Arneth, op. cit., VII, 535.

9. Implementation of the *Urbarium* in Rebro (Bibliog.) and Felhő *et al.* (ed.) (Bibliog.). One copy of Eszterházy's memorandum is in Országos Levéltár (National Archives, Budapest, hereafter OL), I 50

358 ENLIGHTENED ABSOLUTISM

1, fasc.A; another printed in Szabó (Bibliog.) pp. 627–33. For Festetics see Rebro, op. cit., 62–5, and OL P 245 24: letters from Maria Theresa; ibid. 25: letters from Neny. For Balassa see Szabó (Bibliog.), pp. 18–22, 139ff., 167–9, 304ff., 548–81.

10. Quoted Arneth, op. cit., IV, 3; religious background in Evans (Bibliog.), 247ff.

11. Evictions: O. Piszker, *Barokk világ Győregyházmegyében Zichy Ferenc gróf püspöksége idején, 1743–83* (Pannonhalma, 1933) pp. 48ff.; S. Nyíri, *A nagybirtok vallásügyi magatartása a 18. században* (Budapest, 1941). Resettlement: H. von Zwiedineck-Südenhorst in *Archiv für Österreichische Geschichte*, 53 (1875) 457–546; E. Buchinger, *Die 'Ländler' in Siebenbürgen* (Munich, 1980). Towns: I. Révész, *Bécs Debrecen ellen. Vázlatok Domokos Lajos életéből és működéséből* (Budapest, 1966). Catechism: T. Esze in *A Heidelbergi Káté története Magyarországon*, ed. T. Bartha (Budapest, 1965) pp. 169–203.

12. J. H. Schwicker, *Politische Geschichte der Serben in Ungarn* (Budapest, 1880), esp. ch. 4; idem in *Archiv für Österreichische Geschichte*, 52 (1875) 275–400. K. Hitchins, *L'Idée de nation chez les Roumains de Transylvanie, 1691–1849* (Bucharest, 1987), esp. ch. 3.

13. B. Ravasz, *A magyar állam és a protestántizmus Mária Terézia uralkodásának második felében* (Budapest, 1935). Sketch of Batthyány in A. Meszlényi, *A magyar hercegprímások arcképsorozata* (Budapest, 1970) pp. 122–46; and cf. R. J. W. Evans in *Das achtzehnte Jahrhundert und Österreich*, 2 (1985) 9–31, at 28f.

14. Arneth, op. cit., IX, 8–10.

15. F. Krones, *Ungarn unter Maria Theresia und Joseph II* (Graz, 1871), pt 1; A. Csizmadia in *Historický Časopis* (Bratislava), 12 (1964) 215–36; M. Vyvíjalová in Mraz (ed.) (Bibliog.), pp. 125–36; documents in *Der Josephinismus. Quellen zu seiner Geschichte in Österreich*, ed. F. Maass, 5 vols (Vienna, 1951–61), vol. I. On the *Vexatio*: Gy. Concha in *Századok*, 14 (1880) 590–7.

16. J. Tomko, *Die Errichtung der Diözesen Zips, Neusohl und Rosenau, 1776, und das königliche Patronatsrecht in Ungarn* (Vienna, 1968).

17. Mészáros (Bibliog.) prints the text; detailed assessment in Kosáry, *Művelődés* (Bibliog.), pp. 403ff.

18. Schemnitz: J. Vlachovič in *Z Dejin Vied a Techniky na Slovensku*, 3 (1964) 33–95; reheated in German in *Wissenschaftspolitik in Mittel- und Osteuropa . . . im 18. und beginnenden 19. Jahrhundert*, ed. E. Amburger et al. (Berlin, 1976) pp. 206–20; J. Vozár in Mraz (ed.) op. cit., pp. 171–82. Banat: M. Kostić, *Grof Koler kao kulturnoprosvetni reformator kod Srba u Ugarskoi u XVIII veku* (Belgrade, 1932); H. Wolf, *Das Schulwesen des Temesvarer Banats im 18. Jahrhundert* (Baden bei Wien, 1935); P. J. Adler in *Slavic Review*, 33 (1974) 23–45.

19. OL P 245 24: letters from Maria Theresa, 7 May 1765, 18 July 1765, 27 March 1766; ibid. 11, fasc. 12, *passim*. Cf., for the new university

syllabus, *Calendarium regiae universitatis Budensis ad annum . . . 1779* (Buda, 1779).

20. These comments rest largely on autopsy, e.g. of the Batthyány correspondence in OL P 1314, and of Pál Festetics's, ibid., P 245 24 and 25. There is no proper study. But D. Rapant, *K počiatkom mad'arizácie*, 2 vols (Bratislava, 1927) vol. I, contains much good information; and cf. the sounding by C. Michaud in *Dix-huitième Siècle*, 12 (1980) 327–79.

21. For Sonnenfels, cf. below p. 212 and nn.

22. Ember in *Századok*, 69 (1935), at 578ff.; F. A. J. Szabo, 'The social revolutionary conspiracy: the role of Prince W. A. Kaunitz in the policies of Enlightened Absolutism towards Hungary, 1760–1780', paper presented to the Annual Meeting of the Canadian Association of Hungarian Studies, Montreal, 3 June 1985 (and kindly sent to me by James Van Horn Melton, of Emory University).

23. A. F. Kollár, *Vorläufige Ausführung der Rechte des Königreichs Hungarn auf . . . Reußen und Podolien* (Vienna, 1772); Balthasar Kercselich, *De Regnis Dalmatiae, Croatiae, Sclavoniae notitiae praeliminares* (Zagreb, [1770]). On Kercselich/Krčelić, cf. below, ch. 7, n.32.

24. Zips: Arneth, op. cit., vol. VIII, *passim*. Banat: ibid, X, 121–9; J. Szentkláray, *Gróf Niczky Kristóf életrajza* (Pozsony, 1885). Fiume: below, n.36.

25. Arneth, op. cit., vol. X, pp. 131–58, 769. G. A. Schuller, *Samuel von Brukenthal*, 2 vols (Munich, 1967–69), is massively informative (vol. I, pp. 172ff. for the 1765 episode, in which, however, the principality's Magyar Chancellor, Count Gábor Bethlen, played a greater part). Military Frontier: C. Göllner, *Die Siebenbürgische Militärgrenze . . . 1762–1851* (Munich, 1974); Orthodox subjects: M. Bernath, *Habsburg und die Anfänge der rumänischen Nationsbildung* (Leiden, 1972).

26. Arneth, op. cit., IX, 302–5; C. von Hock and H. I. Bidermann, *Der österreichische Staatsrat* (Vienna, 1879) p. 26; Ember in *Századok*, 69 (1935), at 563f.; F. A. J. Szabo, art. cit., 12ff.; Beales (Bibliog.) pp. 213f.

27. OL P 245 24: letter from Maria Theresa, 27 March 1766.

28. Kosáry, *Művelődés*, p. 107.

29. J. Tibenský, *Slovenský Sokrates: život a dielo Adama Františka Kollára* (Bratislava, 1983); idem in *Historický Časopis* (Bratislava), 31 (1983) 371–93. D. Dümmerth, in *Filológiai Közlöny*, 12 (1966) 391–413, agrees with this reading of Kollár, and condemns him for his anti-Magyar animus.

30. A single example is the correspondence of the socialite, Baroness Grass, with the high-flying Balassa, in OL P 1765 59, nos 2946–62.

31. Handy institutional survey by C. Duffy, *The Army of Maria Theresa* (London, 1977); T. M. Barker, *Army, Aristocracy, Monarchy: essays on war, society, and government in Austria, 1618–1780* (Boulder, Colo., 1982), in fact

hardly deals with the period after 1740. Arneth, op. cit., *passim*, describes the wars in detail.

32. The subject needs a serious treatment; a few suggestions in Evans, art. cit., 17.

33. The best-known figures among the Bodyguard-*literati* are Bessenyei, Báróczi and Barcsay. Convenient summaries of their activities in *A magyar irodalom története*, ed. I. Sőtér, 6 vols (Budapest, 1964–66), vol. III, pp. 22–53; Kosáry, *Művelődés*, 301ff., 647ff.; I. Fried and L. Némedi in Mraz (ed.) op. cit., pp. 355–85.

34. This *Lettre* remained unpublished, until edited by I. Vörös (Budapest, 1987). On Fekete see Gy. Morvay, *Galánthai gróf Fekete János* (Budapest, 1903).

35. H. Marczali, *Az 1790/1-diki országgyülés*, 2 vols (Budapest, 1907), vol. II, pp. 82ff.; Morvay, op. cit., pp. 141ff.

36. Fiume: A. Fest in *Századok*, 50 (1916) 239–66; E. H. Balázs in '*Sorsotok előre nézzétek*': *a francia felvilágosodás és a magyar kultúra*, ed. B. Köpeczi and L. Sziklay (Budapest, 1975) pp. 145–61. University: Kosáry, *Művelődés*, pp. 499ff.

37. Maria Theresa quoted from Arneth, op. cit., X, 128. Titles of Mailáth (quoted at vol. V, p. 86), Horváth and Marczali in Bibliog. A further example, from 1826: F. Kazinczy, *Művei*, ed. M. Szauder, 2 vols (Budapest, 1979) vol. II, p. 648.

38. OL P 245 24: letter from György Festetics, 20 July 1781. Teleki: OL P 661 45, nos 2256, 2398.

39. Teleki quoted from OL P 661 45, no. 2200. See Bibliog. for the general literature on Joseph and Hungary.

40. Beales, op. cit., esp. pp. 100–2, 181f., 339, 484, implies this, but without dwelling on the nature of Joseph's attitude to Hungary before his accession there. For Festetics's view of the co-regency see OL P 245 11, fasc. 12, fols 12–19; cf. Beales, op. cit., pp. 135–7, on the general issue.

41. Dickson, op. cit., esp. vol. I, pp. 207ff., 325–9 (quotation at p. 325).

42. For later developments on this front see R. J. W. Evans, 'The Habsburgs and the Hungarian Problem, 1790–1848', forthcoming in *Transactions of the Royal Historical Society*, 5th series 39 (1989), in some ways a companion piece to the present chapter.

7. JOSEPH II AND NATIONALITY IN THE HABSBURG LANDS   *R. J. W. Evans*

1. On the Serbian issue: J. H. Schwicker, *Politische Geschichte der Serben in Ungarn* (Budapest, 1880), ch. viii and *passim*; M. Kostić, *Grof Koler kao kulturnoprosvetni reformator kod Srba u Ugarskoj u 18. veku* (Belgrade,

1932), esp. pp. 162ff.; I. Szeli, *Hajnóczy és a délszlávok* (Novi Sad, 1965) p. 54, etc. For the general impact of the language decree see Mitrofanov (Bibliog.) vol. I, pp. 252ff.; vol. II, pp. 846ff.; and on Hungary especially F. Szilágyi, *A germanizálás történelméből a két magyar hazában II József alatt* (Budapest, 1876), and H. Marczali, *Magyarország története II József korában*, 3 vols. (Budapest, 1882–8) vol. II, pp. 532ff.

2. Mitrofanov, op. cit., vol. I, 81ff. and *passim*. For the origins of the Austrian Enlightenment, cf. R. J. W. Evans in *Das achtzehnte Jahrhundert und Österreich*, 2 (1985) 9–31; for its relation to official policy see Gy. Szekfű in *Századok* 46 (1912) 298–307; F. Valjavec, *Der Josephinismus* (2nd edn, Munich, 1945) p. 7 and *passim*, where the linkage is so loose that the term 'Josephinism' hardly seems the *mot juste* for much of what is described; E. Zöllner in *Österreich und Europa, Festschrift H. Hantsch* (Graz, etc., 1965) pp. 203–19; E. Wangermann, *Aufklärung und staatsbürgerliche Erziehung: Gottfried van Swieten als Reformator des österreichischen Unterrichtswesens, 1781–91* (Vienna, 1978) *passim*.

3. Alois Blumauer, *Beobachtungen über Oesterreichs Aufklärung und Litteratur* (Vienna, 1782).

4. Hungary: Rapant (Bibliog.) pp. 112ff., *passim*; cf. Hans Wolf, *Das Schulwesen des Temesvarer Banats im 18. Jahrhundert* (Baden bei Wien, 1935) pp. 100f., 160f., and *passim*, and D. Kosáry, *Művelődés a 18. századi Magyarországon* (Budapest, 1980) pp. 403ff., esp. 421–4. On the Czech case see the important text by Count Franz Kinsky, scholar, scientist, friend of Joseph II, and chief begetter of the military academy at Wiener Neustadt: *Erinnerungen über einen wichtigen Gegenstand* (Prague, 1773), esp. paras 79–81; also J. Hanzal in *Československý Časopis Historický* 16 (1968) 317–39; Schamschula (Bibliog.), pp. 145ff.; Wangermann, op. cit., p. 23. Some general thoughts in Valjavec, op. cit., pp. 150ff.

5. The most important such newspapers were *Magyar hírmondó* (cf. the selection, ed. Gy. Kókay [Budapest, 1981]) from 1780; V. M. Kramerius's editorship of the *Schönfeldské cís. král. poštovské noviny*, later and better known as the *Krameriusovy cís. král. vlastenské noviny*, from 1786; and the *Prešpurské noviny* from 1783. The early 1790s saw Puglio's *Serbskije novini* and Novaković's *Slaveno-Serbskija Vjedomosti*, both published in Vienna. A. Răduțiu and L. Gyémánt (eds), *Reportoriul actelor oficiale privind Transilvania tipărite în limba română 1701–1847* (Bucharest, 1981), for Rumanian documents. The popular treatises were often too ephemeral to be adequately recorded in standard bibliographies. A good example would be the much-translated works of the Austrian agriculturalist, Johann Wiegand (1707–76), especially his *Handbuch für die österreichische Landjugend zum Unterricht einer wohlgeordneten Feldwirtschaft* (Vienna, 1771, etc.). Other examples in Rapant, op. cit., pp. 136ff., 304ff., 570; Kostić, op. cit., pp. 83ff.; Kosáry, op. cit., pp. 461f., 571ff., esp. 602ff. and 637ff.

6. A. Janša, *Vollständige Lehre von der Bienenzucht* (Vienna, 1775). Czech edns in Z. Tobolka and F. Horák (eds), *Knihopis českých a slovenských tisků*, in progress (Prague, 1939–), nos 3506–10. Cf. *Ottův slovník naučný*,

16 vols (Prague, 1888–1909) vol. VI, pp. 145f.; vol. XXII, p. 477. J. Fándly, *Slovenský včelár* and *O Uhorech aj včeláh rozmlúváňí* (both Trnava, 1802); I. [Piuariu-] Molnár, *Economia stupilor* (Vienna, 1785).

7. On the Puritans cf. R. J. W. Evans in *International Calvinism, 1541–1715*, ed. M. Prestwich (Oxford, 1985) pp. 182ff. and literature there. Bessenyei is quoted from *Rapant*, op. cit., p. 366; for a convenient summary of his views see chapter 6 above, n.33.

8. The German language was perceived thus by, for example, János Batsányi and Ferenc Kazinczy in Hungary (cf. Kosáry, op. cit., pp. 301f.). Compare the changing perceptions of the role of the enlightened circle around K. H. Seibt in Prague, the starting-point for E. Lemberg, *Grundlagen des nationalen Erwachens in Böhmen* (Reichenberg, 1932).

9. Examples of the narrow approach in A. von Arneth, *Geschichte Maria Theresias*, 10 vols (Vienna, 1863–79) vol. IX, p. 244; E. G. von Pettenegg (ed.), *Ludwig und Karl . . . von Zinzendorf* (Vienna, 1879) p. 158; G. Klingenstein in *Wiener Beiträge zur Geschichte der Neuzeit*, 3 (1976) 126–57, and ibid. 5 (1978) 165–204; Wangermann, op. cit. We find the same opinion in an early number of the *Prager Gelehrte Nachrichten* 1 (1772) no. 25, p. 87.

10. I have used the Italian translation of the first edn: *Sull'Amore della Patria*, transl. Baron Antonio Zois (Vienna, 1772), 7ff., *passim*.

11. J. von Sonnenfels, *Ueber die Liebe des Vaterlandes* (2nd edn, Vienna, 1785) pp. 189–91. Sonnenfels cites Pelzl on the Bohemian law [of 1615] which ordered fines for those not using the [Czech] *Landessprache*, and comments: 'How our exchequer would profit from such a law for German nobles!' He still sees only the threat from without, not from within the Monarchy.

12. Myl'nikov (Bibliog.).

13. Cf. Evans, art. cit., 21–3.

14. To de Luca's incomplete work might be added the earlier sketch by F. C. von Khautz, *Versuch einer Geschichte der österreichischen Gelehrten* (Frankfurt/Leipzig, 1755). M. A. Voigt *et al.* (eds), *Effigies virorum eruditorum et artificum Bohemiae et Moraviae*, 2 vols (Prague, 1773–74), expanded as *Abbildungen böhmischer und mährischer Gelehrten und Künstler*, 4 vols (Prague, 1773–82); F. M. Pelzl, *Böhmische, mährische und schlesische Gelehrte und Schriftsteller aus dem Orden der Jesuiten* (Prague, 1786); B. Balbín, *Bohemia Docta*, ed. R. Ungar, 2 vols (Prague, 1776–80); R. Ungar (ed.) *Allgemeine böhmische Bibliothek* (Prague, 1786). P. Bod, *Magyar Athenas . . .* (N.p., 1766); I. Weszprémi, *Succincta medicorum Hungariae et Transylvaniae biographia*, 3 vols (Leipzig/Vienna, 1774–87), much more than just a history of medicine; P. Wallaszky, *Conspectus reipublicae litterariae in Hungaria* (Pressburg/Leipzig, 1785); E. Horányi, *Memoria Hungarorum . . . scriptis editis notorum*, 3 vols (Vienna/Pressburg, 1775–77), and *Nova Memoria Hungarorum . . .* (Pest, 1792); idem, *Scriptores Piarum Scholarum*, 2 vols (Buda, 1808–9) with strongly Hungarian focus.

15. I. de Luca, *Oesterreichische Staatenkunde*, 2 vols (Vienna, 1786–89),

and *Geographisches Handbuch von dem österreichischen Staate*, 6 vols (Vienna, 1790–92), as well as various short-lived periodicals. J. M. Korabinsky, *Geographisch-historisches und Produkten-Lexikon von Ungarn* (Pressburg, 1786); cf. J. Benkő, *Transilvania*, 2 vols (Vienna, 1778), and K. G. von Windisch, *(Neues) Ungrisches Magazin*, 6 vols (Pressburg, 1781–91). J. A. von Riegger (ed.) *Materialien zur ... Statistik von Böhmen*, 12 vols (Prague/Leipzig, 1787–94), and *Archiv der Geschichte und Statistik, insbesondere von Böhmen*, 3 vols (Dresden, 1792–95). Riegger's father of course, in whose footsteps Joseph Anton began, was an important figure for the ideology of Austrian state-building (cf. E. Seifert, *Paul Joseph Riegger* [Berlin, 1973]).

16. *Nationalisten*, in administrative usage, meant especially the Serbs and Rumanians of the Banat: examples in Schwicker, op. cit., pp. 169f., 203, 206, 239, 377 (where 'raizische Nationalisten' seem to be contrasted – as by Bartenstein – with the 'ungarische Nation'); Wolf, op. cit. *passim*; S. Gavrilović and N. Petrović (eds) *Temišvarski Sabor 1790* (Novi Sad, 1972) nos 52, 176, 203, 252, 289; cf. A. Tafferner (ed.) *Quellenbuch zur donauschwäbischen Geschichte*, vol. I (Munich, 1974) pp. xxi, 203, 249, 265.

17. Matěj/Mátyás/Matthias Bél is significant for statistical-topographical developments, for enthusiasm about vernaculars tinged with a quasi-Puritan concern, even to a lesser extent for historical writing. The literature on him is correspondingly diverse: for a Slovak view see J. Oberuč, *Matthieu Bel, un piétiste en Slovaquie au 18ᵉ siècle* (Strasbourg, 1936); Magyar titles are in D. Kosáry, *Bevezetés a magyar történelem forrásaiba és irodalmába*, 3 vols (Budapest, 1951–58) vol. II, pp. 19f. F. Fancev (ed.) (Bibliog.) pp. 273–96 (Derkos).

18. Rapant, op. cit., pp. 273ff., *passim*; cf. ibid., pp. 380ff. for some comments (from Kazinczy, Báróczi, etc.) about the neglect and decay of Magyar. But see also J. Váczy in *Századok*, 48 (1914) 257–76, 370–82, for a different perspective.

19. For the hardly less interesting responses of Westernised Catholic aristocrats, rising officials, and reforming clerics, see comments at the end of chapter 6 above.

20. Kazinczy (Bibliog.); cf. S. Imre in *Budapesti Szemle*, 83 (1895) 161–94, 364–95; J. Váczy, *Kazinczy Ferenc és kora* (Budapest, 1915) 152–95. Berzeviczy: Balázs (Bibliog.) 135ff. Sámuel Teleki: Keresztesi (below, n.22) pp. 108ff., 216ff.; cf. A. Deé Nagy (ed.) *Teleki Sámuel és a Teleki-téka* (Bucharest, 1976). József Teleki: D. F. Csanak, *Két korszak határán: Teleki József* (Budapest, 1983) pp. 273ff. On Benkő cf. Gy. Concha, *A kilencvenes évek reformeszméi és következményei* (Budapest, 1885) pp. 50ff.

21. E. H. Balázs in *Wiener Beiträge zur Geschichte der Neuzeit*, 3 (1976) 251–69 (Podmaniczky and Schlözer); L. Wagner, *Báró Prónay Gábor Pozsony-tankerületi kir. főigazgató* (Pozsony, 1912); Kosáry, *Művelődés* pp. 388, 393, 436, 445, 469 (Vay); I. Révész, *Bécs Debrecen ellen: vázlatok Domokos Lajos életéből és működéséből* (Budapest, 1966).

22. József Keresztesi, *Krónika Magyarország polgári és egyházi közéletéből a 18. század végén* (Pest, 1868) esp. pp. 24ff., 204ff.

23. J. Genersich, *Von der Liebe zum Vaterlande*, i–ii (Vienna, 1793). The conclusions of Csáky (Bibliog.) about the 'Hungarus' ideal as a realistic alternative to rising nationalism rest heavily on the wider loyalties of Hungarian Germans.

24. G. A. Schuller, *Samuel von Brukenthal*, 2 vols (Munich, 1967–69) vol. II, p. 180 (quoted) and *passim*. Cf. I. Markó, *II József és az erdélyi szászok* (Budapest, 1940); C. Göllner and H. Stanescu (eds) *Aufklärung: Schrifttum der Siebenbürger Sachsen und Banater Schwaben* (Bucharest, 1974) esp. pp. 141ff.

25. Institoris: J. Čaplovič in Tibenský (ed.) (Bibliog.) pp. 285–302. Jan Molnár (1757–1818, not to be confused with the contemporary Magyar Jesuit and bibliographer, János Molnár) was the author of a radical Josephinist political tract: *Politisch-kirchliches Manch-Hermaeon* (N.p., 1790); cf. D. Silagi, *Ungarn und der geheime Mitarbeiterkreis Josephs II* (Munich, 1960) pp. 41ff., *passim*.

26. D. Obradović (Bibliog.) pp. 131ff., and introduction, *passim*.

27. Gavrilović and Petrović (eds) op. cit., provide full documentation for the Sabor, including plenty of Serbian grievances, mostly of the traditional kind, though it remains difficult to reconstruct the actual content of the debates or relations between the Serbian leadership and the Habsburg commissioner, Schmidfeld.

28. M. Bernath, *Habsburg und die Anfänge der rumänischen Nationsbildung* (Leiden, 1972), is an extreme version of the manipulative view. Contrast Prodan (Bibliog.) pp. 229ff. and *passim*, and – most recently – I. Lungu, *Şcoala ardeleana* (Bucharest, 1978). There was a similar impact, at a much lower level, on Hungarian Ruthenes in the age of Bishop Andrei Bachyns'kyi; cf. P. R. Magocsi, *The Shaping of a National Identity: Subcarpathian Rus'* (Cambridge, Mass, 1978) pp. 29ff.

29. I leave aside the Poles on the grounds of the special circumstances in newly-acquired Galicia (hence the Uniate Ruthenes are omitted as well) and the very different problems of statehood and nationhood raised. Yet the Habsburg impact on Galicia was substantial (as shown by H. Glassl, *Das österreichische Einrichtungswerk in Galizien 1772–90* (Wiesbaden, 1975)) and cultural linkages were by no means all negative (though barely touched on ibid., pp. 236–46 and *passim*).

30. Some, e.g. J. Strakoš, *Počátky obrozenského historismu v pražských časopisech a M. A. Voigt* (Prague, 1929), and A. Pražák, *České obrození* (Prague, 1948), esp. pp. 9–62, have doubted the genuineness of Bohemian patriotic attachment to Habsburg-directed reform, but rather unconvincingly. Schamschula, op. cit., has curiously little to say about this Austrian dimension. The *Kniha Josefova, sepsaná od jistého spatřujícího osmnácté století, dílem již stclé věci, dílem proroctví* (Prague, 1784) was adapted from a German original. It is reprinted in M. Novotný (ed.) *Kniha Josefova* [etc.] (Prague, 1941) pp. 135–80.

31. Dobrovský's lecture was delivered in the presence of Emperor Leopold II before the 'Private Society', lately renamed Royal Bohemian Society of Sciences, with some passages on the Czech language omitted; the full text was published the same year. F. M. Pelzl, *Akademische Antrittsrede über den Nutzen und die Wichtigkeit der böhmischen Sprache* (Prague, 1793); J. A. Riegger, *Für Böhmen von Böhmen* (Prague, 1794). See in general J. Prokeš, *Počátky České Společnosti Nauk do konce 18. století* (Prague, 1938), and most recently J. Purš and M. Kropilák (eds) *Přehled dějin Československa*, in progress (Prague, 1981–), Vol. I, pt 2; pp. 460ff.

32. Life of Kercselich (Krčelić) in his *Annuae 1748–67*, ed. T. Smičiklas, 2 vols (Zagreb, 1901–2), pp. i–lxvii; cf. above p. 199. Skerlecz (Škrlec): V. Lunaček in *Historijski Zbornik*, 15 (1962) 141–80; his main works have appeared in Magyar translation, ed. P. Berényi (Budapest, 1914); his anonymous *Declaratio* in favour of Croat is printed in Fancev (ed.), op. cit. no. 8. Vrhovac: Josip Horvat, *Kultura Hrvata kroz 1000 godina*, 2 vols (Zagreb, 1939–42) ii, pp. 149ff., 220ff., 225ff.

33. F. M. Pelcl (Pelzl, 1734–1801), *Paměti*, transl. and ed. J. Pán (Prague, 1931) pp. 57ff.; for Skerlecz (1729–99) see previous n. J. Ludvíkovský, *Dobrovského klasická humanita* (Bratislava, 1933), is very perceptive about Dobrovský (1753–1829); for Vrhovac (1752–1827) see previous n.

34. Pogačnik (Bibliog.) pp. 131ff.; but cf. F. Kidrić (ed.) *Zoisova korespondenca 1808–10*, 2 vols (Ljubljana, 1939–41). For the Sonnenfels translation see above, n.10; I have not been able to discover more about this Antonio Zois, who defended theses on *Polizei, Handlung*, and *Finanz* at the Theresianum in 1772.

35. On Kollár see above, ch. 6, nn.15, 29.

36. There is much on Bernolák in Tibenský (ed.) op. cit. For the linguistic side see also A. Bernolák, *Gramatické dielo*, ed. J. Pavelek (Bratislava, 1964); Jozef Ignác Bajza, *René mláď enca príhodi a skusenost'i*, ed. J. Tibenský (Bratislava, 1955), esp. pp. 279ff., 330ff.; Juraj Fándly, *Výber z diela*, ed. J. Tibenský (Bratislava, 1954), and cf. above, n.6.

37. Rapant, op. cit.

38. E. Winter, *Der Josephinismus und seine Geschichte* (Brünn, etc., 1943); F. Maass (ed.) *Der Josephinismus*, 5 vols (Vienna, 1951–61). The current profitable interest in Austrian Jansenism – major contributions are P. Hersche, *Der Spätjansenismus in Österreich* (Vienna, 1977) and Elisabeth Kovács (ed.) *Katholische Aufklärung und Josephinismus* (Munich, 1979) – likewise needs broadening, to recognise the existence of such as Bernolák and Japelj, though Jansenism hardly looks a key issue on the Habsburg periphery, and its correlation with Enlightenment was not, of course, always positive.

39. Able representatives of a Marxist view are Hroch (Bibliog.) and Arató (Bibliog.) esp. vol. I, pp. 13ff., 68ff.; cf. Myl'nikov, *Vznik*, 23ff., and *Épocha*, 140ff. Somewhat similar notions in the second edn of

Winter's *Josephinismus* (Berlin, 1962), p. 361 and *passim*, and in Gogolák (Bibliog.) vol. II, pp. 1–10 and *passim*.

40. For the original comment (it may be apocryphal) see A. Polonsky, *Politics in Independent Poland, 1921–39* (Oxford, 1972) p. 64.

41. The Matica Srpska, founded at Pest in 1826, had numerous imitators. *Matica* means other things too, in various Slavonic languages, but its apian significance seems central to the initiatives which took its name, especially since various nationalist periodicals also used the slogan of the worker bee (*bčela, pčela, včela*).

9. THE DANISH REFORMERS   *Thomas Munck*

1. Claus Bjørn, 'The peasantry and agrarian reform in Denmark', *Scandinavian Economic History Review*, 25 (1977) 117–37; his ' "De danske cahiers": studier i bondereaktionerne på forordningen af 15. april 1768', *Landbohistorisk Tidsskrift*, 5 (1983) 145–70; and his *Bonde, herremand, konge: bonden i 1700–tallets Danmark* (Copenhagen, 1981) *passim*.

2. For a controversial attempt to reconsider some of the assumptions in Danish historical writing about the conditions that led to reform, see Thorkild Kjærgaard, 'The farmer interpretation of Danish history', *Scandinavian Journal of History*, 10 (1985) 97–118.

3. Reventlow's reputation in Danish agrarian history has always been very substantial: see T. Kjærgaard, *Konjunkturer og afgifter: C. D. Reventlows betænkning af 11. februar 1788 om hoveriet* (Copenhagen, 1980) pp. 19–24 and 62ff.

4. Cited in Ole Feldbæk, *Danmarks historie*, vol. IV: *Tiden 1730–1814* (Copenhagen, 1982) p. 164. Translation here and elsewhere is my own.

5. Cited in Hans Jensen, *Dansk Jordpolitik 1757–1919* (Copenhagen, 1936; repr. 1975), vol. I, p. 121.

6. There is no full study of Colbiørnsen; but see the entry on him by Claus Bjørn in *Dansk Biografisk Leksikon*, vol. III (Copenhagen, 1979) pp. 457–62.

7. Bjørn, 'The peasantry', pp. 126–9 and 134–6.

8. On the importance of fiscal incentives for state involvement in agrarian reform in general, and in the protection of tax-paying peasant land as opposed to exempt demesne land in particular, especially in a period of economic growth and in connection with the major reorganisation of state finances attempted in 1785, see Jens Holmgaard, 'Landboreformerne – drivkræfter og motiver', *Fortid & Nutid*, 27 (1977–78) 37–47. Claus Rafner, 'Fæstegårdsmændenes skattebyrder 1660–1802', *Fortid & Nutid*, 33 (1986) 81–94, has recently cast doubts on the oppressiveness of the fiscal burdens on peasant tenants during the eighteenth century, and the crown was no doubt aware of the potential for increasing income from this source.

9. Jensen, op. cit., pp. 175ff.

10. On the protest of the Jutland landowners, see Claus Bjørn, 'den jyske proprietærfejde 1790–91', *Historie – Jyske Samlinger*, 13 (1979) 1–70.

10. FREDERICK THE GREAT   *T. C. W. Blanning*

1. C. B. A. Behrens, 'Enlightened absolutism', *The Historical Journal*, 18:ii (1975); Theodor Schieder, *Friedrich der Grosse: Ein Königtum der Widersprüche* (Frankfurt-am-Main, Berlin and Vienna, 1983), especially pp. 123, 146.

2. C. B. A. Behrens, *Society, Government and the Enlightenment. The experiences of eighteenth-century France and Prussia* (London, 1985) p. 178.

3. Ingrid Mittenzwei, *Friedrich II. von Preussen: Eine Biographie* (East Berlin, 1979) p. 87.

4. Hans Rosenberg, *Bureaucracy, Aristocracy and Autocracy: The Prussian experience 1660–1815* (Cambridge, Mass., 1958) pp. 130–1.

5. Walther Hubatsch, 'Zum Preussenbild in der Geschichte', *Das Preussenbild in der Geschichte*, ed. Otto Büsch (Berlin, 1981) p. 16.

6. Frederick the Great, *The History of My Own Times, Posthumous works of Frederic II King of Prussia*, vol. 1 (London, 1789) pp. 87–8.

7. Wolfgang Neugebauer, *Absolutistischer Staat und Schulwirklichkeit in Brandenburg-Preussen* (Berlin and New York, 1985) pp. 176–80, 632.

8. Ibid., pp. 183, 626.

9. Ibid., p. 630.

10. Gerhard Oestreich, *Friedrich Wilhelm I. Preussischer Absolutismus, Merkantilismus, Militarismus* (Göttingen, 1977) p. 89.

11. Heinrich von Treitschke, *Deutsche Geschichte im 19. Jahrhundert*, 5 vols (Leipzig, 1927) vol. I, p. 43.

12. Otto Hintze, *Die Hohenzollern und ihr Werk*, 8th edn (Berlin, 1916) p. 346.

13. G. B. Volz (ed.) *Die Politischen Testamente Friedrichs des Grossen* (Berlin, 1920) pp. 26, 30. The political testament of 1752 can be found, in German translation, in Otto Bardong (ed.) *Friedrich der Grosse* (Darmstadt, 1982) pp. 174–262.

14. See above, p. 286.

15. Wilhelm Bleek, *Von der Kameralausbildung zum Juristenprivileg. Studium, Prüfung und Ausbildung der höheren Beamten des allgemeinen Verwaltungsdienstes in Deutschland im 18. und 19. Jahrhundert* (Berlin, 1972) p. 69.

16. Karl Demeter, 'Die Herkunft des preussischen Offizierkorps', Otto Büsch and Wolfgang Neugebauer (eds) *Moderne preussische Geschichte*, 3 vols (Berlin and New York, 1981) vol. II, p. 875.

17. Bardong (ed.), *Friedrich der Grosse*, p. 163.

18. Peter Baumgart, 'Zur Geschichte der kurmärkischen Stände im 17. und 18. Jahrhundert', *Ständische Vertretungen in Europa im 17. und 18. Jahrhundert*, ed. Dietrich Gerhard (Göttingen, 1969) p. 148.

19. Gustavo Corni, *Stato assoluto e società agraria in Prussia nell'età di Federico II* (Bologna, 1982) p. 398. For a detailed investigation of the growing problems of the Prussian Junkers, see F. Martiny, *Die Adelsfrage in Preussen vor 1806 als politisches und soziales Problem*, Beiheft 35 of *Vierteljahresschrift für Sozial- und Wirtschaftsgeschichte* (Berlin and Stuttgart, 1938).

20. Gerhard Ritter, *Frederick the Great. An historical profile* (London, 1968) p. 159.

21. Mittenzwei, *Friedrich II*, p. 155. See also his observations on the subject in his political testament of 1768, Volz (ed.) *Die Politischen Testamente*, p. 181.

22. Stephan Skalweit, 'Friedrich Wilhelm I und die preussische Historie', Büsch and Neugebauer, *Moderne preussische Geschichte*, vol. I, p. 125.

23. Behrens, *Society, Government and the Enlightenment*, p. 38.

24. This remark has been variously attributed, most often to Mirabeau. It seems that in fact it was first made by Georg Heinrich von Berenhorst, an illegitimate son of Prince Leopold of Anhalt-Dessau who had served in the Seven Years War as personal adjutant to Frederick and who later became a prominent writer on military affairs – Christian Graf von Krockow, *Warnung vor Preussen* (Berlin, 1981) pp. 105, 213 n.5.

25. Otto Büsch, *Militärsystem und Sozialleben im alten Preussen 1713–1807*, new edn (Frankfurt-am-Main, 1981) p. 1; Ingrid Mittenzwei, *Preussen nach dem Siebenjährigen Krieg. Auseinandersetzungen zwischen Bürgertum und Staat um die Wirtschaftspolitik* (East Berlin, 1979) p. 107; Manfred Messerschmidt, 'Preussens Militär in seinem gesellschaftlichen Umfeld', Hans-Jürgen Puhle and Hans-Ulrich Wehler (eds) *Preussen im Rückblick* (Göttingen, 1980) p. 50; Klaus Schwieger, 'Das Bürgertum in Preussen vor der Französischen Revolution' (unpublished dissertation, University of Kiel, 1971) pp. 116–17.

26. Volz (ed.) *Die Politischen Testamente*, p. 82.

27. Schieder, *Friedrich der Grosse*, p. 27.

28. Instructions for Major Count Adrian Heinrich Borke on the education of the heir to the throne, Bardong (ed.) *Friedrich der Grosse*, p. 168.

29. Martin Greiffenhagen, *Die Aktualität Preussens. Fragen an die Bundesrepublik* (Frankfurt-am-Main, 1981) p. 85.

30. Quoted in Behrens, *Society, Government and the Enlightenment*, p. 60.

31. Comte Jacques-Antoine-Hippolyte de Guibert, *Journal d'un voyage militaire fait en Prusse dans l'année 1787* (Paris, 1790) p. 128.

32. R. Köpcke, *Ludwig Tieck. Erinnerungen aus dem Leben des Dichters nach dessen mündlichen und schriftlichen Mittheilungen*, 2 vols (Leipzig, 1855) vol. I, p. 27.

33. Büsch, *Militärsystem und Sozialleben*, pp. vi–vii.

34. Ibid., pp. 103–4, 164.

35. Ibid., p. 61.

36. Ibid., p. 169.

37. Ernst Consentius, 'Friedrich der Grosse und die Zeitungs-Zensur', *Preussische Jahrbücher*, CXV (1904) 220.

38. Ibid., pp. 230–1.

39. Horst Steinmetz (ed.) *Friedrich II., König von Preussen, und die deutsche Literatur des 18. Jahrhunderts* (Stuttgart, 1985) pp. 50, 290 n.3.

40. Günter Vogler and Klaus Vetter, *Preussen von den Anfängen bis zur Reichsgründung* (Cologne, 1981) p. 41.

41. Ibid., pp. 43, 106.

42. Ingrid Mittenzwei, 'Über das Problem des aufgeklärten Absolutismus', *Zeitschrift für Geschichtswissenschaft*, 18:ix (1970) 1165.

43. Klaus Vetter, *Kurmärkischer Adel und preussische Reformen* (Weimar, 1979) pp. 23–5.

44. Klaus Epstein, *The Genesis of German Conservatism* (Princeton, 1966) pp. 211–13.

45. Rosenberg, *Bureaucracy, Aristocracy and Autocracy*, p. 155.

46. Schieder, *Friedrich der Grosse*, p. 30; Consentius, 'Friedrich der Grosse und die Zeitungs-Zensur', pp. 221–2.

47. Schieder, *Friedrich der Grosse*, p. 28; Mittenzwei, *Friedrich II*, p. 16; Michael Erbe, *Deutsche Geschichte 1713–1790. Dualismus und aufgeklärter Absolutismus* (Stuttgart, 1985) p. 62.

48. Hanns-Martin Bachmann, *Die naturrechtliche Staatslehre Christian Wolffs* (Berlin, 1977) pp. 44–7.

49. Bardong (ed.), *Friedrich der Grosse*, p. 76.

50. For a detailed account of the stormy relationship between Frederick and Voltaire, see Christiane Mervaud, *Voltaire et Frédéric II: une dramaturgie des lumières 1736–1778* (Oxford, 1985) *passim*.

51. Schieder, *Friedrich der Grosse*, p. 128.

52. Adolf von Harnack, *Geschichte der königlich preussischen Akademie zu Berlin*, 3 vols in 4 (Berlin, 1900) vol. I, pp. 248–9, 293–318, 358–61.

53. Quoted in Peter Gay, *Voltaire's Politics* (New York, n.d.) p. 171.

54. Honoré Gabriel Victor Riquetti comte de Mirabeau, *De la monarchie prussienne sous Frédéric le Grand*, 7 vols (London, 1789) vol. V, p. 348.

55. 'Instruction König Friedrich Wilhelms I. für seinen Nachfolger', *Acta Borussica: Die Behördenorganisation und die allgemeine Staatsverwaltung Preussens im 18. Jahrhundert*, vol. III, ed. G. Schmoller, D. Krauske and V. Loewe (Berlin, 1901) pp. 441–67.

56. Volker Hentschel, *Preussens streitbare Geschichte 1594–1945* (Düsseldorf, 1980) p. 54; Oestreich, *Friedrich Wilhelm I*, p. 49.

57. See Frederick's remarks on his father's prudence, above, p. 267.

58. 'Essai sur les formes de gouvernement et sur les devoirs des souverains', *Oeuvres*, vol. IX (Berlin, 1848) p. 208.

59. Ibid.

60. Bardong (ed.), *Friedrich der Grosse*, p. 73.

61. See above, pp. 284–5.

62. Ewald Friedrich Graf von Hertzberg, 'Mémoire sur le troisième année du règne de Frédéric Guillaume II, et pour prouver que le gouvernement prussien n'est pas despotique', *Mémoires de l'Académie Royale des Sciences et Belles-Lettres depuis l'avénement de Frédéric Guillaume II au throne. Août 1786 jusqu'à la fin de 1787. Avec l'histoire pour le même temps* (Berlin, 1792) pp. 645–9.

63. Friedrich Meinecke, *Machiavellism. The doctrine of raison de'état and its place in modern history* (London, 1957) pp. 277–9.

64. Bardong (ed.), *Friedrich der Grosse*, p. 227; Frederick, *The History of My Own Times*, p. 48.

65. Volz (ed.), *Die Politischen Testamente*, pp. 184–5.

66. Frederick, *The History of My Own Times*, vol. I, p. 92.

67. Volker Sellin, 'Friedrich der Grosse und der aufgeklärte Absolutismus', U. Engelhardt, Volker Sellin and Horst Stuke (eds) *Soziale Bewegung und politische Verfassung. Beiträge zur Geschichte der modernen Welt* (Stuttgart, 1976) pp. 97–9.

68. 'Essai sur les formes de gouvernement', p. 205.

69. Ibid., p. 206.

70. See above, p. 274.

71. See above, p. 272.

72. Corni, *Stato assoluto e società agraria*, p. 415; Behrens, *Society, Government and the Enlightenment*, pp. 88, 140, 148, 182–3.

73. Schieder, *Friedrich der Grosse*, p. 218.

74. See above, pp. 269–72.

75. Günter Birtsch, 'Religions- und Gewissensfreiheit in Preussen von 1780 bis 1817', *Zeitschrift für historische Forschung*, vol. XI, 2 (1984) p. 184.

76. Bardong (ed.), *Friedrich der Grosse*, p. 542.

77. Gay, *Voltaire's politics*, pp. 278–80.

78. Hans Reiss (ed.) *Kant's political writings* (Cambridge, 1970) pp. 58–9.

79. Reinhold Koser, *Geschichte Friedrichs des Grossen*, 2 vols (Stuttgart and Berlin, 1914) vol. I, p. 13. In theory, torture could still be employed in cases of mass murder and high treason but in practice it was never used again. It was abolished without qualification in 1755.

80. Bardong (ed.), *Friedrich der Grosse*, p. 495.

81. Koser, *Geschichte Friedrichs des Grossen*, vol. I, p. 12.

82. Hintze, *Die Hohenzollern und ihr Werk*, pp. 349–52. Hintze gives the best easily accessible account of the progress of Cocceji's reforms.

83. Behrens, *Society, Government and the Enlightenment*, p. 105.

84. There is a good account in ibid., pp. 111–14.

85. Hermann Conrad, *Die geistigen Grundlagen des Allgemeinen Landrechts für die preussischen Staaten von 1794*, Arbeitsgemeinschaft für Forschung des Landes Nordrhein-Westfalen, LXXVII (1958) pp. 13–14.

86. Reinhart Koselleck, *Preussen zwischen Reform und Revolution. Allgeme-*

ines Landrecht, Verwaltung und soziale Bewegung von 1791 bis 1848, 3rd edn (Stuttgart, 1981) p. 23.

87. Mittenzwei, *Friedrich II*, pp. 193–5.

88. Bardong (ed.) *Friedrich der Grosse*, p. 454.

89. Volz (ed.) *Die Politischen Testamente*, p. 2.

90. Otto Hintze, 'Preussens Entwicklung zum Rechtsstaat', *Forschungen zur Brandenburgischen und Preussischen Geschichte*, XXXII (1920) p. 406.

91. Baumgart, 'Zur Geschichte der kurmärkischen Stände', p. 159.

92. Behrens, *Society, Government and the Enlightenment*, pp. 105–6.

93. Ibid., p. 106; William W. Hagen, 'The Junkers' faithless servants: peasant insubordination and the breakdown of serfdom in Brandenburg-Prussia 1763–1811', Richard J. Evans and W. R. Lee (eds) *The German Peasantry: conflict and community in rural society from the eighteenth to the twentieth centuries* (London, 1985) pp. 77–86.

94. Behrens, *Society, Government and the Enlightenment*, p. 113.

95. Quoted in Mittenzwei, *Friedrich II*, p. 192.

96. Peter Baumgart, 'Wie absolut war der preussische Absolutismus?', *Preussen – Versuch einer Bilanz* (5 vols, Hamburg, 1981), vol. II: *Preussen – Beiträge zu einer politischen Kultur*, ed. Manfred Schlenke, p. 103.

97. Mittenzwei, *Friedrich II*, p. 193.

98. Daniel Jenisch, 'Denkschrift auf Friedrich den Zweiten', in Steinmetz (ed.) *Friedrich II und die deutsche Literatur*, p. 244.

99. Behrens, *Society, Government and the Enlightenment*, p. 115.

100. See above, p. 273.

101. Eckhart Hellmuth, *Naturrechtsphilosophie und bürokratischer Werthorizont. Studien zur preussischen Geistes- und Sozialgeschichte des 18. Jahrhunderts* (Göttingen, 1985) pp. 27–9, 280–1.

102. Horst Möller, 'Königliche und bürgerliche Aufklärung', Schlenke (ed.) *Preussen – Versuch einer Bilanz*, vol. II, pp. 127–9. Cf. H. B. Nisbet, '*Was ist Aufklärung?* The concept of Enlightenment in eighteenth-century Germany', *Journal of European Studies*, XII (1982) 90.

103. Horst Möller, *Aufklärung in Preussen. Der Verleger, Publizist und Geschichtsschreiber Friedrich Nicolai* (Berlin, 1974) p. 266.

104. Ibid., p. 252.

105. Behrens, *Society, Government and the Enlightenment*, p. 187.

106. J. R. Seeley, *Life and Times of Stein, or Germany and Prussia in the Napoleonic age*, 3 vols (Cambridge, 1878) vol. I, p. 40.

107. See above, p. 273.

108. Quoted in Rudolf Vierhaus, 'Deutschland vor der französischen Revolution. Untersuchungen zur deutschen Sozialgeschichte im Zeitalter der Aufklärung' (unpublished Münster dissertation, 1961) p. 280.

109. John Moore, *A View of Society and Manners in France, Switzerland and Germany*, 2 vols (4th edn, Dublin, 1789) vol. II, p. 130.

110. Reiss (ed.) *Kant's political writings*, p. 59.

111. Ibid., p. 58.

112. Jacques Droz, *L'Allemagne et la Révolution française* (Paris, 1949) p. 80.

11. CATHERINE THE GREAT   *Isabel de Madariaga*

1. N. Hans, 'Dumaresq, Brown and some early educational projects of Catherine II', *Slavonic and East European Review*, 40 (1961) 481–91.

2. For a discussion of the use of the terms autocracy and absolutism, see Isabel de Madariaga, 'Autocracy and sovereignty', *Canadian American Slavic Studies*, 16 (1982) 369–87.

3. D. Griffiths in 'Catherine II, the republican empress', *Jahrbücher für Geschichte Osteuropas*, NF, 21 (1973) 323–4, is guilty of an anachronistic judgment in an otherwise very balanced article. So also are V. S. Nersesiants and S. I. Shtamm, 'La politica e il diritto penali nella Russia della seconda metà del XVIII secolo', in *La "Leopoldina"*: *Criminalità e giustizia criminale nelle riforme del settecento europeo*, Siena, 1986, preprint, II, pp. 1731–44 at p. 1740.

4. See e.g. I. Fedossov, 'Europe Centrale et Orientale et la Russie' in *L'absolutisme éclairé*, eds B. Köpecki, A. Soboul, E. H. Balàzs and D. Kosàry (Budapest and Paris, 1985) pp. 233–54, at p. 244; see also D. Ransel, *The Politics of Catherinian Russia: The Panin Party* (New Haven, Conn., 1973) p. 179 and n. These distortions all go back to the misreading of Montesquieu not only by these authors, but by their common source, F. V. Taranovsky, in 'Politicheskaya doktrina Nakaza' in *Sbornik statey po istorii prava posvyashchennyy M. F. Vladimirskomu Budanovu* (Kiev, 1904) pp. 44–86.

5. Catherine commissioned a Russian translation of Blackstone prepared by the Russian jurist who studied at Glasgow University, S. I. Desnitsky, which was published in 1780–1782.

6. The origins of the Conscience Court are much debated and have been sought in the *Encyclopédie* and also in the Ukrainian court of arbitration.

7. See J. M. Hartley, 'Catherine's Conscience Court – an English equity court?' in *Russia and the West in the Eighteenth Century*, ed. A. G. Cross (Newtonville, Mass., 1983) pp. 306–18. The Conscience Courts continued to function until the 1840s.

8. See R. E. Jones, *The Emancipation of the Russian Nobility* (Princeton, 1973) p. 230; J. P. LeDonne, *Ruling Russia: politics and administration in the Age of Absolutism* (Princeton, 1984) p. 93.

9. M. Raeff, *The Origins of the Russian Intelligentsia* (New York, 1966) p. 125.

10. But see J. M. Hartley, 'The Boards of Social Welfare and the financing of Catherine II's state schools', *Slavonic and East European Review* 67 (1989), 211–27.

11. Marcel Le Clère, *Histoire de la Police* (Paris, 1964) pp. 32–4.

12. The fact that nobles did not pay the poll-tax should not surprise one since it was designed for the maintenance of the armed forces, and most of the nobles served in the armed forces; even if all males had paid it, however, the total raised would not have been more than 50,000 roubles – a drop in the bucket of Russian military expense.

13. M. Raeff, 'The Empress and the Vinerian Professor: Catherine II's projects of government reforms and Blackstone's *Commentaries*', *Oxford Slavonic Papers*, New Series, 7 (1974) pp. 18–41.

14. LeDonne argues (op. cit., p. 16) that since 'the state is synonymous with the ruling class . . . it has no separate existence'.

15. G. Marker, *Publishing, Printing and the Origins of Intellectual Life in Russia, 1700–1800* (Princeton, 1985) p. 91.

16. Op. cit., p. 108.

17. For a modern account of the satirical journals, see W. Gareth Jones, 'The polemics of the 1769 Journals: a reappraisal' in *Canadian American Slavic Studies*, 16 (1982) 432–3.

18. *Svodnyy katalog russkoy knigi grazhdanskoy pechati XVIII veka 1725–1800*, IV, entry 110, pp. 104–5.

19. See for instance Gabriel Ardant, 'I have not mentioned Catherine II on purpose because she cannot qualify as a reformer. She duped a certain number of great European writers . . . but she cannot fool history', in 'Financial policy and economic infrastructure of modern states and nations' in C. Tilly (ed.) *The Formation of National States in Western Europe* (Princeton, 1975) at p. 204, n.34.

20. It counted as 'little treason' in England and the penalty was still burning alive in the eighteenth century.

21. See his introduction to the Memoirs of Princess Dashkova, repr. in Yekaterina Dashkova, *Zapiski 1743–1810* ed. G. N. Moiseyeva (Leningrad, 1985) at p. 214.

22. This point is very effectively made in M. Raeff, *The Well-Ordered Police State. Social and institutional change through law in the Germanies and Russia, 1600–1800* (New Haven, Conn., 1983) p. 233.

23. *Krest'yane v tsarstvovaniye Yekateriny II* (2 vols, St Petersburg, 1901).

Notes on Contributors

M. S. ANDERSON is Professor Emeritus of International History in the University of London, where he taught for many years at the London School of Economics. His principal publications include: *Britain's Discovery of Russia 1553–1815* (London, 1958), *Europe in the Eighteenth Century* (London, 1961; 3rd edn, 1987), *The Eastern Question 1774–1923* (London, 1966), *The Ascendancy of Europe* (1972; 2nd edn, 1986), *Peter the Great* (London, 1978), *Historians and Eighteenth-Century Europe* (Oxford, 1979), and *War and Society in Europe of the Old Regime, 1618–1789* (London, 1988).

DEREK BEALES is Professor of Modern History in the University of Cambridge, where he is a Fellow of Sidney Sussex College. He has recently published the first volume of a major life of *Joseph II: In the shadow of Maria Theresa 1741–1780* (Cambridge, 1987); his earlier publications include *England and Italy 1859–60* (London, 1961), *From Castlereagh to Gladstone 1815–85* (London, 1969), *The Risorgimento and the Unification of Italy* (London, 1971), and, edited with Geoffrey Best, *History, Society and the Churches: essays in honour of Owen Chadwick* (Cambridge, 1985).

T. C. W. BLANNING is Reader in Modern European History at the University of Cambridge, where he is a Fellow of Sidney Sussex College. His principal publications are: *Joseph II and Enlightened Despotism* (London, 1970), *Reform and Revolution in Mainz 1743–1803* (Cambridge, 1974), *The French Revolution in Germany* (Oxford, 1983), *The Origins of the French Revolutionary Wars* (London, 1986), and *The French Revolution* (London, 1987).

R. J. W. EVANS is University Lecturer in the Modern History of East Central Europe at the University of Oxford and a Senior Research Fellow of Brasenose College. He is the author of *Rudolf II and His World* (Oxford, 1973), *The Wechel Presses* (Oxford, 1975) and *The Making of the Habsburg Monarchy 1550–1700* (Oxford, 1979) and of numerous articles on the intellectual and social history of the Habsburg Lands in the early modern period. A fellow of the British Academy, he is also Joint-Editor of the *English Historical Review*.

375

C. W. INGRAO is Professor of History at Purdue University, Indiana, USA. He is the author of *In Quest and Crisis: Emperor Joseph I and the Habsburg Monarchy* (West Lafayette, Indiana, 1979) and *The Hessian Mercenary State: ideas, institutions and reform under Frederick II, 1760–1785* (Cambridge, 1987) and of numerous articles on eighteenth-century Austrian and German history.

ISABEL DE MADARIAGA is Professor Emeritus of Russian Studies in the University of London, where she taught for many years at the School of Slavonic and East European Studies. Her publications include *Britain, Russia and the Armed Neutrality of 1780* (London, 1962), *Opposition* (with G. Ionescu) (London, 1968), and *Russia in the Age of Catherine the Great* (London, 1981), and numerous articles on eighteenth-century Russian history.

KENNETH MAXWELL is Director of the Camões Center for the Portuguese Speaking World at Columbia University in New York. A specialist in eighteenth-century Portuguese and Brazilian history, he is the author of *Conflicts and Conspiracies: Brazil and Portugal 1750–1808* (Cambridge, 1973) and of a forthcoming study of Pombal's ministry.

THOMAS MUNCK is a Lecturer in Modern History at the University of Glasgow. An expert on early modern Danish history, he is the author of *The Peasantry and the Early Absolute Monarchy in Denmark 1660–1708* (Copenhagen, 1979) and of *Seventeenth-century Europe 1598–1700: State, Conflict and the Social Order* (Macmillan, 1990).

CHARLES C. NOEL is Senior Lecturer in Humanities at Ealing College of Higher Education, London. He has published several articles on eighteenth-century Spanish History and is preparing a major study of Charles III's reign.

H. M. SCOTT is a Lecturer in Modern History at the University of St Andrews. He is the author of *The Rise of the Great Powers 1648–1815* (with Derek McKay) (London, 1983), *British Foreign Policy in the Age of the American Revolution 1763–1783* (Oxford University Press, forthcoming), and of articles on eighteenth-century international history.

Index

Note: individuals or places mentioned incidentally in one chapter are not generally indexed.